D0916004

Autophagy in Immunity and Infection

Edited by
Vojo Deretic

Related Titles

Frosch, M., Maiden, M. C. J. (eds.)

Handbook of Meningococcal Disease

Infection Biology, Vaccination, Clinical Management

2006
ISBN 3-527-31260-9

Lutz, M. B., Romani, N., Steinkasserer, A. (eds.)

Handbook of Dendritic Cells

Biology, Diseases and Therapies

2006
ISBN 3-527-31109-2

Frelinger, J. A. (ed.)

Immunodominance

The Choice of the Immune System

2006
ISBN 3-527-31274-9

Pollard, K. M. (ed.)

Autoantibodies and Autoimmunity

Molecular Mechanisms in Health and Disease

2006
ISBN 3-527-31141-6

Kropshofer, H., Vogt, A. B. (eds.)

Antigen Presenting Cells

From Mechanisms to Drug Development

2006
ISBN 3-527-31108-4

Hamann, A., Engelhardt, B. (eds.)

Leukocyte Trafficking

Molecular Mechanisms, Therapeutic Targets, and Methods

2005
ISBN 3-527-31228-5

Autophagy in Immunity and Infection

A Novel Immune Effector

Edited by
Vojo Deretic

WILEY-VCH

WILEY-VCH Verlag GmbH & Co. KGaA

The Editor

Prof. Dr. Vojo Deretic
Department of Molecular Genetics
& Microbiology & Physiology
University of New Mexico Health Sciences Center
915 Camino de Salud, NE
Albuquerque, NM 87131-001
USA

Library of Congress Card No.: applied for

British Library Cataloguing-in-Publication Data
A catalogue record for this book is available from the British Library.

Bibliographic information published by Die Deutsche Bibliothek
Die Deutsche Bibliothek lists this publication in the Deutsche Nationalbibliografie; detailed bibliographic data is available in the Internet at <http://dnb.ddb.de>

Printed in the Federal Republic of Germany
Printed on acid-free paper

Composition K+V Fotosatz GmbH, Beerfelden
Printing betz-druck GmbH, Darmstadt
Bookbinding Litges & Dopf Buchbinderei GmbH, Heppenheim
Cover Design Adam Design, Weinheim

ISBN-13: 978-3-527-31450-8
ISBN-10: 3-527-31450-4

Contents

Autophagy in Immunity and Infection. Edited by Vojo Deretic
Copyright © 2006 WILEY-VCH Verlag GmbH & Co. KGaA, Weinheim
ISBN: 3-527-31450-4

Preface

This book was inspired by a stream of nearly simultaneous reports in 2004 and 2005 demonstrating that the fundamental biological process of autophagy, primarily known for its role in cytoplasmic maintenance, represents a previously unrecognized innate and adaptive immunity mechanism that functions as a defense against intracellular pathogens and probably has other roles within the immune system. Although hints to the role of autophagy in immune defenses and other roles in immunity have existed in the literature, the most recent burst of publications made a compelling and definitive case for the importance of autophagy in immunity. A further motivation for this project came from the opportunity to merge these new findings with the superb recent progress on genetics, biochemistry and cell biology of autophagy. The product is a book covering the basic aspects of autophagy as a cytoplasmic maintenance process playing a role in cell survival and death, its role in health and disease in general, and the new cutting edge – the role of autophagy in immunity. Using this book, the reader can find a full range of information on autophagy in one place covering both its fundamental molecular mechanisms and its many physiological roles.

Autophagy is a homeostatic intracellular mechanism, whereby a cell digests parts of its own cytoplasm for removal or turn-over, as eloquently summarized in the Foreword by P. Seglen. The term autophagy represents a set of distinct yet related pathways. These range from the robust process of macroautophagy to a rather subtle process of chaperone mediated autophagy, as detailed in Chapter 1 by J. Legakis and D. Klionsky, which also provides the fundamentals of autophagy based on the powerful genetics in yeast and other organisms. Macroautophagy sequesters significant portions of the cytosol or whole organelles into a characteristic double membrane vacuole termed the autophagosome, for eventual degradation in autolysosomes, covered extensively in Chapter 2 by S. Tooze and colleagues and Chapter 3 by N. Mizushima. Chaperone mediated autophagy, covered in some detail in Chapter 4 by A. Cuervo and colleagues and touched upon in Chapter 12 by D. Schmid and C. Münz, is a degradative pathway whereby individual proteins are imported directly into the lysosomes. In macroautophagy, or its variant manifestation of microautophagy, the trapped cytosol or organelles are eventually delivered to degradative compartments (in mammalian cells – autolysosome) for digestion and removal. In its probably

Autophagy in Immunity and Infection. Edited by Vojo Deretic
Copyright © 2006 WILEY-VCH Verlag GmbH & Co. KGaA, Weinheim
ISBN: 3-527-31450-4

most common presentation, autophagy recycles stable cytosolic macromolecules, such as proteins with long half-lives, to supply nutrients and maintain essential cellular anabolic needs and viability under starvation conditions. The organelle removal function of autophagy is a just as important housekeeping function, by controlling the pool of peroxisomes or removing compromised mitochondria, in the latter case potentially protecting cells from unscheduled apoptosis. Although autophagy is a cell maintenance mechanism, under certain conditions, excessive autophagy can cause non-apoptotic programmed cell death, covered in Chapter 5 by Y. Debnath and C. Fung. Autophagy has been implicated in cancer, degenerative disorders, such as Huntington, Parkinson, and Alzheimer diseases, normal development, and aging, covered in detail in Chapter 4 by A. Cuervo and colleagues.

A number of very precise studies on anti-viral action of autophagy have been the true forerunner of our present more general understanding of the role of autophagy in defense against intracellular pathogens, as covered in Chapter 13 by B. Levine. More recent studies demonstrate that autophagy is also an innate immunity effector against intracellular bacteria, a central theme of the second half of this book, encompassing: Chapter 6 on *Mycobacterium tuberculosis* elimination by autophagy (Harris et al.); Chapter 7 by T. Yoshimori and A. Amano on autophagic elimination of streptococci if they invade host cells and find themselves in the cytosol; Chapter 8 on the role of autophagy in capturing the intracellular *Shigella* and its ability to escape this process; and Chapter 9 by K. Rich and P. Webster on *Listeria*. Some highly evolved pathogens have mechanisms for harnessing autophagy to their own benefit, as suggested in Chapter 10 by M. Gutierrez and M. Colombo and discussed in Chapter 11 by M.-P. Stein and C. Roy. The duality of effects of autophagy is also reflected in the Addendum to B. Levine's Chapter 13 provided by J. Sparks and M. Denison. Significantly, autophagy has a strong impact on MHCII presentation (Chapter 12 by D. Schmid and C. Münz) and is controlled by cytokines (Chapters 6 and 13) clearly extending the role of autophagy to adaptive immunity.

The goal of this volume was to provide the reader not only with the applications of autophagy in infectious diseases and immunity, but also to generate a definitive text for autophagy in general. In other words, a reader who is interested primarily in the fundamental principles and broad biological aspects of autophagy, should find this book an indispensable companion and a comprehensive source of information. For those who are primarily interested in the burgeoning field of autophagy in innate and adaptive immunity, the chapters covering the basic principles of autophagy are just as important to understand fully the underlying processes.

The book starts with a foreword by Professor Per Seglen, a doyen in the field of autophagy, who has defined many biochemical and cell biological features of autophagy and has also produced both classical and contemporary highly cited papers in this field. A careful reader of the foreword will extract many useful concepts on autophagosomes, amphisomes and autolysosomes, and precious cautionary notes on interpretations of cause and effect in diseases and in cell survival vs. cell-death promoting faces of autophagy. The editor is indebted to Per for his willingness to write a foreword to this volume and give the reader both his sage advice on general aspects of autophagy and sum it all up including the latest developments in the context of defense against intracellular pathogens.

Furthermore, the reader is a true beneficiary of the combination of excitement and enthusiasm that pervades the field of autophagy research, and enormous expertise of the contributing authors in this area. The editor of this book is indebted immensely to all contributing authors. The chapters by Drs. Ana Maria Cuervo, Daniel Klionsky, Beth Levine, Sharon Tooze and Naboru Mizushima, taken together, can give a textbook on autophagy as a standalone product. Likewise, the chapters that link autophagy with innate and adaptive immunity by Drs. Christian Münz, Chichiro Sasakawa, Tamotsu Yoshimori, and others summarize the new breakthroughs in immunological applications of autophagy. They also define the nidus for the developing field of immunophagy, a term used by the Editor of this book in a recent review in Current Opinion in Immunology to describe collectively these processes.

I acknowledge the excellent coordination and open lines of communication with the publisher including the gentle prompts from Andreas Sendtko, importance of NIH funding (AI45148 and AI42999) for all my scientific activities including this one, great support and understanding at home beyond what a person can expect or deserves, and above all the collective responsiveness and enthusiasm for this book by the main protagonists in the field of autophagy. My great personal and professional respect for many of the contributors to this book has been reaffirmed in the process.

Placitas (between Albuquerque and Santa Fe), April 2006   Vojo Deretic

Foreword

Autophagy, the mechanism by which cells envelop and degrade their own cytoplasm, plays a dual role in cellular physiology. On the one hand, autophagy serves vital functions such as the supply of essential amino acids during nitrogen starvation, the mobilization of iron from intracellular stores, the sequestration of aggregated (and potentially harmful) abnormal proteins that cannot be digested by the proteasomes, and the containment and degradation of infectious organisms. On the other hand, autophagy is frequently turned on during programmed cell death, complementing the apoptotic caspases in the orderly liquidation of the cell. In certain cases, particularly if the major caspases are somehow incapacitated, autophagy can, by itself, complete the death process. Autophagy may thus either support or prevent cell survival, depending on the biological context.

In a pathological setting, this autophagic duality may cause problems of interpretation. Many diseases are accompanied by alterations in cellular membrane fluxes, often causing massive accumulations of intracellular vacuoles of varying morphologies. Do these changes represent an attempt to combat the disease or do they contribute to disease progression (or both – or neither)? What is the nature of the vacuoles that are the affected steps in the vacuolar dynamics and in what direction are they altered?

As a first step in the analysis, the observed vacuoles need to be identified; however, unfortunately, this is not a straightforward matter. In addition to the three major types of autophagic vacuoles, i.e. autophagosomes, amphisomes and (auto)lysosomes, endosomes may contain cellular material derived from disintegrated surrounding cells or, in late, multivesicular endosomes, internalized by invagination of the endosomal delimiting membrane. In certain diseases, such as Alzheimer's, both the endocytic and autophagic pathways are afflicted, causing the accumulation of an extremely heterogeneous array of vacuoles. It should be noted that a prolonged disturbance of vacuole fluxes may induce the formation of unusual vacuoles, which may be difficult to classify by morphological criteria.

Autophagosomes, the autophagic vacuoles formed when the sequestering membrane cisternae (the phagophores) have completed the enclosure process, can be recognized in the electron microscope as areas of absolutely normal cytoplasm, circumscribed by delimiting membranes, but not deviating morphologi-

Autophagy in Immunity and Infection. Edited by Vojo Deretic
Copyright © 2006 WILEY-VCH Verlag GmbH & Co. KGaA, Weinheim
ISBN: 3-527-31450-4

cally from their surroundings. The delimiting membrane can sometimes be seen as a double-membrane (cisternal) structure, sometimes as a thick, osmiophilic layer and sometimes (artificially) as an open, electron-lucent cleft. However, since later autophagic vacuoles may contain sequestered membraneous elements closely apposed to their (single) delimiting membrane, an apparent "double membrane" is a less reliable diagnostic criterion for an autophagosome than its contents of unaltered cytoplasm.

Amphisomes, the products of fusion between endosomes and autophagosomes, quickly get their contents denatured due to acidification by the proton pump brought in by the endosomal fusion partner. The denaturation is visible in the electron microscope as a darkening and a somewhat altered morphology relative to the cytoplasmic surroundings. Multiple inputs from both autophagy and endocytosis often make the amphisomes large and complex. The contents will usually serve to distinguish amphisomes from autophagosomes, but they cannot be reliably distinguished from early autolysosomes by morphological criteria alone, particularly because the endosomal fusion partner contributes small amounts of lysosomal enzymes that may initiate degradation of the amphisomal contents. With more advanced degradation, autolysosomes usually become distinctive.

Organelle markers can make the identification of autophagic and endocytic vacuoles considerably easier. Few markers are entirely specific, but by using them in combination, information can be obtained both from the presence and the absence of a marker. In immunogold labeling studies, a relatively degradation-resistant cytosolic enzyme such as superoxide dismutase (SOD) can be used to mark autophagic vacuoles (autophagosomes, amphisomes and autolysosomes), an endocytosed, gold-conjugated protein like bovine serum albumin (BSA) can be used to mark endosomes, amphisomes and lysosomes, and a lysosomal membrane protein, e.g. LGP120, can be used to mark lysosomes. The combination of positive and negative markers will then identify endosomes ($SOD^-/BSA^+/LGP^-$), autophagosomes ($SOD^+/BSA^-/LGP^-$), amphisomes ($SOD^+/BSA^+/LGP^-$) and autolysosomes ($SOD^+/BSA^\pm/LGP^+$). A similar approach can be used in light microscopic studies, using, for example, the lipidated mammalian Atg8 analogue, LC3-II, as a marker for all types of autophagic vacuoles, in combination with a marker of acidic vacuoles, e.g. monodansylcadaverine and suitable endosomal and lysosomal markers.

Markers may also give information about flux perturbations. The accumulation of a specific vacuolar organelle is not necessarily the result of an increased rate of its formation, but may equally well reflect a reduced rate of its disappearance due to a defective fusion step. For example, a microtubule poison like vinblastine will block all vacuole transport and fusion, and cause autophagosomes and endosomes to accumulate. An inhibitor of intralysosomal protein degradation, like leupeptin, will not only increase the size and visibility of autolysosomes, but the impaired fusion capacity of the congested lysosomes will also cause amphisomes to pile up. Similar changes in cellular vacuole populations may occur as a result of pathological alterations in vacuolar fusion rates. Even moderate fusion defects may have large morphological consequences if they persist over long periods of time, as may be the case in many of the slowly progressing autophagy-related diseases.

Experimental interruption of the autophagic-lysosomal flux offers useful ways of measuring flux rates. Since inhibition of intralysosomal protein degradation has been shown not to affect autophagic sequestration on a short-term basis, the intravacuolar accumulation of an autophagocytosed cytosolic enzyme after leupeptin treatment provides a precise measure of the autophagic sequestration rate (an autophagic membrane marker like LC3-II is less suitable for this purpose, because its vacuolar dynamics are influenced by factors other than the rates of sequestration and intralysosomal degradation). However, by blocking the autophagic flux altogether with a sequestration inhibitor such as 3-methyladenine (3-MA), the flux rate can be calculated, e.g. as the 3MA-sensitive part of the degradation of long-lived cellular protein. The effectiveness of this inhibitor also makes it useful in assessing the secondary effects of autophagy: if a cellular response is *in*sensitive to 3-MA, an autophagic causation can be excluded. In contrast, 3-MA *sensitivity* is compatible with an involvement of autophagy, although it does not prove it (as is the case with inhibitors in general).

Although most disease-related alterations in autophagic-lysosomal traffic are likely to be secondary, they can be the primary causes of some pathological conditions, most notably the lysosomal storage diseases. In these diseases, a deficiency in a single lysosomal enzyme will cause a massive intralysosomal accumulation of undegradable material that eventually disrupts all lysosomal functions, resulting in complex cellular and pathological alterations. Autophagic and endocytic influxes to the lysosome will gradually slow down, and prelysosomal autophagic and endocytic vacuoles will accumulate, their contents of undegraded material representing a spreading of the storage syndrome beyond the lysosomes. In the closely related Danon disease, lysosomes have apparently become fusion-incompetent due to a mutation in the lysosomal membrane protein, LAMP-2, resulting in reduced influxes to the lysosome, and an accumulation of amphisomes and autophagosomes. During aging, the gradual intralysosomal accumulation of undegradable lipofuscin inclusions will similarly disturb lysosomal function and has been shown to cause a reduced chaperone-mediated lysosomal protein uptake as well as a reduced flux through the autophagic pathway.

In many neurodegenerative diseases, mutant proteins that somehow escape proteasomal degradation may instead become autophagocytosed and gradually form undegradable aggregates inside lysosomes. The resulting lysosomal storage syndrome, including the accumulation of prelysosomal autophagic vacuoles, may in the long run impair the autophagic sequestration of toxic protein aggregates and thus contribute to progression of the disease. Events that take place in the piled-up amphisomes may exacerbate the situation: in Alzheimer's disease, toxic peptides seem to be generated by intra-amphisomal proteolysis. If exocytic recycling from amphisomes takes place, it could possibly be involved in the formation of the extracellular aggregates (plaques) characteristic of several neurodegenerative diseases. Preciously little is known about amphisome physiology; hopefully, a better understanding of this pivotal organelle, strategically located at the junction between the autophagic and endocytic pathways, may shed some light on the complex pathology of degenerative diseases.

In relation to infectious pathogens, autophagy has been shown to play a dual role. On the one hand, autophagy is a part of the innate and adaptive immune defense, participating in the generation of antigenic peptides for MHC class II presentation as well as in the sequestration, containment and degradation of bacteria like *Streptococcus*, *Shigella* and *Mycobacterium*. The bacteria fight back by attempting to suppress autophagic activity. On the other hand, bacteria like *Coxiella*, *Legionella* and several RNA viruses enter cells by a phagocytic route, but eventually become autophagocytosed intracellularly and take up residence inside autophagic vacuoles. In these cases, autophagy may promote infectivity.

Can better knowledge about autophagy help to combat autophagy-related diseases? Clearly, the slow progression of many of the degenerative diseases should leave a lot of room for therapeutic intervention. A stimulation of autophagic activity by intermittent amino acid starvation is one obvious strategy that seems to work well in mice (which can prolong their lifespan by fasting), but adequate data for humans are lacking. Autophagy-stimulatory drugs, such as rapamycin, represent another possibility that has shown considerable promise in several degenerative disease models. Conversely, in the case of infections or programmed cell death promoted by autophagy, autophagy suppressants like 3-MA have been demonstrated to be protective under experimental conditions. The development of autophagy modifiers that are pharmacologically acceptable and effective *in vivo* would seem like a promising therapeutic avenue.

Consideration should also be given to the possibility of overcoming or circumventing lysosomal dysfunctions. Some improvement has been reported with lysosomal enzyme replacement therapy to lysosomal storage disease patients, but the endocytic delivery of missing enzymes to lysosomes may be hampered by poor lysosomal uptake or fusion capacity. The amphisomes should, at least at an early stage, be more accessible. By supplying amphisomes, through the endocytic pathway, with lytic enzymes and other factors required for efficient degradation of problematic substrates, these organelles could possibly be turned into artificial lysosomes, tailored for a specific purpose. Hopefully, additional therapeutic strategies will be suggested by future research into the inner workings of the autophagic–endocytic-lysosomal network.

Oslo, December 1st, 2005   Per O. Seglen

List of Contributors

Atsuo Amano
Department of Oral Frontier Biology
Osaka University Graduate School
of Dentistry
1–8 Yamadaoka
Suita-Osaka 565-0871
Japan

Edmond Chan
Cancer Research UK
London Research Institute
44 Lincoln's Inn Fields
London WC2A 3PX
UK

María I. Colombo
Laboratorio de Biología Celular y
Molecular
Instituto de Histología y Embriología
(IHEM)-CONICET
Facultad de Ciencias Médicas
Universidad Nacional de Cuyo
Casilla de Correo 56
5500 Mendoza
Argentina

Ana Maria Cuervo
Department of Anatomy
and Structural Biology
Marion Bessin Liver Research Center
Albert Einstein College of Medicine
1300 Morris Park Ave
Bronx, NY 10461
USA

Jayanta Debnath
Department of Pathology
and Comprehensive Cancer Center
University of California San Francisco
513 Parnassus Ave HSW 511
San Francisco, CA 94143-0511
USA

Sergio De Haro
Departments of Molecular Genetics &
Microbiology and Cell Biology
and Physiology
University of New Mexico Health
Sciences Center
915 Camino de Salud, NE
Albuquerque, NM 87131-001
USA

Mark R. Denison
Department of Pediatrics
Vanderbilt University Medical Center
D-7235 Medical Center North
Nashville, TN 37232-2581
USA

Autophagy in Immunity and Infection. Edited by Vojo Deretic
Copyright © 2006 WILEY-VCH Verlag GmbH & Co. KGaA, Weinheim
ISBN: 3-527-31450-4

Vojo Deretic
Departments of Molecular Genetics &
Microbiology and Cell Biology &
Physiology
University of New Mexico Health
Sciences Center
915 Camino de Salud, NE
Albuquerque, NM 87131-001
USA

Christopher Fung
Department of Pathology
and Comprehensive Cancer Center
University of California San Francisco
513 Parnassus Ave HSW 511
San Francisco, CA 94143-0511
USA

Maximiliano G. Gutierrez
Laboratorio de Biología Celular y
Molecular
Instituto de Histología y Embriología
(IHEM)-CONICET
Facultad de Ciencias Médicas
Universidad Nacional de Cuyo
Casilla de Correo 56
5500 Mendoza
Argentina

James Harris
Departments of Molecular Genetics &
Microbiology and Cell Biology &
Physiology
University of New Mexico Health
Sciences Center
915 Camino de Salud, NE
Albuquerque, NM 87131-001
USA

Susmita Kaushik
Department of Anatomy
and Structural Biology
Marion Bessin Liver Research Center
Albert Einstein College of Medicine
1300 Morris Park Ave
Bronx, NY 10461
USA

Daniel J. Klionsky
Department of Molecular, Cellular
and Developmental Biology
University of Michigan
210 Washtenaw Ave
Ann Arbor, MI 48109-2216
USA

Robert Köchl
Cancer Research UK
London Research Institute
44 Lincoln's Inn Fields
London WC2A 3PX
UK

Julie E. Legakis
Life Sciences Institute,
Department of Molecular, Cellular
and Developmental Biology
University of Michigan
210 Washtenaw Ave
Ann Arbor, MI 48109-2216
USA

Beth Levine
Department of Internal Medicine
University of Texas Southwestern
Medical Center
5323 Harry Hines Blvd.
Dallas, TX 75390-9113
USA

Marta Martinez-Vicente
Department of Anatomy
and Structural Biology
Marion Bessin Liver Research Center
Albert Einstein College of Medicine
1300 Morris Park Ave
Bronx, NY 10461
USA

Noboru Mizushima
Department of Bioregulation
and Metabolism
Tokyo Metropolitan Institute of
Medical Science
3-18-22 Honkomagome, Bunkyo-ku
Tokyo 113-8613
Japan

Christian Münz
Laboratory of Viral Immunobiology
and Christopher H. Browne Center
for Immunology and
Immune Diseases
The Rockefeller University
New York, NY 10021
USA

Michinaga Ogawa
Department of Microbiology
and Immunology
Institute of Medical Science
University of Tokyo
4-6-1, Shirokanedai, Minato-ku
Tokyo 108-8639
Japan

Kathryn A. Rich
House Ear Institute
2100 W Third Street
Los Angeles, CA 90057
USA

Craig R. Roy
Section of Microbial Pathogenesis
Boyer Center for Molecular Medicine
Yale University School of Medicine
295 Congress Ave
New Haven, CT 06536
USA

Chihiro Sasakawa
Department of Microbiology
and Immunology
Institute of Medical Science
University of Tokyo
4-6-1, Shirokanedai, Minato-ku
Tokyo 108-8639
Japan

Dorothee Schmid
Laboratory of Viral Immunobiology
and Christopher H. Browne Center
for Immunology
and Immune Diseases
The Rockefeller University
New York, NY 10021
USA

Jennifer Sparks
Department of Pediatrics
Vanderbilt University Medical Center
D-7235 Medical Center North
Nashville, TN 37232-2581
USA

Mary-Pat Stein
Section of Microbial Pathogenesis
Boyer Center for Molecular Medicine
Yale University School of Medicine
295 Congress Ave
New Haven, CT 06536
USA

Sharon A. Tooze
Cancer Research UK
London Research Institute
44 Lincoln's Inn Fields
London WC2A 3PX
UK

Paul Webster
House Ear Institute
2100 W Third Street
Los Angeles, CA 90057
USA

Tamotsu Yoshimori
Department of Cell Genetics
National Institute of Genetics
Sokendai, Yata 1111
Mishima, Shizuoka-ken 411-8540
Japan

Color Plates

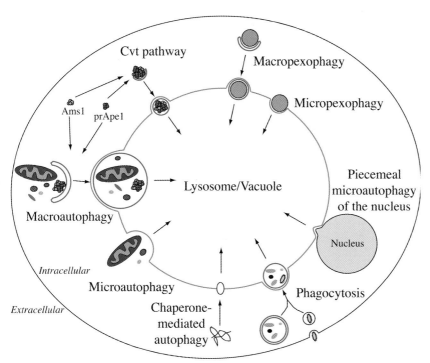

Fig. 1.1 Schematic representation of various transport routes to the lysosome/vacuole. There exist a number of pathways by which substrates are delivered to the lysosome/vacuole. Some of the sequestration events occur at the organelle membrane, these are denoted by the prefix "micro". In other cases, the enclosure of the substrate occurs spatially away from the lysosome/vacuole membrane. These pathways begin with the prefix "macro". Macro- and microautophagy are nonspecific degradation pathways, which include a variety of cargoes, depending on the organism and the particular stress conditions or stage of development. Selective degradation of peroxisomes, small parts of the nucleus or foreign pathogens occurs via macropexophagy, micropexophagy, piecemeal microautophagy of the nucleus or phagocytosis, respectively. Chaperone-mediated autophagy is a receptor-driven degradative pathway that is a secondary response to starvation conditions. The biosynthetic Cvt pathway is a method of delivery for at least two vacuolar hydrolases.

Autophagy in Immunity and Infection. Edited by Vojo Deretic
Copyright © 2006 WILEY-VCH Verlag GmbH & Co. KGaA, Weinheim
ISBN: 3-527-31450-4

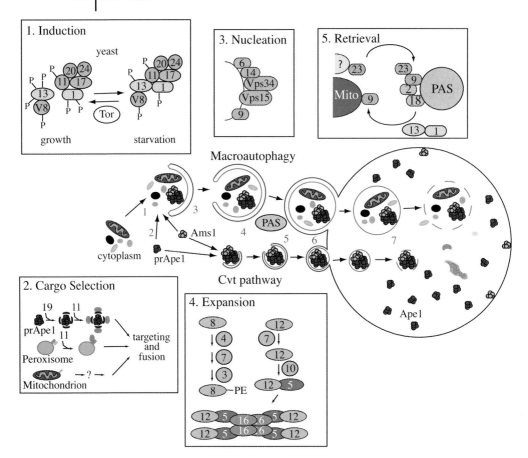

Fig. 1.2 Autophagy and the Cvt pathway. Autophagy and the Cvt pathway can be depicted as a series of separate steps. The roles of Atg and other proteins, shown to participate in different parts of the pathway, are depicted. The proteins classified by only a number are the corresponding Atg gene product. Otherwise, the protein name is specified, except for Vac8, which is indicated as "V8". "P" denotes phosphorylation of the indicated protein. (1) *Induction.* TOR kinase becomes inactivated upon nutrient limitation, eliciting a series of events, which result in the induction of autophagy. These include partial dephosphorylation of Atg13, which alters its association with Atg1. Atg1 is thought to play a key role in the switch between growth and starvation. Autophagy-specific proteins are shown in blue, whereas Cvt-specific proteins are depicted in purple.

(2) *Cargo selection and packaging.* Examples of specific autophagy include the Cvt pathway, pexophagy and possibly mitophagy. During growth, the Cvt pathway is active. The cargo, prApe1, is synthesized as an inactive precursor and rapidly oligomerizes. Atg19, the cargo receptor, binds to the oligomer, followed by Atg11 binding to the complex. Upon induction of pexophagy, the peroxin, Pex3, is degraded, thus exposing the docking protein, Pex14. Although it is not proven, Atg11 is proposed to bind to the newly exposed Pex14. The mechanism of mitophagy is unknown. Once these binding events occur, the cargo are enwrapped by a double-membrane vesicle and delivered to the lysosome/vacuole. (3) *Vesicle nucleation.* Membrane is acquired from an unknown location and the cargo associates with the forming vesicle. Membrane formation re-

quires the PI3K complex I; the components of this complex are shown in Step 3. The PI3-phosphate (PI3P) generated by this complex recruits a number of Atg proteins to the PAS, including Atg18, Atg20, Atg21, and Atg24 [24]. (4) *Vesicle expansion and completion*. There are two sets of Atg proteins, which participate in a series of ubiquitin-like (Ubl) conjugation reactions. These generate Atg12–Atg5–Atg16 and Atg8–PE (see text for details). The functions of these proteins are not known but they are needed for expansion and completion of the sequestering vesicle. (5) *Retrieval*. As most of the Atg proteins are not included in the completed vesicle, there must be a mechanism to release and return these components back to their original site. Atg9 and Atg23 have been shown to be cycling proteins, moving between the PAS and other punctate structures. Atg9 has been shown to cycle betweent he mitochondria and the PAS. The non-PAS localizations of Atg23 are as yet unidentified. These two proteins may aid in the recovery of Atg components, allowing them to be reused for another round of delivery. (6) *Targeting, docking and fusion of the vesicle with the lysosome/vacuole*. The docking and fusion of the completed vesicle requires a number of components (see text for details). The fusion event results in a single-membrane vesicle within the lumen of the lysosome/vacuole. (7) *Breakdown of the vesicle and its contents*. Once inside the lysosome/vacuole, the autophagic or Cvt body must be degraded in order for the cargo to be released. The lipase responsible for vesicle lysis is thought to be Atg15. Upon release into the lumen, the cargoes of pexophagy and bulk autophagy are broken down for re-use in the cell, while the cargoes of the Cvt pathway carry out their function as hydrolases.

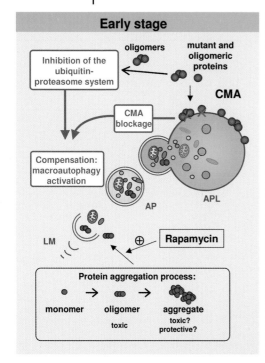

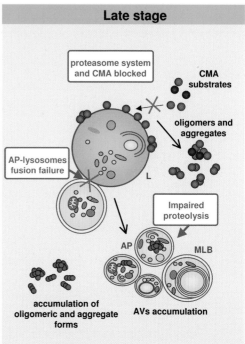

Fig. 4.1 Autophagy in protein conformational disorders. Protein conformational disorders result from abnormal conformational changes in particular proteins (due to mutations or post-translational modifications) that make them prone to aggregation. In the early stages of the disorder, the abnormal proteins often block the activity of proteolytic systems normally responsible for the degradation of soluble proteins (proteasome and CMA by the lysosome), resulting in compensatory activation of macroautophagy to eliminate the oligomeric toxic forms. As the diseases progress (late stage), a macroautophagic failure often occurs, probably due to problems in the clearance of the autophagocytosed materials, leading to the accumulation of AVs with partially degraded contents and eventually to cell death. Abbreviations: L=lysosome; LM=limiting membrane; AP=autophagosome; APL=autophagolysosome; MLB=multilamelar bodies.

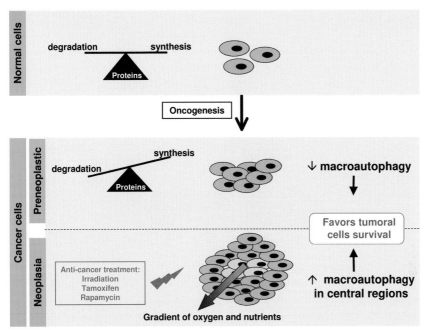

Fig. 4.2 Dual role of macroautophagy in cancer. In normal cells, there is a balance between protein synthesis and degradation, which is deranged in cancer cells. However, in early stages of carcinogenesis (preneoplastic cells), low levels of macroautophagy would serve as a prosurvival mechanism for the cells. In contrast, in late stages (neoplasia) upregulation of macroautophagy in the hypoxic, low-nutrient center regions of tumors would be advantageous for cell survival. Possible therapeutic attempts try to activate macroautophagy by means of γ-irradiation, tamoxifen or rapamycin (red callout). However, these treatments at later stages of tumorigenesis, during which macroautophagy activity is required for survival can, in fact, promote tumor cell proliferation rather than cell death.

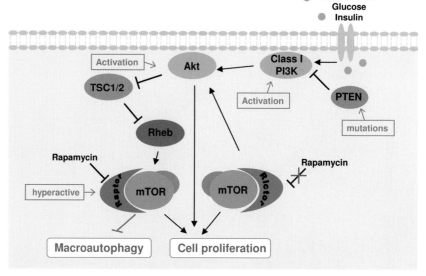

Fig. 4.3 A model for the PI3K–Akt–mTOR pathway. Akt, a serine/threonine kinase, downstream of class I PI3K regulates cell proliferation and cell survival, and also activates mTOR indirectly, in response to extrinsic stimuli such as nutrients and growth factors. PTEN, a lipid phosphatase, dephosphorylates and inactivates class I PI3K. mTOR forms complexes with two cytosolic proteins: raptor (rapamycin-sensitive) and rictor (rapamycin-insensitive). mTOR inhibits macroautophagy and upregulates protein synthesis. In various cancers, mutations in PTEN, amplifications of both PI3K and Akt, and hyperactivation of the mTOR have been observed, resulting in the blockage of macroautophagy, thus providing an anabolic advantage (shown in red callouts) to the cancer cells. Abbreviations: mTOR = mammalian target of rapamycin; PTEN = phosphatase and tensin homolog deleted on chromosome 10; TSC = tuberous sclerosis complex.

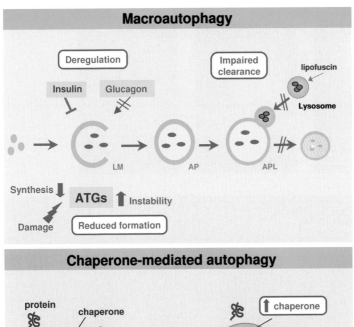

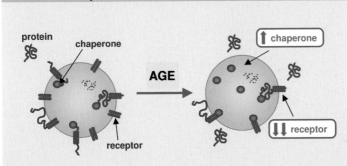

Fig. 4.4 Changes in autophagy during aging. The activity of both macroautophagy and CMA decreases with age. (Top) Impaired macroautophagic activity results from the combined effect of reduced formation of autophagic vesicles, impaired clearance and deregulation of the hormonal control. (Bottom) Levels of LAMP-2A, a receptor for CMA substrates at the lysosomal membrane, decrease with age. The decrease in levels of the receptor is initially compensated for by an increase in the levels of the chaperone that assist in substrate uptake. However, at advanced ages, the levels of the receptor decrease to a point for which compensation is no longer possible and failure of CMA becomes evident. Abbreviations: ATG = autophagy-related proteins; LM = limiting membrane; AP = autophagosome; APL = autophagolysosome.

A

B

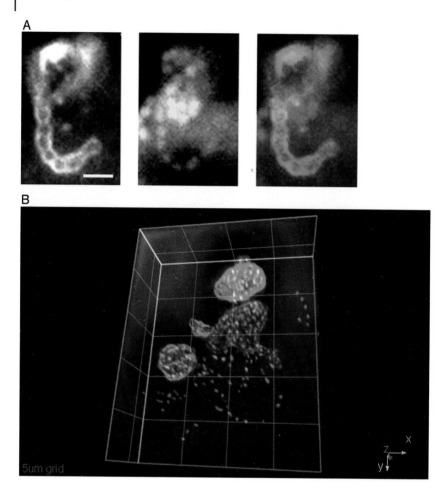

Fig. 7.2 GcAVs. (A) LC3-positive compartments (a left panel and green in a right panel) sequestering intracellular GAS (a middle panel and magenta in a right panel) in HeLa cells expressing enhanced GFP (EGFP)–LC3 at 1 h post-infection [40]. The bacteria were visualized by staining bacterial DNA with propidium iodide. Bar=2 μm. (B) Three-dimensional image of a large GcAV at 3 h post-infection. The image was made by Shunsuke Kimura, Research Institute of Microbial Diseases, Osaka University, Japan. Grid=5 μm.

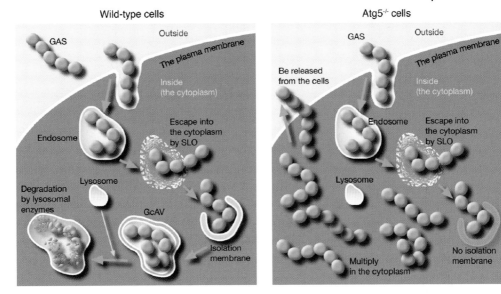

Fig. 7.5 Model for the fate of GAS in host cells with ("Wild-type cells") or without ("Atg5$^{-/-}$ cells") autophagic activity.

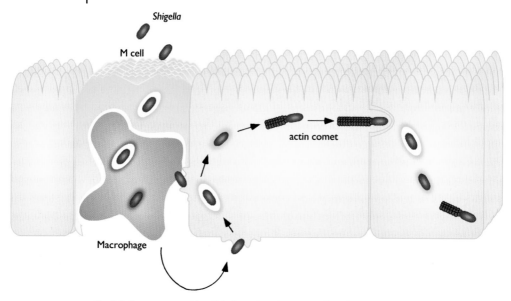

Fig. 8.1 Strategy unsed by *Shigella* to invade intestinal epithelial cells. Simplified illustration of infection of colonic epithelial cells by *Shigella*. *Shigella* are able to multiply in the cytosol of host cells and move into neighboring cells.

A

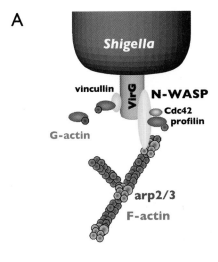

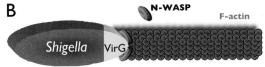

B

Fig. 8.2 The VirG–N-WASP–Arp2/3 complex formed at one pole of the bacterium mediates local actin nucleation and elongation.

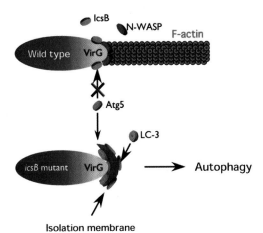

Fig. 8.4 Proposed model for the camouflage against autophagic recognition during *Shigella* infection. In the presence of IcsB (wild type), IcsB binds to VirG and competitively inhibits the binding of Atg5 to VirG. In the absence of IcsB (*icsB* mutant), Atg5 can bind to the VirG accumulated at one pole of the bacterium, and intracellular Shigella are recognized by autophagy

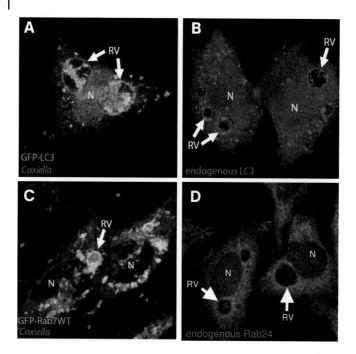

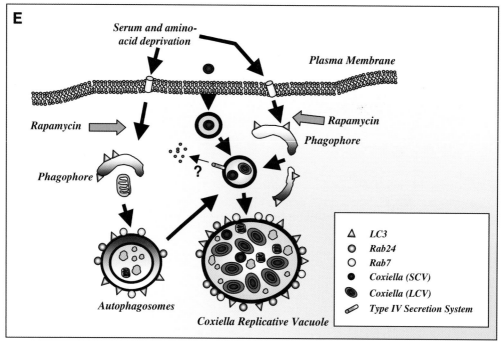

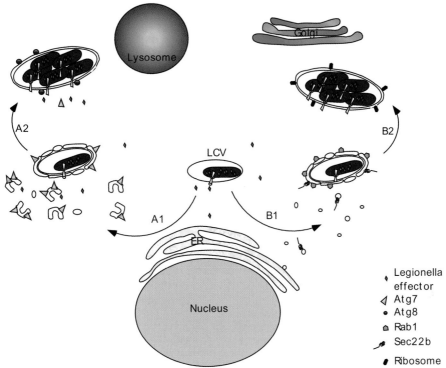

Legionella
effector
◁ Atg7
• Atg8
◮ Rab1
🖋 Sec22b
🍠 Ribosome

Fig. 11.1 Model depicting two alternative pathways for the intracellular trafficking of *Legionella*. (A1) *Legionella* induces autophagy by injecting effector proteins into the host cell cytoplasm using the type IV Dot/Icm secretion apparatus. Autophagic membranes labeled with Atg7 and Atg8 are recruited to the LCV, remodeling the limiting membrane into an autophagosome-like compartment where *Legionella* replication takes place. (A2) After a long delay of 16–24 h after infection, autophagosomes containing *Legionella* fuse with lysosomes. (B1) *Legionella* inject effec-

tor proteins into the host cell cytoplasm using the type IV Dot/Icm secretion apparatus to inhibit transport to lysosomes and direct the recruitment of ER-derived vesicles to the LCV limiting membrane. Remodeling of the LCV involves the recruitment of Rab1- and Sec22b-containing ER-derived vesicles to the LCV. (B2) Remodeling of the LCV membrane to resemble the ER, including thinning of the membrane and attachment of ribosomes, facilitates bacterial growth and replication.

Fig. 10.2 *C. burnetii* RVs interact with the autophagic pathway. (A) Stably transfected CHO cells overexpressing GFP–LC3 (green) were infected with *C. burnetii* phase II for 48 h. Cells were fixed and the *Coxiella* (red) were detected by indirect immunofluorescence with a specific antibody. (B) CHO cells were infected as indicated in (A), fixed and subjected to indirect immunofluorescence to detect endogenous LC3 (red) using a specific antibody. (C) Stably transfected CHO cells

overexpressing GFP–Rab7 (green) were infected with *Coxiella* and processed as indicated in (A). (D) HeLa cells were infected with *C. burnetii* phase II for 72 h, fixed and subjected to indirect immunofluorescence to detect endogenous Rab24 (red) using a mouse polyclonal antibody against Rab24. Confocal images are depicted. (E) A model showing the interaction between the *Coxiella*-containing vacuole and the autophagic pathway.

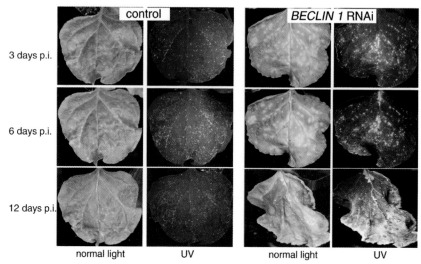

Fig. 13.5 Beclin 1 limits tobacco mosaic virus replication and limits the spread of tobacco mosaic virus-induced programmed cell death in tobacco plants. UV images show sites of replication of a GFP-expressing TMV and normal light images show areas of cell death (yellow) in nonsilenced (control) and *BECLIN 1*-silenced (*BECLIN 1* RNAi) plants. p.i. = post-infection. Adapted with permission from Ref. [24].

Part I
Introduction to Autophagy

Autophagy in Immunity and Infection. Edited by Vojo Deretic
Copyright © 2006 WILEY-VCH Verlag GmbH & Co. KGaA, Weinheim
ISBN: 3-527-31450-4

1
Overview of Autophagy

Julie E. Legakis and Daniel J. Klionsky

1.1
Overview of Autophagy

Just as cells must manufacture necessary components for proper function, so must they break down damaged or unnecessary organelles and other cellular constituents. In order to maintain this balance, the cells employ two primary degradative pathways – the proteasome, which is responsible for the breakdown of most short-lived proteins, and autophagy, a process induced by nutrient limitation and cellular stress, which governs the degradation of the majority of long-lived proteins, protein aggregates and whole organelles. It enables cells to survive stress from the external environment, such as nutrient deprivation, as well as internal stresses like accumulation of damaged organelles and pathogen invasion. Autophagy is induced by starvation in all eukaryotic systems examined, including several species of fungi, plants, slime mold, nematodes, fruit flies, mice, rats and humans [1]. By degrading superfluous intracellular components and reusing the breakdown products, these organisms are able to survive periods of scarce nutrients [1, 2]. Along these lines, autophagy aids in maintenance of homeostasis in cellular differentiation, tissue remodeling, growth control [3, 4] and a type of programmed cell death separate from apoptosis [5–7]. There exist multiple types of autophagy, which differ mainly in the site of cargo sequestration and in the type of cargo. These include micro- and macroautophagy, chaperone-mediated autophagy, micro- and macropexophagy, piecemeal microautophagy of the nucleus, and the cytoplasm-to-vacuole targeting (Cvt) pathway (Fig. 1.1) (reviewed in Ref. [8]). This chapter will focus on the process known as macroautophagy, which will be referred to as autophagy from this point on.

Autophagy is induced during certain developmental states, in response to various hormones, under conditions of nutrient deprivation or by other types of stress. This process involves the sequestration of bulk cytoplasm within a cytosolic double-membrane vesicle termed the autophagosome, which ultimately fuses with the lysosome (or the vacuole in yeast). Fusion results in the release

Autophagy in Immunity and Infection. Edited by Vojo Deretic
Copyright © 2006 WILEY-VCH Verlag GmbH & Co. KGaA, Weinheim
ISBN: 3-527-31450-4

of the inner vesicle, now termed an autophagic body, into the lysosome lumen. Within the lysosome the engulfed material is degraded and the products are recycled. Autophagy has been implicated in a number of human diseases and conditions, including cancer, neurodegenerative disorders, certain myopathies, aging and defense against pathogens. The potential ability to control autophagy for therapeutic intervention will require a better understanding of the mechanistic details of this degradative process.

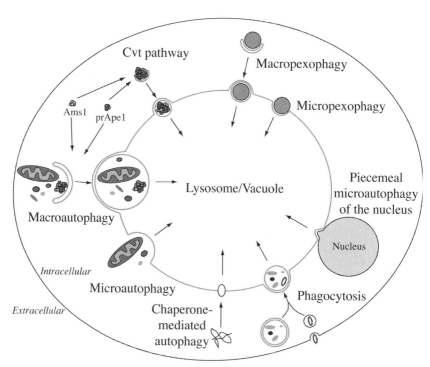

Fig. 1.1 Schematic representation of various transport routes to the lysosome/vacuole. There exist a number of pathways by which substrates are delivered to the lysosome/vacuole. Some of the sequestration events occur at the organelle membrane, these are denoted by the prefix "micro". In other cases, the enclosure of the substrate occurs spatially away from the lysosome/vacuole membrane. These pathways begin with the prefix "macro". Macro- and microautophagy are nonspecific degradation pathways, which include a variety of cargoes, depending on the organism and the particular stress conditions or stage of development. Selective degradation of peroxisomes, small parts of the nucleus or foreign pathogens occurs via macropexophagy, micropexophagy, piecemeal microautophagy of the nucleus or phagocytosis, respectively. Chaperone-mediated autophagy is a receptor-driven degradative pathway that is a secondary response to starvation conditions. The biosynthetic Cvt pathway is a method of delivery for at least two vacuolar hydrolases. (This figure also appears with the color plates).

1.2
The Discovery of Autophagy

The first description of autophagy was published almost 50 years ago. For nearly four decades, studies of the mammalian lysosome were primarily pharmacological, biochemical and morphological in nature. Nonetheless, many of the questions raised, and conclusions drawn, from those investigations are still valid today. In more recent years, the discovery of autophagy in yeast, allowing the application of genetic and molecular genetic techniques, led to a greater understanding of the machinery required for this essential cellular process. In particular, the systematic isolation and characterization of autophagy-related (*ATG*) genes in the yeasts *Saccharomyces cerevisiae*, *Pichia pastoris* and *Hansenula polymorpha* has allowed identification of 27 gene products that appear to be specific to autophagy (Tab. 1.1) [9]. Accordingly, the proposed functions of lysosomes and the accepted cellular roles of autophagy have evolved as a more detailed characterization of this degradative pathway has been achieved. This chapter will serve to highlight the progression of our knowledge concerning autophagy, give a general introduction to the process and set the stage for a discussion of its role in cellular immunity.

Lysosomes were first identified in rat liver, recognizable in electron microscopy images by their intense acid phosphatase staining [12]. It was soon demonstrated that these "particles" harbored additional hydrolases. So-called "dense bodies" were identified during attempts to purify lysosomes; from the initial studies, it was unclear whether these structures were distinct from lysosomes, as they shared many of the same properties [13]. Subsequently, these intracellular structures were identified as compartments similar to, but distinct from, lysosomes and they were named autophagic vacuoles (AVs; also referred to as autophagosomes, particularly in yeast). One of the first clues as to the degradative capacity of the AVs was demonstrated in newborn mouse kidney cells, where it was shown that some vacuole-like structures contained dense, amorphous material and even whole organelles, including mitochondria [14]. Similar investigations continued, attempting to determine lysosomal function and the origin of the AVs through electron microscopy.

Two aspects of autophagy which have been intensely studied since the original studies mentioned above are the membrane source for the nascent AV and the induction of autophagy. Despite the focus of research on these facets of the process, they are still not completely known or understood. Early studies provided evidence supporting both the Golgi and the endoplasmic reticulum (ER), as well as an area of the cell termed GERL (Golgi endoplasmic reticulum lysosomes), as the source of the AV membrane (reviewed in Ref. [15]), although none of this evidence was conclusive. Later investigations similarly were unable to reach a consensus, although many of the conclusions implicated the ER [16–18]. The identity of the donor membrane is still not known with certainty, there being evidence implicating the Golgi, the plasma membrane, as well as the ER as the source for the AV (reviewed in Ref. [19]).

Table 1.1 Atg proteins, their orthologs and putative roles in autophagy-related processes

Atg protein	Orthologs in other species	Putative function or component of	Step involved in
1	Sp, Nc, At, Dd, Ce, Dm, Mm, Hs	protein kinase	induction, retrieval
2	Sp, Pp, Ce, Dm, Hs	Atg9 cycling	retrieval
3	Sp, Nc, At, Ce, Dm, Mm, Hs	Atg8 conjugation	expansion
4	Sp, At, Ce, Dm, Mm, Hs	Atg8 conjugation	expansion
5	Sp, Dd, At, Ce, Dm, Mm, Hs	Atg12 conjugation	expansion
6	Sp, Nc, Dd, At, Ce, Dm, Mm, Hs	PI3K complex	nucleation
7	Sp, Pp, Nc, At, Dd, Ce, Dm, Mm, Hs	Atg8 and Atg12 conjugation	expansion
8	Sp, Nc, Dd, At, Ce, Dm, Mm, Hs	Atg8 conjugation	expansion
9	Sp, Nc, At, Ce, Dm, Hs	membrane delivery	expansion
10	At, Ce, Mm, Hs	Atg12 conjugation	expansion
11	Sp, Pp	cargo specificity	cargo selection
12	Sp, Nc, Dd, At, Ce, Dm, Mm, Hs	Atg12 conjugation	expansion
13	Sp, Nc	regulates Atg1 activity	induction, retrieval
14		PI3K complex	nucleation
15	Sp, Nc	lipase	vesicle breakdown
16	At, Ce, Dm, Mm, Hs	Atg12 conjugation	expansion
17	Sp, Nc	Atg1–Atg13 complex	induction, formation
18	Sp, Nc, At, Ce, Dm, Mm, Hs	Atg9 cycling	retrieval
19		Cvt receptor	cargo selection
20		Atg1–Atg13 complex	induction
21		PI3P binding	
22	Sp, Nc	transmembrane protein	
23		cycling protein	formation, expansion
24		Atg1–Atg13 complex	induction
25[a]	Hp	coiled-coil protein	regulation
26	Pp	glucosyltransferase	
27		PI3P binding	

These proteins were first identified in the yeast *S. cerevisiae*. The species with known orthologs are abbreviated as follows: *At, Arabidopsis thaliana*; *Ce, Caenorhabditis elegans*; *Dd, Dictyostelium discoideum*; *Dm, Drosophila melanogaster*; *Hp, Hansenula polymorpha*; *Hs, Homo sapiens*; *Mm, Mus muscularis*; *Nc, Neurospora crassa*; *Pp, Pichia pastoris*; *Sp, Schizosaccharomyces pombe*. The orthologs were compiled from the references listed below and from Homologene (http://www.ncbi.nlm.nih.gov/entrez/query.fcgi?DB=homologene). The information presented in this table is a compilation of information from Refs. [1, 9–11], as well as those cited throughout the text.

a) This protein has not been identified in *S. cerevisiae*, only in the species of yeast indicated.

Despite their inherent limitations, the morphological studies provided important information on the basic membrane dynamics involved in autophagy. For example, these studies suggested that AVs must acquire their resident enzymes through fusion with mature lysosomes [20]. These and more recent analyses have led to a model in which AVs develop into mature degradative autophagolysosomes in a series of discrete steps: (1) following induction, an initial isolation membrane or phagophore forms in a nucleation step; (2) this membrane expands into an AV; (3) the AV fuses with an endosome to form an intermediate structure known as an amphisome; (4) the amphisome acidifies; (5) fusion with a lysosome allows the AV/amphisome to acquire hydrolases [18, 21, 22].

Another mechanistic aspect of autophagy that has been the focus of much research concerns regulation and, in particular, induction. Initially it was not known whether autophagy was a random process, indiscriminately enwrapping and degrading cytoplasmic components, or whether it was a more directed action, selecting substrates according to the cellular needs at that particular moment. At present, we know that autophagy occurs at a basal level, but that in many cell types it is also an inducible process. In addition, although autophagy is generally considered to be nonspecific, there are various examples of specific types of autophagy. As the control of autophagic induction is important for defense against extracellular pathogens [22, 23], it is addressed in more detail later in this chapter.

Many of the questions that were raised shortly after the discovery of autophagy are still relevant today. The source of the sequestering membrane is still not known and it is possible that the forming AV may derive its membrane from multiple sources within the cell. Some of the molecular components that induce autophagy are now known, but the precise manner in which they act to bring about autophagic degradation is still unclear. Finally, the steps involved in the maturation of the newly formed AV are partially understood, but there are still aspects of this process that need to be clarified. As mentioned previously, over 20 genes have been identified that are involved in some form of autophagy, but the functions of the gene products are still largely unknown; however, based on recent genetic, molecular genetic and biochemical studies, along with new morphological analyses, a model describing a series of discrete steps and the components involved in these steps has been postulated.

1.3
Mechanistic Aspects of Autophagy

Autophagy is an evolutionarily conserved process in which the basic components are similar from unicellular (i.e. yeast) to multicellular eukaryotes. Very few of the autophagy proteins have motifs that provide insight into their function. Accordingly, the role of most of the Atg proteins is unknown; however, their interacting partners and order of action have been determined through various studies. Although autophagy is a dynamic process, the pathway has been delineated into

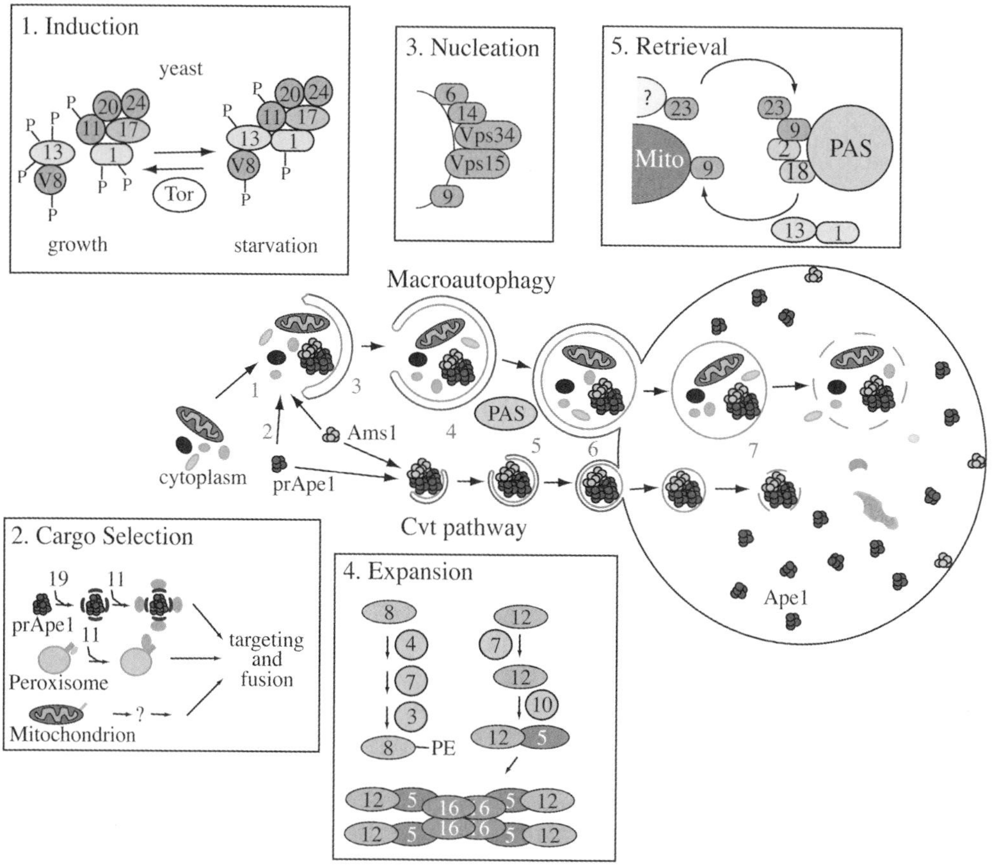

1. Induction

yeast

P 20 24
P P 11 17
P 13 1
P V8 P P
Tor
P
growth

P 20 24
P 11 17
13 1
V8
P
starvation

3. Nucleation

6
14
Vps34
Vps15
9

5. Retrieval

? 23
Mito 9
23
9
2
18
PAS
13 1

Macroautophagy

1
2
3
Ams1
PAS
4
5
6
7

cytoplasm
prApe1

Ape1

Cvt pathway

2. Cargo Selection

19 11
prApe1 11
Peroxisome
Mitochondrion → ? →
targeting and fusion

4. Expansion

8
4
7
3
8—PE

12
7
12
10
12 5

12 5 16 6 5 12
12 5 16 6 5 12

Fig. 1.2 Autophagy and the Cvt pathway. Autophagy and the Cvt pathway can be depicted as a series of separate steps. The roles of Atg and other proteins, shown to participate in different parts of the pathway, are depicted. The proteins classified by only a number are the corresponding Atg gene product. Otherwise, the protein name is specified, except for Vac8, which is indicated as "V8". "P" denotes phosphorylation of the indicated protein. (1) *Induction*. TOR kinase becomes inactivated upon nutrient limitation, eliciting a series of events, which result in the induction of autophagy. These include partial dephosphorylation of Atg13, which alters its association with Atg1. Atg1 is thought to play a key role in the switch between growth and starvation. Autophagy-specific proteins are shown in light gray, whereas Cvt-specific proteins are depicted in dark gray. (2) *Cargo selection and packaging*. Examples of specific autophagy include the Cvt pathway, pexophagy and possibly mitophagy. During growth, the Cvt pathway is active. The cargo, prApe1, is synthesized as an inactive precursor and rapidly oligomerizes. Atg19, the cargo receptor, binds to the oligomer, followed by Atg11 binding to the complex. Upon induction of pexophagy, the peroxin, Pex3, is degraded, thus exposing the docking protein, Pex14. Although it is not proven, Atg11 is proposed to bind to the newly exposed Pex14. The mechanism of mitophagy is unknown. Once these binding events occur, the cargo are enwrapped by a double-membrane vesicle and delivered to the lysosome/vacuole. (3) *Vesicle nucleation*. Membrane is acquired from an unknown location and the cargo associates with the forming vesicle. Membrane formation re-

several static steps for the convenience of description: (1) induction, (2) cargo selection and packaging, (3) vesicle nucleation, (4) vesicle expansion and completion, (5) retrieval, (6) targeting, docking and fusion of the vesicle with the lysosome/vacuole, and (7) breakdown of the vesicle and its contents (Fig. 1.2) [1, 25].

1.3.1
Induction

During vegetative growth, autophagy operates at a basal level both in yeast and mammalian cells. In addition, during growth in yeast, a second, more specific pathway operates, the Cvt pathway, which mediates delivery of the resident vacuolar hydrolase aminopeptidase I (Ape1) [26]. Upon a change in nutrient status, or other stress conditions, autophagy is induced. Therefore there must be some intracellular stimulus signaling the need to degrade intracellular components. As stated earlier, the mechanism of induction is not precisely understood, but some of the molecules involved are known. The best characterized regulatory component in yeast is the protein kinase target of rapamycin (Tor) [27]. Tor either directly or indirectly controls a putative protein complex that is sometimes called "the switching complex". This complex includes Atg1, a serine/threonine protein kinase, Atg13, a protein that modulates Atg1 kinase activity, and pathway-specific proteins including the autophagy-specific protein Atg17, and Cvt-specific factors Atg11, Atg20, Atg24 and Vac8 [28, 29]. Although the associations among these proteins have been demonstrated as bimolecular interactions, it is not known whether all of these proteins are ever present in a single complex.

---

quires the PI3K complex I; the components of this complex are shown in Step 3. The PI3-phosphate (PI3P) generated by this complex recruits a number of Atg proteins to the PAS, including Atg18, Atg20, and Atg21, Atg24 and Atg27 [24]. (4) *Vesicle expansion and completion.* There are two sets of Atg proteins, which participate in a series of ubiquitin-like (Ubl) conjugation reactions. These generate Atg12–Atg5–Atg16 and Atg8–PE (see text for details). The functions of these proteins are not known but they are needed for expansion and completion of the sequestering vesicle. (5) *Retrieval.* As most of the Atg proteins are not included in the completed vesicle, there must be a mechanism to release and return these components back to their original site. Atg9 and Atg23 have been shown to be cycling proteins, moving between the PAS and other punctate structures. Atg9 has been shown to cycle between he mitochondria and the PAS. The non-PAS localizations of Atg23 are as yet unidentified. These two proteins may aid in the recovery of Atg components, allowing them to be reused for another round of delivery. (6) *Targeting, docking and fusion of the vesicle with the lysosome/vacuole.* The docking and fusion of the completed vesicle requires a number of components (see text for details). The fusion event results in a single-membrane vesicle within the lumen of the lysosome/vacuole. (7) *Breakdown of the vesicle and its contents.* Once inside the lysosome/vacuole, the autophagic or Cvt body must be degraded in order for the cargo to be released. The lipase responsible for vesicle lysis is thought to be Atg15. Upon release into the lumen, the cargoes of pexophagy and bulk autophagy are broken down for re-use in the cell, while the cargoes of the Cvt pathway carry out their function as hydrolases. (This figure also appears with the color plates).

Nutrient limitation results in inactivation of Tor and induction of autophagy, whereas during vegetative growth conditions, Tor is active. Tor may mediate the activity of autophagy directly or indirectly: Tor activity alters the phosphorylation state of Atg13, thereby changing its binding affinity for Atg1; Tor also controls global transcription and translation through various downstream effectors [29–31]. Regulation through additional factors in yeast is not well understood. For example, protein kinase A is another negative regulatory component, but it is not known whether this protein acts downstream of TOR or in parallel [32]. In higher eukaryotes, mammalian (m) TOR is controlled through a phosphatidylinositol-3-kinase (PI3K) pathway that includes phosphoinositide-dependent kinase 1 (PDK1) and Akt/protein kinase B [33].

1.3.2
Cargo Selection and Packaging

Just as a signal must exist to dictate which of the various types of autophagy are functioning at a given moment, there also must be components that confer specificity on the cargo to be sequestered. The biosynthetic Cvt pathway involves the specific sequestration of the resident vacuolar hydrolases Ape1 and α-mannosidase (Ams1), and their subsequent delivery to the vacuole [26]. This is a receptor-mediated route of transport, although the primary cargo, precursor Ape1 (prApe1), is not concentrated via binding to Atg19, the Cvt receptor [34]; the ability of prApe1 to assemble into a large oligomeric complex appears to be an inherent property of the precursor protein. None of the other Atg proteins are needed for the interaction between Atg19 and the prApe1 propeptide [26]; however, in the absence of Atg11 the prApe1–Atg19 complex (termed the Cvt complex) is not localized at the pre-autophagosomal structure (PAS), the presumed site of Cvt vesicle and autophagosome formation. Therefore, Atg11 is thought to act in part as a tethering factor. Following delivery to the PAS, Atg19 binds Atg8 conjugated to phosphatidylethanolamine (Atg8–PE) that is present on the PAS membrane. Atg11 is not found in the completed Cvt vesicle, so it is thought to leave the Cvt complex at this time [35]. The interaction between Atg19 and Atg8–PE may trigger completion of the sequestering vesicle.

Upon fusion with the vacuole, the inner vesicle is released into the lumen and is now termed a Cvt body. Precursor Ape1 is activated by removal of the propeptide [36]. Although this pathway has only been reported in the yeast *S. cerevisiae*, it provides an ideal model system to analyze specific autophagy-related pathways, variations of which are likely to operate in all cell types. For example, the degradation of peroxisomes (pexophagy) is another form of specific autophagy, which has been detected in yeast as well as in plants and mammalian cells [37]. When yeasts are grown on carbon sources that require peroxisome function, this organelle proliferates. Shifting to a preferred carbon source results in rapid elimination of the now superfluous organelle. This is a highly specific process – the sequestering vesicles contain solely peroxisomes and it is presumed that this is also receptor driven, although a receptor similar to Atg19 has not been identified [38]. The peroxisomal tag for pexophagy in yeast appears to be Pex14 [39]. Other examples of selective autophagy involve exclusive packaging and delivery of ER and mitochondria, as well as certain cytosolic proteins and even the nucleus [40–43]. As investigations continue to reveal these different forms of selective autophagy, they serve to highlight the point that autophagy is not only a random process, but can also be highly discriminatory, capable of degrading only the specific components necessary at a given time.

1.3.3
Vesicle Nucleation

The putative site for vesicle formation is the PAS [44, 45]. This is the structure believed to be the organizing center for the assembly and organization of the autophagic machinery. Very little is understood about this process, but it seems that autophagic vesicles may form *de novo*, meaning that they are not generated in one step by segregation of membrane from a pre-existing organelle. Rather, the sequestering vesicles appear to begin at some nucleation site and then appropriate additional intracellular membrane to form a cup-shaped structure around the cargo. The intermediate structure is termed a phagophore, or isolation membrane. Formation of the double-membrane cup requires PI3K activity, which is mediated by a complex containing Atg6/Vps30, Atg14 (in yeast), Vps15 and the PI3K Vps34 [46, 47]. Atg5 is one of the first Atg proteins that can be visualized on this structure, but whether it is the first protein to arrive at the PAS is unknown [47].

1.3.4
Vesicle Expansion and Completion

In order to completely enclose the cargo, the membrane must undergo an expansion step. Involved in this process are two groups of Atg proteins that include some ubiquitin-like (Ubl) proteins, which participate in conjugation reactions [48]. One set of proteins is involved in the covalent attachment of the Ubl Atg12 to Atg5. Atg5 binds Atg16 noncovalently and the subsequent tetramerization of Atg16 forms a large multimeric complex. Atg8 is another Ubl important for membrane expansion. This protein is proteolytically cleaved at its C-terminus to expose a glycine residue and is then used as a modifier of PE [49]. In mutants lacking Atg8, autophagosomes can still be generated but they are of reduced size [50]. These and the other proteins depicted in Fig. 1.2 are thought to be delivered to the forming autophagosome, and possibly control the size and curvature of the nascent vesicle; however, the exact function of these proteins is not known.

1.3.5
Retrieval

Protein targeting pathways generally utilize components that are reused, enabling them to be used for multiple rounds of substrate delivery. Atg8 and Atg19 are the only components that are known to be included in the completed autophagosome, suggesting that all of the proteins involved in the previously described steps must dissociate from the forming vesicle before completion. This is particularly problematic for integral membrane proteins such as Atg9 [51], which cannot simply dissociate from the vesicle. Retrieval of Atg proteins has been recently demonstrated for two factors, Atg9 and Atg23, which have been shown to cycle between punctate cytosolic structures and the PAS [52, 53]. Interestingly, the Atg9-containing structures have been identified as corresponding at least in part to mitochondria [54]; it remains to be determined whether this is also the case for Atg23. Atg1 and Atg13 are required for cycling of both of these proteins (although higher Atg1 kinase activity is needed for Atg23), whereas Atg2 and Atg18 are required only for cycling of Atg9 [52]. The function of Atg9 and Atg23 cycling is unknown – Atg9- and Atg23-containing structures may contribute membrane to the expanding autophagosome or these proteins may serve to mediate delivery of other necessary components to the PAS.

1.3.6
Targeting, Docking and Fusion of the Vesicle with the Lysosome/Vacuole

There must be some mechanism for preventing fusion of the incomplete Cvt vesicle or autophagosome with the lysosome or vacuole. This may be achieved by the presence of coat proteins that sterically interfere with the interaction of soluble *N*-ethylmalemide-sensitive fusion protein (NSF) attachment receptor (SNARE) proteins; however, the presence of coat proteins has not been clearly demonstrated. Once the vesicle is complete, it fuses with the degradative organelle. As noted previously, in mammalian cells the initial fusion step may involve an endosome. The proteins required for fusion appear to be common to those involved in other lysosomal/vacuolar fusion events including SNARE proteins, NSF, soluble NSF attachment protein (SNAP) and GDP dissociation inhibitor (GDI) homologs, a Rab protein, and the class C Vps/HOPS complex [55]. After fusion of the double-membrane vesicle with the lysosome/vacuole, the inner vesicle is released into the organelle's lumen.

1.3.7
Breakdown of the Vesicle and its Cargo

The outer membrane of the sequestering vesicle becomes continuous with the limiting membrane of the lysosome/vacuole. This membrane may be removed through a microautophagic process. The membrane of the Cvt or autophagic body must be broken down within the lumen to release the contents. The lipase thought to be responsible for breakdown of the membrane is Atg15 [56]. In the Cvt pathway, lysis of the Cvt body allows release and subsequent activation of prApe1. For pexophagy and nonspecific autophagy, release of the vesicle cargo results in its subsequent degradation by resident vacuolar hydrolases. These macromolecular components are then made available for reuse in the cell. These processes are only depicted in general in Fig. 1.2; a more detailed summary can be found in other reviews [1, 25, 55].

The details of the steps outlined above have been best characterized in yeast, but many of the components also have homologs in higher eukaryotes cells (Tab. 1.1). In addition, the development of novel techniques is allowing investigators to determine if the mammalian components function similar to their yeast counterparts.

1.4
Autophagy and Immunity

One of the most important functions of autophagy appears to be its role in the host defense against cellular pathogens. In general, bacterial pathogens enter the cell via an endocytosis-like pathway, enclosed within a vesicle called a phagosome that ultimately fuses with and is degraded by the lysosome [57]. This was demonstrated to be the method cells use to avoid infection by *Streptococcus pyogenes* [58]. In contrast, it has recently been shown that certain bacteria undermine the autophagic machinery to promote their replication and survival [22, 59, 60]. This evasion is accomplished in different ways by different bacteria. In the case of *Listeria monocytogenes, Shigella* and certain other bacteria, the microbes induce lysis of the phagosome, causing their release into the cytoplasm and enabling them to replicate in that environment [61, 62]. Other invasive bacteria including *Mycobacterium tuberculosis* modify the phagosome in which they are contained, to prevent fusion with the lysosome [63]. Still other pathogens such as *Legionella pneumophila* induce the autophagic pathway and replicate within autophagosome-like compartments [59, 64]. In organisms such as *L. pneumophila* and *Coxiella burnetii*, induction of autophagy enhances the replication and survival of the invading bacteria [65].

The role of autophagy in defense against viral infection is also dual in nature, both protecting against infection and being exploited to promote viral invasion. For example, induction of autophagy increases MHC class II antigen presentation and contributes to the immune response by aiding in antigen processing of certain viruses [66, 67]. Like bacteria, some viruses can also use the autophagic machinery to their advantage. For example, coronavirus replication and viability is increased by autophagy [68], and stimulation of autophagy increased the yield of poliovirus [69].

With these recent discoveries, it is now clear that autophagy can aid in defense against pathogens, but the microbes can also employ this pathway to promote their viability. This is similar to the role of autophagy in cancer and other diseases – depending on the progression of the disease, autophagy may be a protective mechanism, eliminating damaged organelles or even damaged cells, or it may have harmful effects by causing cell death or in the case of cancer by promoting the survival of tumor cells under limiting nutrient conditions [70]. These varying effects of autophagy only serve to emphasize the importance of being able to control the activity of this degradative pathway if it is ever to be used therapeutically. In addition, as will become evident in the following chapters, the various analyses of autophagy and its role in immunity have led to a series of intriguing new questions. For example, is the mechanism of induction of autophagy the same during pathogen infection as it is during nutrient deprivation? What signals allow bacterial pathogens to induce autophagy and how do pathogens prevent maturation or fusion of autophagosomes? Is the sequestration of these pathogens specific, enclosing only the bacteria or virus, or is it a bulk degradation process, which includes other cytoplasmic components? Continued investigations of autophagy will aid in our understanding of this important process, and hopefully lead to treatments for the human conditions and diseases in which autophagy is implicated.

References

1 Levine, B., D. J. Klionsky. **2004**. Development by self-digestion: molecular mechanisms and biological functions of autophagy. *Dev Cell 6*, 463–477.

2 Yoshimori, T. **2004**. Autophagy: a regulated bulk degradation process inside cells. *Biochem Biophys Res Commun 313*, 453–458.

3 Tanida, I., T. Ueno, E. Kominami. **2004**. LC3 conjugation system in mammalian autophagy. *Int J Biochem Cell Biol 36*, 2503–2518.

4 Meijer, A. J., P. Codogno. **2004**. Regulation and role of autophagy in mammalian cells. *Int J Biochem Cell Biol 36*, 2445–2462.

5 Edinger, A. L., C. B. Thompson. **2004**. Death by design: apoptosis, necrosis and autophagy. *Curr Opin Cell Biol 16*, 663–639.

6 Lockshin, R. A., Z. Zakeri. **2004**. Apoptosis, autophagy, and more. *Int J Biochem Cell Biol 36*, 2405–2419.

7 Debnath, J., E. H. Baehrecke, G. Kroemer. **2005**. Does autophagy contribute to cell death? *Autophagy 1*, 66–74.

8 D. J. Klionsky (Ed.). **2004**. *Autophagy*. Georgetown, TX: Landes Bioscience.

9 Klionsky, D. J., J. M. Cregg, W. A. Dunn, Jr., S. D. Emr, Y. Sakai, I. V. Sandoval, A. Sibirny, S. Subramani, M. Thumm, M. Veenhuis, Y. Ohsumi. **2003**. A unified nomenclature for yeast autophagy-related genes. *Dev Cell 5*, 539–545.

10 Mizushima, N., Y. Ohsumi, T. Yoshimori. **2002**. Autophagosome formation in mammalian cells. *Cell Struct Funct 27*, 421–429.

11 Reggiori, F., D. J. Klionsky. **2002**. Autophagy in the eukaryotic cell. *Eukaryot Cell 1*, 11–21.

12 De Duve, C., B. C. Pressman, R. Gianetto, R. Wattiaux, F. Appelmans. **1955**. Tissue fractionation studies. 6. Intracellular distribution patterns of enzymes in rat-liver tissue. *Biochem J 60*, 604–617.

13 Novikoff, A. B., E. Podber, J. Ryan, E. Noe. **1953**. Biochemical heterogeneity of the cytoplasmic particles isolated from rat liver homogenate. *J Histochem Cytochem 1*, 27–46.

14 Clark, S. L. J. **1957**. Cellular differentiation in the kidneys of newborn mice studied with the electron microscope. *J Biophys Biochem Cytol 3*, 349–364.

15 De Duve, C., R. Wattiaux. **1966**. Functions of lysosomes. *Annu Rev Physiol 28*, 435–492.

16 Novikoff, A. B., W. Y. Shin. **1978**. Endoplasmic reticulum and autophagy in rat hepatocytes. *Proc Natl Acad Sci USA 75*, 5039–5042.

17 Marzella, L., J. Ahlberg, H. Glaumann. **1982**. Isolation of autophagic vacuoles from rat liver: morphological and biochemical characterization. *J Cell Biol 93*, 144–154.

18 Dunn, Jr., W. A. **1990**. Studies on the mechanisms of autophagy: formation of the autophagic vacuole. *J Cell Biol 110*, 1923–1933.

19 Fengsrud, M., M. L. Sneve, A. Overbye, P. O. Seglen. **2004**. Structural aspects of mammalian autophagy. In: *Autophagy*, D. J. Klionsky (Ed.). Georgetown, TX: Landes Bioscience, pp. 11–25.

20 Deter, R. L., C. De Duve. **1967**. Influence of glucagon, an inducer of cellular autophagy, on some physical properties of rat liver lysosomes. *J Cell Biol 33*, 437–449.

21 Dunn, Jr., W. A. **1990**. Studies on the mechanisms of autophagy: maturation of the autophagic vacuole. *J Cell Biol 110*, 1935–1945.

22 Levine, B. **2005**. Eating oneself and uninvited guests: autophagy-related pathways in cellular defense. *Cell 120*, 159–162.

23 Shintani, T., D. J. Klionsky. **2004**. Autophagy in health and disease: a double-edged sword. *Science 306*, 990–995.

24 Nice, D. C., T. K. Sato, P. E. Stromhaug, S. D. Emr, D. J. Klionsky. **2002**. Cooperative binding of the cytoplasm to vacuole targeting pathway proteins, Cvt13 and Cvt20, to phosphatidylinositol 3-phosphate at the pre-autophagosomal structure is required for selective autophagy. *J Biol Chem 277*, 3098–3207.

25 Klionsky, D. J. **2005**. The molecular machinery of autophagy: unanswered questions. *J Cell Sci 118*, 7–18.

26 Shintani, T., W.-P. Huang, P. E. Stromhaug, D. J. Klionsky. **2002**. Mechanism of cargo selection in the cytoplasm to vacuole targeting pathway. *Dev Cell 3*, 825–837.

27 Carrera, A. C. **2004**. TOR signaling in mammals. *J Cell Sci 117*, 4615–4616.

28 Scott, S. V., D. C. Nice, III., J. J. Nau, L. S. Weisman, Y. Kamada, I. Keizer-Gunnink, T. Funakoshi, M. Veenhuis, Y. Ohsumi, D. J. Klionsky. **2000**. Apg13p and Vac8p are part of a complex of phosphoproteins that are required for cytoplasm to vacuole targeting. *J Biol Chem 275*, 25840–25849.

29 Kamada, Y., T. Funakoshi, T. Shintani, K. Nagano, M. Ohsumi, Y. Ohsumi. **2000**. Tor-mediated induction of autophagy via an Apg1 protein kinase complex. *J Cell Biol 150*, 1507–1513.

30 Talloczy, Z., W. Jiang, H. W. Virgin, IV D. A. Leib, D. Scheuner, R. J. Kaufman, E.-L. Eskelinen, B. Levine. **2002**. Regulation of starvation- and virus-induced autophagy by the eIF2a kinase signaling pathway. *Proc Natl Acad Sci USA 99*, 190–195.

31 Abeliovich, H. **2004**. Regulation of autophagy by the target of rapamycin (Tor) proteins. In: *Autophagy*, D. J. Klionsky (Ed.). Georgetown, TX: Landes Bioscience, pp. 60–69.

32 Budovskaya, Y. V., J. S. Stephan, F. Reggiori, D. J. Klionsky, P. K. Herman. **2004**. Ras/cAMP-dependent protein kinase signaling pathway regulates an early step of the autophagy process in *Saccharomyces cerevisiae*. *J Biol Chem 279*, 20663–20671.

33 Codogno, P., A. J. Meijer. **2004**. Signaling pathways in mammalian autophagy. In: *Autophagy*, D. J. Klionsky (Ed.). Georgetown, TX: Landes Bioscience, pp. 26–47.

34 Scott, S. V., J. Guan, M. U. Hutchins, J. Kim, D. J. Klionsky. **2001**. Cvt19 is a receptor for the cytoplasm-to-vacuole targeting pathway. *Mol Cell 7*, 1131–1141.

35 Yorimitsu, T., Klionsky D. J. **2005**. Atg11 links cargo to the vesicle-forming machinery in the cytoplasm to vacuole targeting pathway. *Mol Biol Cell 16*, 1593–1605.

36 Klionsky, D. J., R. Cueva, D. S. Yaver. **1992**. Aminopeptidase I of *Saccharomyces cerevisiae* is localized to the vacuole independent of the secretory pathway. *J Cell Biol 119*, 287–299.

37 Habibzadegah-Tari, P., W. A. Dunn, Jr. **2004**. Glucose-induced pexophagy. In: *Autophagy*, D. J. Klionsky (Ed.). Georgetown, TX: Landes Bioscience, pp. 107–114.

38 Kiel, J. A. K. W., M. Veenhuis. **2004**. Selective degradation of peroxisomes in the methylotrophic yeast *Hanensula polymorpha*. In: *Autophagy*, D. J. Klionsky (Ed.). Georgetown, TX: Landes Bioscience, pp. 140–156.

39 Bellu, A. R., M. Komori, I. J. van der Klei, J. A. Kiel, M. Veenhuis. **2001**. Peroxisome biogenesis and selective degradation converge at Pex14p. *J Biol Chem 276*, 44570–44574.

40 Hamasaki, M., T. Noda, M. Baba, Y. Ohsumi. **2005**. Starvation triggers the delivery of the endoplasmic reticulum to the vacuole via autophagy in yeast. *Traffic 6*, 56–65.

41 Kissova, I., M. Deffieu, S. Manon, N. Camougrand. **2004**. Uth1p is involved in the autophagic degradation of mitochondria. *J Biol Chem 279*, 39068–39074.

42 Onodera, J., Y. Ohsumi. **2004**. Ald6p is a preferred target for autophagy in yeast, *Saccharomyces cerevisiae*. *J Biol Chem 279*, 16071–16076.

43 Roberts, P., S. Moshitch-Moshkovitz, E. Kvam, E. O'Toole, M. Winey, D. S. Goldfarb. **2003**. Piecemeal microautophagy of nucleus in *Saccharomyces cerevisiae*. *Mol Biol Cell 14*, 129–141.

44 Kim, J., W.-P. Huang, P. E. Stromhaug, D. J. Klionsky. **2002**. Convergence of multiple autophagy and cytoplasm to vacuole targeting components to a perivacuolar membrane compartment prior to *de novo* vesicle formation. *J Biol Chem 277*, 763–773.

45 Suzuki, K., T. Kirisako, Y. Kamada, N. Mizushima, T. Noda, Y. Ohsumi. **2001**. The pre-autophagosomal structure organized by concerted functions of *APG* genes is essential for autophagosome formation. *EMBO J 20*, 5971–5981.

46 Kihara, A., Y. Kabeya, Y. Ohsumi, T. Yoshimori. **2001**. Beclin–phosphatidylinositol 3-kinase complex functions at the *trans*-Golgi network. *EMBO Rep 2*, 330–335.

47 Mizushima, N., A. Yamamoto, M. Hatano, Y. Kobayashi, Y. Kabeya, K. Suzuki, T. Tokuhisa, Y. Ohsumi, T. Yoshimori. **2001**. Dissection of autophagosome formation using Apg5-deficient mouse embryonic stem cells. *J Cell Biol 152*, 657–668.

48 Ohsumi, Y. **2001**. Molecular dissection of autophagy: two ubiquitin-like systems. *Nat Rev Mol Cell Biol 2*, 211–216.

49 Huang, W.-P., S. V. Scott, J. Kim, D. J. Klionsky. **2000**. The itinerary of a vesicle component, Aut7p/Cvt5p, terminates in the yeast vacuole via the autophagy/Cvt pathways. *J Biol Chem 275*, 5845–5851.

50 Abeliovich, H., W. A. Dunn, Jr., J. Kim, D. J. Klionsky. **2000**. Dissection of autophagosome biogenesis into distinct nucleation and expansion steps. *J Cell Biol 151*, 1025–1034.

51 Noda, T., J. Kim, W.-P. Huang, M. Baba, C. Tokunaga, Y. Ohsumi, D. J. Klionsky. **2000**. Apg9p/Cvt7p is an integral membrane protein required for transport vesicle formation in the Cvt and autophagy pathways. *J Cell Biol 148*, 465–480.

52 Reggiori, F., K. A. Tucker, P. E. Stromhaug, D. J. Klionsky. **2004**. The Atg1–Atg13 complex regulates Atg9 and Atg23 retrieval transport from the pre-autophagosomal structure. *Dev Cell 6*, 79–90.

53 Tucker, K. A., F. Reggiori, W. A. Dunn, Jr., D. J. Klionsky. **2003**. Atg23 is essential for the cytoplasm to vacuole targeting pathway and efficient autophagy but not pexophagy. *J Biol Chem 278*, 48445–48452.

54 Reggiori, F., T. Shintani, U. Nair, D. J. Klionsky. **2005**. Atg9 cycles between mitochondria and the pre-autophagosomal structure in yeasts. autophagy *1*, 101–109.

55 Wang, C. W., D. J. Klionsky. **2003**. The molecular mechanism of autophagy. *Mol Med 9*, 65–76.

56 Teter, S. A., K. P. Eggerton, S. V. Scott, J. Kim, A. M. Fischer, D. J. Klionsky. **2001**. Degradation of lipid vesicles in the yeast vacuole requires function of Cvt17, a putative lipase. *J Biol Chem 276*, 2083–2087.

57 Kirkegaard, K., M. P. Taylor, W. T. Jackson. **2004**. Cellular autophagy: surrender, avoidance and subversion by microorganisms. *Nat Rev Microbiol 2*, 301–314.

58 Nakagawa, I., A. Amano, N. Mizushima, A. Yamamoto, H. Yamaguchi, T. Kamimoto, A. Nara, J. Funao, M. Nakata, K. Tsuda, S. Hamada, T. Yoshimori. **2004**. Autophagy defends cells against invading group A *Streptococcus*. *Science 306*, 1037–1040.

59 Dunn, Jr., W.A., B.R. Dorn, A. Progulske-Fox. **2004**. Trafficking of bacterial pathogens to autophagosomes. In: *Autophagy*, D. J. Klionsky (Ed.). Georgetown, TX: Landes Bioscience, pp. 233–240.

60 Gorvel, J.P., C. de Chastellier. **2005**. Bacteria spurned by self-absorbed cells. *Nat Med 11*, 18–19.

61 Rich, K.A., C. Burkett, P. Webster. **2003**. Cytoplasmic bacteria can be targets for autophagy. *Cell Microbiol 5*, 455–468.

62 Ogawa, M., T. Yoshimori, T. Suzuki, H. Sagara, N. Mizushima, C. Sasakawa. **2005**. Escape of intracellular *Shigella* from autophagy. *Science 307*, 727–731.

63 Gutierrez, M.G., S.S. Master, S.B. Singh, G.A. Taylor, M.I. Colombo, V. Deretic. **2004**. Autophagy is a defense mechanism inhibiting BCG and *Mycobacterium tuberculosis* survival in infected macrophages. *Cell 119*, 753–766.

64 Amer, A.O., M.S. Swanson. **2005**. Autophagy is an immediate macrophage response to *Legionella pneumophila*. *Cell Microbiol 7*, 765–778.

65 Gutierrez, M.G., C.L. Vazquez, D.B. Munafo, F.C. Zoppino, W. Beron, M. Rabinovitch, M.I. Colombo. **2005**. Autophagy induction favours the generation and maturation of the *Coxiella*-replicative vacuoles. *Cell Microbiol 7*, 981–993.

66 Dengjel, J., O. Schoor, R. Fischer, M. Reich, M. Kraus, M. Muller, K. Kreymborg, F. Altenberend, J. Brandenburg, H. Kalbacher, R. Brock, C. Driessen, H.G. Rammensee, S. Stevanovic. **2005**. Autophagy promotes MHC class II presentation of peptides from intracellular source proteins. *Proc Natl Acad Sci USA 102*, 7922–7927.

67 Paludan, C., D. Schmid, M. Landthaler, M. Vockerodt, D. Kube, T. Tuschl, C. Munz. **2005**. Endogenous MHC class II processing of a viral nuclear antigen after autophagy. *Science 307*, 593–596.

68 Prentice, E., W.G. Jerome, T. Yoshimori, N. Mizushima, M.R. Denison. **2004**. Coronavirus replication complex formation utilizes components of cellular autophagy. *J Biol Chem 279*, 10136–10141.

69 Jackson, W.T., T.H. Giddings, Jr., M.P. Taylor, S. Mulinyawe, M. Rabinovitch, R.R. Kopito, K. Kirkegaard. **2005**. Subversion of cellular autophagosomal machinery by RNA viruses. *PLoS Biol 3*, e156.

70 Furuya, N., X.H. Liang, B. Levine. **2004**. Autophagy and cancer. In: *Autophagy*, D. J. Klionsky (Ed.). Georgetown, TX: Landes Bioscience, pp. 241–255.

2
Cell Biology and Biochemistry of Autophagy

Edmond Y. W. Chan, Robert Köchl and Sharon A. Tooze

2.1
Introduction

Autophagy, or self-digestion, is widely recognized as an important process in normal and pathological cell physiology, for both cell survival and cell death (see Chapter 5). Macroautophagy, or what is commonly referred to now as simply autophagy, was first identified in mammalian cells during a morphological analysis of the kidney of newborn mice [1] and the term "cellular autophagy" was coined by DeDuve in 1963 (for a review of early literature, see Ref. [2]). The early morphological investigations were primarily focused on the relationship of autophagosomes to lysosomes, largely because autophagosomes stained for acid phosphatase, a popular histochemical stain, and it was recognized that the process was distinct from heterophagy, also known as phagocytosis. As early as 1962, the origin of the membranes of autophagosomes was the subject of speculation and research. Several hypotheses were discussed including the possibility that the membranes were entirely unique or formed *de novo* [3, 4], or were derived from existing smooth endoplasmic reticulum (ER) [5, 6] or endocytic vacuoles [2]. While the controversy remains, the current most popular view favors the ER as the primary source of membrane for mammalian autophagosomes (for a recent review, see Ref. [7]). On the other hand, recent studies in yeast show that autophagosomes originate from a unique compartment called the pre-autophagosomal structure (PAS) [8, 9].

Other unresolved issues concern the molecular components which are downstream of the initial autophagy induction signal and the signaling pathways which they impinge upon. Perhaps the best understood aspect of autophagy is the final maturation process into an autolysosome and lysosome, where the morphological results are informative and where the parallels with yeast (in which autophagosomes fuse with the yeast vacuole) are robust enough to provide a clear indication of the molecular details.

The focus of this chapter is a summary of our current understanding of the autophagic process in mammalian cells. In particular, our discussion focuses on

Autophagy in Immunity and Infection. Edited by Vojo Deretic
Copyright © 2006 WILEY-VCH Verlag GmbH & Co. KGaA, Weinheim
ISBN: 3-527-31450-4

the relationship between autophagy and the normal compartmentalization of a mammalian cell, in addition to the signaling pathways that regulate the initiation of the process. Finally, we summarize the wide range of current cell biological approaches currently used to study autophagy in mammalian cells.

2.2
Autophagic Pathway

2.2.1
Formation of Autophagosomes

Starvation or other inductive signals (see Section 2.3) give rise to almost flat, double-membrane structures or individual "saccules" (defined as irregularly shaped flattened vesicles) of membranes in the cytosol. During the early formation process these membranes increase in size and area, changing shape to form a cup-shaped structure (0.5–1.5 μm in diameter) that can envelop substantial portions of cytosol and subcellular organelles. These early membranes were originally called phagophores and, more recently, isolation membranes [7] and their origin is still unclear (see below). A major advance was the development and use of Green Fluorescent Protein (GFP)-tagged Atg5 (a gene first identified in yeast which is essential for autophagy) expressed in mouse embryonic stem cells. Here the isolation membrane was clearly characterized by electron microscopy using immunogold labeling for GFP as a cup-shaped structure which developed from a small crescent-shaped membrane [10]. Importantly, the same group generated Atg5-deficient embryonic stem cells, and showed that the Atg5–Atg12 complex is required for autophagy and for the elongation of the isolation membrane. How the isolation membrane identified in this study relates to the original description of the phagophore is not yet clear.

In theory, an increase in size and area of the isolation membrane can be achieved by either (a) expansion of the original membrane through fusion of the isolation membrane saccules together, (b) extension (tubulation) of existing compartments or (c) expansion by incorporation of additional lipid (Fig. 2.1 and see Ref. [11] for a related discussion). Expansion by the addition of lipid could occur by three different mechanisms: delivery of lipid monomers by phospholipid transfer proteins, delivery by vesicular transfer or through *de novo* lipid synthesis on the isolation membrane. The phagophore (isolation membrane) surrounds and sequesters cytosol and/or organelles, finally closing to form an autophagic vacuole (AV). It is not understood how the membrane bends to become a spherical structure. The sequestration of cytosolic enzymes has been shown to be nonselective [12] and, while there is no direct evidence to the contrary, sequestration of organelles is also assumed to be nonselective, although targeted removal of damaged organelles (such as mitochondria) through autophagy remains an attractive hypothesis.

The closure of the cup-shaped membrane to form an AV must occur through a bilayer–bilayer fusion, although this has never been directly demonstrated. Fu-

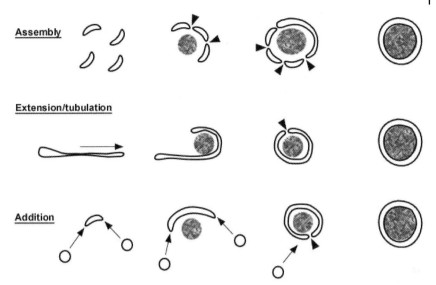

Fig. 2.1 Three possible models for the expansion of the isolation membrane or phagophore to form an AV. Expansion can occur by assembly (top panel) of individual isolation membranes through fusion of the ends (arrowheads), extension (tubulation) of existing membranes (middle panel), followed by fusion at the ends, or by addition of membrane (bottom panel), e.g. by fusion of vesicles with the ends of the isolation membranes (arrows) followed by closure of the membrane at the ends (arrowhead). Note that in the first two examples only homotypic fusion events occur, while in the last example (addition) both heterotypic fusion (between isolation membrane and incoming vesicles) and homotypic fusion, occurs.

sion of the rims of the cup-shaped structure to finally seal the AV is thought to occur at a single point, although again this also has not been experimentally demonstrated. The final closure entails fusing the curved edge of the cisternae such that the outer and inner membrane bilayers are now continuous, generating two vesicles. The content between the inner and outer vesicle is derived from the lumen of the original cisternae, the content of which is not known, and the lumen of the inner vesicle contains cytosolic components.

Formation of AVs can occur in the absence of ongoing protein synthesis [13], which supports the concept that the membrane is derived from pre-existing membranes. The pre-existing membrane could be a compartment such as the ER or Golgi. It could also be a unique structure such as the phagophore, although as Seglen points out, recycling of components would be necessary to maintain the unique structure [14]. Early morphological experiments using OsO_4 as a fixative showed that the phagophore stains particularly well with osmium [15], which can be explained by its high content of unsaturated fatty acids [16]. It has also been shown that the phagophore membrane contains no detectable cholesterol [17]. Surprisingly, freeze-fracture studies suggested the phagophores and autophagosomes were devoid of transmembrane proteins [18].

The most widely accepted view is that the phagophore and autophagosomes are derived from the smooth ER, a conclusion reinforced by immunoperoxidase cytochemistry with antisera against membrane proteins of the rough ER proteins [19]. This view was originally developed following a comprehensive histochemical and subcellular fractionation study performed in 1968 that demonstrated two markers of the ER, inosine diphosphate and glucose-6-phosphate in AVs [6], while no markers for Golgi membranes were detectable. However, acid phosphatase staining of *cis*-Golgi membranes [20] suggested that because the phagophore also stains for acid phosphatase [21], the latter may be derived from the Golgi. Additional evidence for a post-Golgi origin was obtained using lectin cytochemistry, which demonstrated that the isolation membrane contained complex-Golgi-derived carbohydrates [22]. However, data obtained from an analysis of purified phagophores and autophagosomes using immunoblotting techniques have demonstrated that neither markers for the ER nor Golgi were enriched in the purified membranes [23]. Further independent evidence suggests that an intact Golgi is not required for autophagosome formation, as treatment with Brefeldin A, a compound which inactivates ADP-ribosylation factor and leads to Golgi disassembly, does not impair AV formation [24]. Finally, the morphological studies using GFP–Atg5 mouse cells clearly demonstrated that the small crescent shaped structures identified by electron microscopy (see above) are not connected or contiguous with any other membrane structure [10].

Most recently, several informative mouse lines with manipulations in autophagy genes have been generated: transgenic mice expressing GFP–microtubule-associated protein 1 light chain 3 (LC3) (Atg8) [25], mice with heterozygous and homozygous disruption of beclin 1 (Atg6) [26], mice with a homozygous knockout of Atg5 [27], and mice with a conditional knockout for Atg7 [28] (see Chapter 3). As expected, in both the Atg5 and Atg7 knockout mice there are no autophagosomes detectable under conditions in which AV formation is normally seen, i.e. immediately after birth or after starvation. These mouse models will provide additional tools to allow a further understanding of the formation process.

In conclusion, while we have a firm morphological description of the autophagosomal structures, our lack of knowledge about the molecular machinery needed to form autophagosomes in mammalian cells remains a stumbling block towards understanding where and how the cell is able to produce AVs, and subsequently carry out regulated autophagy. While the genes required in yeast (see Chapter 1) have provided important tools and considerable increases in our knowledge, a great deal remains unknown.

2.2.2
Fusion of AV with Endocytic Pathways and Maturation

Newly formed AVs are called initial or immature autophagosomes (AVi), in part to reflect the fact that their content is not yet degraded. In the final stages of autophagy, the AVs, which for the most part correspond to a lysosomal compartment, are degradative AVs (AVd) (Figs. 2.2 and 2.3). The most important func-

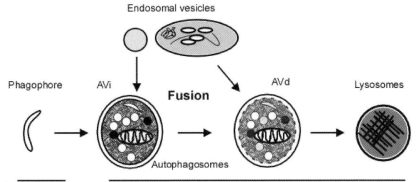

Endosomal vesicles

Phagophore   AVi   AVd   Lysosomes

Fusion

Autophagosomes

Sequestration   **Maturation**

Fig. 2.2 Maturation of an autophagosome. Autophagosomes sequester cytosol and close (see Fig. 2.1) to form an AVi. Endosomal vesicles and endosomes then fuse with AVis to generate AVds. The intermediate stage between an AVi and an AVd is called an amphisome (AVi/d). The final stages of the mature AVd are called autolysosomes and are virtually indistinguishable from lysosomes.

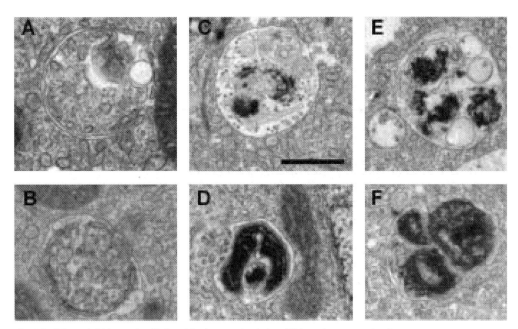

Fig. 2.3 AVis and AVds can be distinguished by electron microscopy. AVis (A and B) with a double membrane contain sequestered cytosol and organelles which appear virtually identical to those present in cytosol. AVds (C–F) have a more heterogenous content which appears degraded and is often densely stained. In addition, the content is often composed of several discrete regions, which may have been derived from distinct individual AVis which have subsequently fused together (F). Bar corresponds to 1 μm. Electron micrographs are courtesy of Dr. X. Hu, Cancer Research UK, London, UK.

tional consequences of the maturation of an AVi to an AVd are the acidification of the AV lumen and delivery of lysosomal hydrolases. Both are required for degradation of the sequestered protein content to amino acids. A recent review has covered this topic and the reader is referred to it for more detail [29].

Structures during intermediate stages in the maturation process are called amphisomes [30] (AVi/ds), reflecting their transient intermediate nature. Several fusion steps are believed to be required to achieve a fully functional AVd, including fusion with early endosomes, late endosomes and lysosomes. Fusion with endocytic vesicles has also been demonstrated, although the precise identity of these endocytic vesicles is not known. The confusion in the literature about whether endosomal carrier vesicles, late endosomes and multivesicular bodies (MVBs) represent equivalent compartments in terms of membrane composition and content [31] suggests it may be hard to precisely define them. While this multistep fusion process is similar to the fusion steps which a phagosome (produced by the internalization of particles in leukocytes, cells capable of phagocytosis) may undergo [32], multistep fusion has not been demonstrated to be an obligate sequence of events for AVs. For example, the AVi may possibly fuse directly with a lysosome and in one step acquire full functionality.

Morphologically, the amphisome (AVi/d) has the hallmarks of both an AV and endocytic compartment. They usually have a more complex morphology than the AVi which typically simply reflects the composition of the cell cytosol (see Fig. 2.3). Amphisomes often have multiple morphologically discrete regions which could only have arisen by coalescence through fusion with other organelles such as endocytic vesicles, late endosomes, MVBs or other AVis. This morphological data suggests that the AVi can participate in heterotypic, as well as homotypic, fusion events.

Heterotypic fusion events are required for all biosynthetic vesicular transport from the ER to the plasma membrane, e.g. when an ER-derived vesicle fuses with an early Golgi compartment, as well as during endocytosis from the plasma membrane. Homotypic fusion events are less frequent, but are also essential during organelle biogenesis, e.g. during immature secretory granule biogenesis [33] and endocytosis [34]. The cytoplasmic machinery which ensures the specificity of membrane fusions events in general, as well as the actual bilayer fusion events, is relatively well understood. In addition, much of the protein machinery required for fusion is conserved between yeast and mammals. Membrane fusion occurs in three distinct steps, tethering, docking and finally fusion of the bilayers. Tethering occurs over relatively long ranges (greater than 25 nm), while docking brings the membranes in proximity (less than 5–10 nm) for fusion. Many of the tethering and docking events are facilitated by multi-subunit protein complexes (for a review see, Ref. [35]) and the specificity of the events is provided by GTPases belonging to the Rab family, which interact in their GTP-bound conformation with membranes [36]. Fusion is driven by assembly of cognate pairs of membrane proteins called soluble *N*-ethylmalemide-sensitive fusion protein (NSF) attachment receptors (SNAREs) on each fusion partner.

While there has not been any detailed analysis of whether AVis or AVds can undergo homotypic fusion, there is abundant data demonstrating that AVis and

AVds can undergo heterotypic fusion. In the context of the need for accurate and specific targeting, docking and fusion, a major question arises as to how the AVis or AVi/ds acquire the SNAREs to allow them to fuse with multiple members of the endocytotic pathway. We do know, however, that in yeast fusion of the AV with the vacuole requires the SNARE complex including Vam3 [37] and Vti1 [38]. A similar question of specificity arises when one considers the role of Rab family members. It has been established that Ypt7 (the yeast equivalent of mammalian Rab7) is required for yeast autophagosome fusion with the vacuole [39] and, recently, Rab7 has been shown to be required for fusion of AVis to late endosomes [40, 41].

During maturation there is conversion of an AVi, which has a neutral pH internal compartment and contains no hydrolytic enzymes, to an AVd, which has an acidic pH and active, mature hydrolyases. Additionally, lysosomal membrane proteins such as lysosome-associated membrane protein types 1 and 2 (LAMP-1 and -2) are characteristic components of AVds. Maturation has been proposed to occur in a stepwise process [42]. First, delivery of endosomal vesicles to the AVi provides lysosomal membrane proteins and proton pumps. This allows acidification to occur. Next, fusion of the AVi/d occurs with late endosomes and MVBs (and possibly lysosomes), which delivers hydrolases and produces an AVd [43]. Based on this model, one can assume that the vesicles involved in the first step (delivery of lysosomal membrane proteins and proton pumps) could also deliver the machinery required for the second fusion step, e.g. SNARE proteins. How the endosomal vesicles in the first step can identify the AVi as a fusion partner remains unknown. The precise definition of the donor compartment (i.e. the vesicles or the late endosomes/MVBs) also remains vague. This is in part due to the complexity of the entire endocytic pathway, both in morphological and functional terms, which has led to a plethora of descriptions and names for various endocytic compartments, and a lively debate about how endocytosed material flows from the early endosome to the late endosome and lysosome.

2.2.3
Endosomes and Lysosomes

Understanding of the function and composition of endocytic compartments has a direct impact on our understanding of autophagosome maturation. The process of autophagy is ultimately to degrade self-constituents, be it for the purpose of survival or suicide. To do this the AV must acquire hydrolases from the endocytic pathway and essentially become a lysosome, and thus a clear understanding of lysosome biogenesis will help to understand autophagosome maturation.

Lysosomes were identified in the middle of the 20th century [44]. They are morphologically heterogenous organelles that contain acid hydrolases and are characterized by highly glycosylated membrane proteins called LAMPs, LGPs or LIMPs [45]. How the cell maintains such terminal degradative compartments has been a question that has attracted much investigation and these investigations have produced several models to explain lysosome biogenesis. Lysosomes receive traffic from both the biosynthetic and endocytic pathways, and the inputs almost all funnel through late endosome compartments. The biosynthetic pathway is responsible for the delivery of newly synthesized lysosomal membrane proteins, and hydrolases bound and sorted by the mannose-6-phophate receptor (M6PR). M6PR and the glycoproteins of the lysosome have intrinsic sorting signals that direct them to the lysosome [46]. The biosynthetic route originates in the Golgi complex where proteins such as LAMPs and MP6PRs are incorporated into transport vesicles and targeted to late endosome.

Several models exist to explain lysosome biogenesis from endocytic compartments and it is likely that different cells use one or all of these processes [47]. Basically, the distinction lies between maturation versus vesicular transport models. Lysosome maturation models are based on the conversion of an initial compartment (such as an early endosome) to a lysosome, via an intermediate (a late endosome), facilitated by vesicle-mediated delivery and remodeling. In contrast, vesicle transport models rely on the hypothesis that the early endosomes and late endosome are stable compartments that communicate (for the purpose of delivering materials) via endocytic carrier vesicles. The late endosome can then mature into, or form, a lysosome. Late endosomes and lysosomes can also transiently fuse to exchange components (kiss and run) or stably fuse to create a hybrid organelle [48]. The hybrid organelle then can form a lysosome and a residual body.

Although the endocytic pathway is complex, it is generally agreed that the early endosomes contain molecules internalized from the plasma membrane, including cell surface receptors. Early endosomes, which arise from clathrin-coated pits formed from the plasma membrane, can be classified into two types: sorting endosomes, from which the major route would be to late endosomes, and recycling endosomes, from which the major route would be back to the plasma membrane. Late endosomes are often characterized as multivesicular compartments; however, a distinction has been drawn to distinguish a MVB from a late endosome. The functional significance of this distinction is the identification of a complex sorting machinery (the ESCRT complex) which serves to generate the internal vesicles of an MVB, and is crucial in the process of downregulating cell surface receptors using HRS and Annexin-II, and ubiquitination [31].

Both early and late endosomes can be described as acidic compartments containing M6PRs and their bound hydrolases, while lysosomes do not contain M6PR. A major compositional difference between early and late endosomes is the Rab proteins associated with each compartment. Generally, Rab4, Rab5 and Rab11 are associated with early compartments, while Rab7 and Rab9 are associated with late endocytic compartments. As markers for these compartments, Rab proteins have proven invaluable in many studies. Lipids also make a major contribution to generating differences between the compartments, in particular the phosphorylated inositols, phosphatidylinositol-3-phosphate (PI3P) and PI-3,5-bisphosphate, which are produced by lipid kinases such as PI-3-kinases (PI3Ks), in particular VPS34, and PI3P-5-kinases such as PIKfyve [49], the mammalian homolog of Fab1 [50].

Do the current models of how the endocytic compartments form impact upon our ideas about how autophagosomes form and mature? All the current data about the formation of the initial AVs refutes any possibility that AVs are formed in any way analogous to the formation of early endosomes via clathrin-mediated internalization from the plasma membrane or that this endocytic compartment contributes to the isolation membrane. However, it is worth considering the new data about the phagocytic compartment, traditionally thought to form by invagination or enwrapping of the plasma membrane, and the controversy they have generated in the field. Using proteomic approaches it was shown that isolated latex-bead phagosomes contain markers of the ER [51]. This was explained by the identification of a transient connection between the phagocytic cup at the plasma membrane and the ER, and was supported by *in vitro* data which demonstrated that the ER SNAREs can pair with plasma membrane SNAREs [52]. However, recent models now suggest that the phagocytosed particle may in fact enter the ER and form a phagosome from the ER [32]. This is clearly an area of intense speculation and controversy which challenges the well-established models [53]. The dramatic shifts in models for phagosome maturation serve to illustrate the need for a biochemical analysis to support morphological data. As our knowledge of the isolation membrane and phagophore is in fact largely morphological, more accurate compositional data about the isolation membrane, or phagophore, is required to advance the existing model for AV formation.

More parallels between the autophagic and endocytic pathway are found when one considers AV maturation. An essential aspect of AV maturation is fusion of the AVi and AVd with endocytic compartments to form an amphisome (AVi/d), which would be analogous to a hybrid organelle. Lysosomes, and residual bodies, form from hybrid organelles and it is possible that the same structures form during autophagosome maturation. However, it has been suggested that before fusion to generate a hybrid organelle in the autophagic pathway, there is vesicular transport for the delivery, for example, of proton pumps [43], although the delivery of the pumps to a double membrane structure poses a potential topological problem. Another key issue is the question of vesicular transport out of the AVs, e.g. to recycle components from the AVi or AVd. Can components be recycled from the maturing AV in a similar manner as they are postulated to be from the endosomes?

2.3
Regulation of Mammalian Autophagy by Amino Acids and Hormones

The primary purpose of autophagy may be to degrade cellular proteins and organelles in response to nutrient starvation, since this application is present in yeast to higher organisms. In light of this conserved function, much interest has been directed at understanding the mechanisms linking nutrient starvation and the activation of autophagy. In the following sections, we review our current knowledge regarding the mechanisms that control autophagy, ranging from well-established ideas to newly proposed models based on recent findings.

2.3.1
Amino Acids

To stimulate autophagy, the most common type of nutrient starvation is deprivation of amino acids. One of the first demonstrations of this effect was performed by Woodside and Mortimore in 1972 [54], who measured free amino
acid that was released after protein degradation in a perfused rat liver system. A
high level of basal degradation in the liver was observed when using standard
perfusate (roughly a 50:50 mixture of defibrinated rat blood and Krebs–Ringer
bicarbonate buffer). However, the high levels of protein degradation were
quickly inhibited when a cocktail of amino acids was added into the perfusate
(Fig. 2.4). This inhibition of autophagy in rat liver by amino acids was later
quantitatively confirmed using electron microscopy [55, 56]. The ability of amino acids to negatively regulate autophagy is robust since it can be also recapitu-
lated in rat hepatocytes that have been freshly isolated or maintained in culture
for days [57–60]. The regulation of autophagy by amino acids could be universal
since it has been documented in isolated rat diaphragm [61], perfused rat heart
[62], and, recently, a number of cultured cells including HeLa [63], HT-29 [64],
C2C12 myotubes [65], mouse embryonic stem cells [10] and 293 human embryonic kidney cells [59]. The universality appears to apply across organisms
since this amino acid-sensitive response mechanism is conserved from yeast,
Dictyostelium [66] and *Drosophila* [67] to humans. Recently, our understanding of
autophagy regulation across various tissues has been developed to greater detail
by analyses in the transgenic mice expressing GFP–LC3 (see Chapter 3).

The autophagic response to amino acid starvation is characteristically rapid.
In rat liver, increases in the amount of nascent autophagosomes (AVis) could be
observed within 5 min after perfusion with amino acid-depleted buffer and maximal rates of autophagosome formation appear to be reached within 20 min
using electron microscopy [42, 55]. Findings from the perfused liver system indicate that the sensor for amino acid changes is also highly sensitive [55, 60,
68]. When the perfusion buffer contained amino acids at the normal plasma
concentration, low levels of basal autophagy and protein degradation were observed. This basal activity was robustly increased when all amino acids were removed. Conversely, the low levels of basal autophagy could be further inhibited
by inclusion of higher concentrations of amino acids (Fig. 2.4). These data suggest that autophagy, in liver and possibly other tissues, is constantly maintained
at a low, but active, basal level that can rapidly respond to small decreases in
plasma amino acid levels. These data imply that autophagy in certain tissues
(such as liver and muscle) might comprise a part of the metabolic homeostasis
mechanism that helps maintain plasma amino acid levels within a small range.
A critical metabolic role has been supported by recent data showing bursts of
autophagic activity in mouse tissues within hours of birth, suggestive of a response to the first neonatal nutritional challenges [27]. Consistent with this
metabolic role, autophagy-impaired knockout mice that lack Atg5 or Atg7 show
lower levels of plasma amino acids and die early after birth (see Chapter 3).

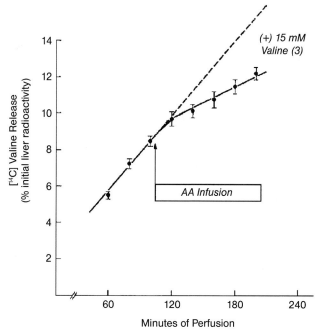

Fig. 2.4 Inhibition of autophagy in perfused rat liver by amino acid. The graph has been extracted from Woodside and Mortimore [54]. Autophagy was assessed by measuring protein degradation in a perfused liver that had been radiolabeled with [^{14}C]valine. Initially, the livers were perfused in a buffer consisting of defibrinated blood and Krebs solution, which essentially produced amino acid starvation conditions and a high "basal" rate of degradation observed as [^{14}C]valine release. Extrapolation of this initial slope expresses this initial rate (dotted line). Addition of an amino acid mixture (AA infusion) into the perfusion reduced the rate of degradation and maximal inhibition was reached within roughly 20 min. The residual degradation is a background rate that is amino acid-independent.

In the pioneering work that used perfused livers and isolated hepatocytes, it was determined that certain amino acids were especially potent at inhibiting autophagy. In the lists of "regulatory amino acids" identified using the two systems, Leu, Phe, Tyr, Trp, His and Gln were common [60]. Furthermore, Leu alone appeared to be the most potent inhibitor of protein degradation in liver and other tissues [61, 62]. Based on the primacy of this particular amino acid, a number of studies have gone on to postulate that the strong inhibitory effect of Leu is initiated from a receptor at the plasma membrane. For example, the inhibitory effect of the Leu analog isovaleryl-L-carnitine on protein degradation is nearly identical to that of Leu [70]. Since isovaleryl-L-carnitine is rapidly hydrolysed once inside the cell, it seems more plausible that the compound elicited its effects before transport into the cytoplasm. To strengthen their argument, these authors coupled eight Leu moieties to multiple antigen peptide (MAP), a dense branched lysine core that is used during antibody generation to support a covalently bound peptide antigen. Leu8MAP, cannot be transported into the cell, could inhibit autophagic protein degradation as effectively as Leu [58, 70]. The identity of the putative plasma membrane Leu receptor remains to be identified. Yet, the especially robust property of Leu alone to inhibit autophagy has been especially useful in recent work that attempts to dissect the intracellular signaling mechanisms controlling the process (further discussed below). However, it is important to note that Leu alone is generally not as effective as a mixture of regulatory amino acids at inhibiting autophagy [57, 58, 68]. Thus, it is possible that signals from multiple amino acid receptors converge to provide synergistic regulation.

2.3.2
Hormones

To complement the role of amino acids, it had been recognized early that hormones can robustly and rapidly regulate autophagy. Ashford and Porter first noted by electron microscopy an increased numbers of cytoplasmic "lysosome-like bodies" in livers that were perfused with a buffer containing glucagon [3]. Thus, it was demonstrated that autophagy in the liver could be simulated by increased levels of the pancreatic starvation hormone, glucagon. Conversely, insulin, which is synthesized in the pancreas typically in response to high blood glucose, has been observed to rapidly inhibit liver autophagy in injected whole rats and perfusion-based systems [71, 72]. These effects are consistent with a metabolic role of autophagy in response to starvation conditions in the context of a whole animal. For example, in situations of low blood glucose, glucagon may be stimulating protein degradation in the liver and the production of free amino acids, which can be further metabolized to generate ATP.

There appears to be mechanisms that integrate the multiple positive and negative signals controlling autophagy, but the dominant regulator is amino acids (Fig. 2.5). For example, in isolated hepatocytes, Blommaart et al. could demonstrate that addition of glucagon to an incubation medium containing regulatory amino acids at the plasma concentration could robustly stimulate protein degradation to near maximal levels [57]. However, the strong stimulatory effect of glucagon could not be observed in the context of higher amino acid concentrations. Thus, in the hepatocyte, the inhibitory signal arising from high levels of regulatory amino acids dominates over a simultaneous stimulatory signal from glucagon. Similarly, the inhibitory effects of insulin were clearly apparent when studied in the context of normal concentrations of regulatory amino acids, but almost completely masked in amino acid-deficient medium. Thus, the inhibitory effect of insulin can be overridden by the stimulatory effects of total amino acid starvation. Clearly, the animal employs multiple metabolic cues to regulate autophagic protein degradation. In later sections, we discuss some of the intracellular molecular signaling mechanisms that have been proposed to transmit the hormonal and amino acid stimulations. Regarding the prevalence of such mechanisms, different cell types will undoubtedly possess varying hormonal sensitiv-

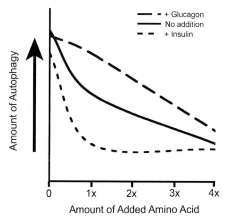

Fig. 2.5 Coordinate regulation of autophagy by amino acid, glucagon and insulin. The graph was derived from data presented in Blommaart et al. [57] and supported by data from [55, 57]. In their experiment, degradation of long-lived proteins (labeled with radioactive valine) in isolated rat hepatocytes was used to measure autophagy. High levels of autophagy were observed when the cells were incubated in Krebs buffer alone (without any additions or amino acid). When a mixture of amino acids was added back to concentrations of 1× normal plasma levels, autophagy was inhibited and this inhibition was more pronounced when 4× levels of amino acid were added. Co-addition of glucagon can stimulate autophagy, but the effect is most robust in the context of amino acid levels near 1× plasma concentration. Similarly, insulin co-addition inhibits autophagy, but its effect is best in the context of 1× amino acid. The cell's intracellular signaling machinery integrates these signals to coordinately regulate autophagy.

ities, but evidence has already been presented to suggest that the regulation of autophagy by insulin is conserved in both *Caenorhabditis elegans* and *Drosophila* [73, 74].

2.3.3
Longer-term Regulation

Amino acids, insulin and glucagon are all able to modulate autophagic protein degradation within minutes. As such, these stimuli have been proposed to trigger signal transduction mechanisms that primarily employ post-translational modifications. In contrast, it is now apparent that autophagy can also be activated over longer timecourses (from 3 to 48 h) by a wide range of treatments. As discussed in Chapters 6–13, infection by different types of bacteria and viruses leads to increased autophagy, possibly as a defense response. C2-ceramide treatment of U-373 malignant glioma cells also leads to several hallmarks of autophagy within 24 h, along with correlative increased protein levels of LC3 and the pro-death Ccl-2 interacting protein, BNIP3 [76]. Since BNIP3 overexpression alone can stimulate autophagy in these cells, much of the C2-ceramide effects seem to be mediated through BNIP3 upregulation. In HT-29 cells, C2-ceramide has also been reported to stimulate an upregulation of Atg6/beclin levels and autophagy within 3 h of treatment [77]. Furthermore, the ability of the estrogen receptor antagonist tamoxifen to stimulate upregulation of Atg6/beclin and autophagy in MCF-7 cells has been proposed to be mediated through endogenous ceramide second messengers. These cases all demonstrate that increased expression of particular genes, manifested over a relatively wider time frame, can provide an alternative mechanism for activating autophagy.

Autophagy is also a category of programmed cell death (termed type II) that is elicited in response to a variety of cellular stresses, purportedly as an alternative to type I programmed cell death, i.e. apoptosis. For example, it has been shown that inhibition of caspases using zVAD in several cell types activates type II autophagic cell death [78]. As discussed in Chapter 5, there appears to be a dynamic balance between suicide in type I or II programmed cell death. This balance is forcibly tipped in embryonic fibroblasts derived from Bax/Bak$^{-/-}$ double-knockout mice, which are resistant to entering type I programmed cell death. Interestingly, treatment of these cells with etoposide, which typically activates overt type I death, appears to shunt the death program into type II death, and this bypass is associated with increased levels of Atg5, Atg6 and Atg12 [79]. Thus, altered expression of autophagy genes might be another means for the cell to control the balance between alternate forms of cell suicide. The signal transduction mechanisms that coordinate cell death, gene expression and autophagy remain largely uncharted. Although the links remain to be made, it is likely that these slower-acting additional layers of signaling mediate their effects by modulating the molecular machinery that is understood to regulate autophagy in response to acute stimuli. The increased complexity arising from these novel mechanisms is intriguing, but further investigation is required to define the details.

2.3.4
The Nutrient Sensor Target of Rapamycin (TOR)

In response to amino acid starvation, one of the best appreciated molecules implicated in autophagy regulation is TOR, which, along with the downstream molecule p70S6 kinase (p70S6K), forms the core of a signaling cascade commonly termed the nutrient sensor. The original link between autophagy and TOR–p70S6K signaling was provided in 1995 by Blommaart et al., who observed increased phosphorylation on a 31-kDa protein upon re-addition of an amino acid-rich medium to rat hepatocytes [57]. This amino acid-stimulated phosphoprotein turned out to be ribosomal protein S6, which was being phosphorylated by p70S6K. These authors went on to investigate the relationship between levels of autophagy and p70S6K activity in isolated rat hepatocytes. Using a large number of experimental conditions involving combinations of amino acid starvation, insulin treatment, glucagon treatment and osmolarity shock, a strong correlation was constructed between increased p70S6K activity and autophagic inhibition (Fig. 2.6).

Another breakthrough was when Noda and Ohsumi could show in yeast that the drug rapamycin (discussed in more detail below) could stimulate autophagy,

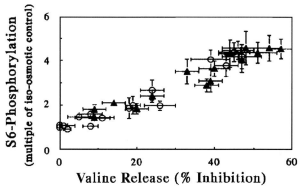

Fig. 2.6 Correlation between autophagy and decreased activity of p70S6K. Figure extracted from Blommaart et al. [57]. Shown are multiple data points obtained in the isolated hepatocyte system described in the legend of Fig. 2.5. Each point represents a different combination of amino acid starvation, insulin treatment, glucagon treatment and osmolarity stress (refer to original report for further details). Autophagy is expressed in this graph as a percentage of inhibition in (radiolabeled) valine release. Activities of p70S6K were assessed by measuring levels of phosphorylation on ribosomal protein S6 and expressed as fold change relative to the control (no additions in iso-osmotic conditions). The correlation between decreased autophagy (inhibition of valine release) and higher p70S6K activity is clear.

even under nutrient-rich conditions, by inhibiting the signaling through TOR and p70S6K [80]. Although there are variations in the magnitudes of response, inhibition of TOR via rapamycin treatment has been shown to activate autophagy in a wide variety of mammalian cell types such as macrophages [81], COS7 cells [82], 293 cells [59] and rat hepatocytes [57]. Overall, an inhibitory role on autophagy is conceptually consistent with the other important roles of activated TOR signaling in regulating cell growth (reviewed in Refs. [83, 84]). Under amino acid-rich conditions, TOR signals through both downstream molecules p70S6K and eukaryotic translation initiation factor 4E-binding protein (4E-BP1) to stimulate protein translation. Under amino acid-deprived situations (when synthesis of new proteins is repressed), it is reasonable that evolution has provided the cell with a mechanism to coordinately increase autophagy, degrade existing cellular proteins and recycle free amino acids into the cytosol. Interestingly, there are data indicating that Leu alone can be a robust activator of TOR [85], further bolstering the correlation between high TOR function and autophagy inhibition. There is some controversy, however, regarding the extent of autophagy regulation by TOR and rapamycin. Despite the accumulated data to implicate rapamycin as an autophagy inducer, we and others have also observed that certain cells, such as hepatocytes, can exhibit a degree of rapamycin-insensitive signaling that controls autophagy [58, 59].

The molecular pathways upstream and downstream of TOR that regulate autophagy are beginning to be defined in greater detail. Indeed, the steady progression of our understanding of TOR in autophagy likely stems from the considerable interest in TOR function within the contexts of cell growth, immune response and tumorigenesis. Although the evidence supporting a role of TOR and p70S6K in autophagy is strong, many complexities and controversies currently exist. Before discussing these in more detail, it is helpful to review the salient features of TOR biochemistry. The reader is referred to Fig. 2.7, which summarizes much of the following discussion.

Mammalian TOR (mTOR) is a protein kinase that can exist in two distinct protein complexes, of which only one is involved in autophagy. TOR complex 1 (TORC1), which is comprised of mTOR, GβL and raptor (regulatory associated protein of mTOR), is characterized by its rapamycin sensitivity and central role within the nutrient sensor [86–89]. The exact roles of raptor and GβL remain unclear, but some findings suggest that these subunits regulate mTOR by stabilizing the binding with or modulating kinase activity towards the substrates p70S6K and 4E-BP1. The immunosuppressant rapamycin is structurally similar

---

Fig. 2.7 Model showing proposed intracellular pathways that regulate autophagy. More details of the model are presented in the text. Amino acids are one of the most robust regulators of autophagy and it has been proposed that leucine (Leu) alone mediates the majority of the effects. Lower levels of extracellular Leu are detected by a putative Leu receptor (Leu-R) in the plasma membrane which signals into the cytoplasmic pathways via an unknown mechanism (black box). This mechanism likely also responds to changes in intracellular Leu levels. Lower Leu levels result in decreased activity of the PI3K-III complex [consisting of Atg6, PI3K-III (also known as hVps34p) and the protein kinase p150). As such, the signaling upstream of Leu is depicted to stimulate PI3K-III. One of the effects of activated PI3K-III appears to be positive regulation of TORC1 (which consists of TOR, raptor and GL). Lower cellular energy levels (low ATP) also signal to TORC1 by activating AMPK, which phosphorylates the TSC1–TSC2 complex on TSC2. Within the complex, TSC2 has GAP activity and stimulates conversion of Rheb-GTP to Rheb-GDP. Since Rheb-GTP binding is required for activation of TORC1, AMPK effectively is a negative regulator of TORC1. Amino acids such as Leu have been proposed to regulate TORC1 by promoting the binding of Rheb to TORC1. Insulin (Ins) can also signal into this pathway by first activating its tyrosine kinase receptor in the plasma membrane, which goes on to activate PI3K-I. Second messengers produced by PI3K-I activate the protein kinase AKT, which then phosphorylates TSC2. Phosphorylation by AKT promotes instability of TSC2 and decreases its overall GAP activity. As such, Ins positively regulate TORC1. The lipid phosphatase PTEN antagonizes this pathway by degrading the products of PI3K-I. When added to cells, rapamycin (Rap) binds to FKBP and, while complexed together, inhibits TORC1. When in an activated state, TORC1 phosphorylates p70S6K at residue Thr389, which creates a docking site for another protein kinase, PDK. PDK then phosphorylates p70S6K on a site within the activation loop of the catalytic domain, thereby fully activating p70S6K activity. Once activated, p70S6K phosphorylates ribosomal protein S6 to positively regulate ribosome biogenesis and protein translation. Activated TORC1 also phosphorylates 4E-BPa to regulate protein translation. In the context of autophagy, it has been proposed that TORC1 regulates Atg1 via an unknown pathway (depicted here to also include p70S6K although this may not be the case). In this mechanism, TORC1 activity promotes hyperphosphorylation of Atg13, which results in lower complex formation with Atg1. Since Atg13 binding is required for Atg1 to stimulate autophagy, TORC1 activity inhibits autophagy in this model, which is consistent with existing experimental data.

to FK506 and, when added to cells, is bound by the FK506 binding protein (FKBP) to form an active-drug–protein toxin complex (for a review, see Ref. [84]). Within this complex, rapamycin binds to a specific domain of around 100 amino acids within mTOR just N-terminal to the kinase catalytic domain, thereby inhibiting the ability to signal downstream, although the molecular mechanism for this remains unclear. In contrast, TORC2, which is composed of

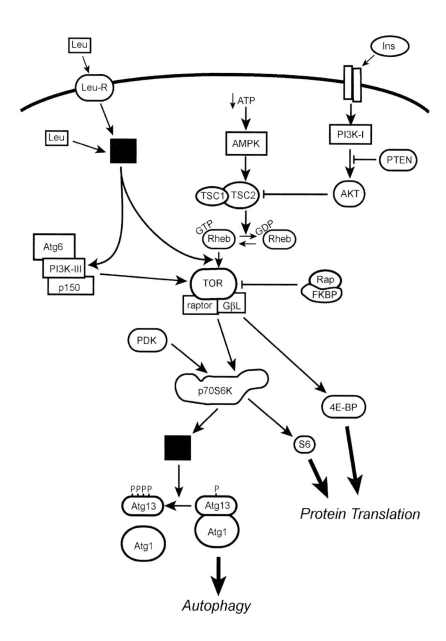

mTOR, GβL and rictor (rapamycin-insensitive companion of mTOR), does not bind FKBP–rapamycin and does not signal to p70S6K. TORC2 appears to have a distinct role in actin cytoskeleton regulation [89, 90].

The overall regulation of p70S6K is complex, involving phosphorylation on at least eight different sites that coordinately modulate activity via a multistep mechanism [91]. Within p70S6K, four functional domains can be delimited. An acidic region in the N-terminus is followed by the catalytic domain, a linker region and then a basic C-terminal domain that contains an inhibitory pseudosubstrate. In the inactive state, the acidic N-terminus interacts with the basic C-terminal region and stabilizes an inhibitory interaction between the kinase domain and the pseudosubstrate. Activation of p70S6K is thought to initiate with phosphorylation at four sites within the C-terminus (Ser411, Ser418, Ser421 and Thr424), which would promote a conformational change that is prerequisite for phosphorylation at Thr389. TORC1 directly phosphorylates p70S6K at Thr389, thereby creating a proposed docking site for another kinase, PDK1 [92, 95]. PDK1 then performs the final phosphorylation of the activation sequence at Thr229, which is within the p70S6K activation loop [94]. It is important to note, however, that despite the ability of this model to account for a large collection of data, other findings suggest additional mechanisms. In addition to inhibiting phosphorylation of p70S6K at Thr389 by TORC1, rapamycin treatment also promotes rapid dephosphorylation at this site, suggesting an activated phosphatase activity [95]. Other kinases and autophosphorylation have also been proposed as pathways leading to phosphorylation at Thr389. On a practical note, it has become routine to assay activation in the TORC1–p70S6K module by exploiting robust phosphorylation-specific antibodies towards Thr389 of p70S6K. However, it is becoming apparent that Thr389 phosphorylation does not always reflect the activity of TORC1 nor the final activation status of p70S6K. These concerns suggest that multiple methods may be required in conjunction to get a truer picture of TORC1 and p70S6K activity. In later sections, the signaling downstream of TORC1 and p70S6K that regulates autophagy will be revisited.

2.3.5
Upstream of TOR

The role of TORC1 and p70S6K in autophagy has recently been corroborated by evidence (from largely autophagy-independent research) of several upstream molecular pathways that can link in signals from known regulatory stimuli. Data from multiple recent studies can be condensed into a model that puts the protein complex containing TSC1 and TSC2 at a nexus where upstream signals converge towards TORC1. TSC1 (also termed hamartin) and TSC2 (also termed tuberin) are both tumor suppressors, and mutations in either can lead to the pediatric autosomal dominant disease tuberous sclerosis (TSC). This condition is associated with benign tumors known as hamartomas that can form in a variety of tissues in addition to other debilitating symptoms (for a review, see Ref. [96]).

TSC2, when heterodimerized with TSC1, has activity as a GTPase-activating protein (GAP) for the small GTP-binding protein Rheb (Ras homolog enriched in brain). Rheb binds to a portion of the mTOR catalytic domain, but this binding appears to be largely independent of the type of nucleotide bound. However, GTP-bound Rheb potently activates TORC1 to phosphorylate p70S6K (and 4E-BP1) [97]. Conversely, TORC1 bound to mutant forms of Rheb that are nucleotide deficient are almost catalytically inactive. Negative regulation of TORC1 by the TSC complex could be clearly observed in rodent cells with a homozygous disruption in either *TSC1* or *TSC2*, which displayed abnormally high levels of TORC1 activity [98]. Together, these findings define a pathway where the GAP activity of TSC1–TSC2 stimulates the conversion of Rheb-GTP to Rheb-GDP, thereby negatively regulating TORC1.

Interestingly, the TSC1–TSC2 complex provides a potential mechanism to explain how insulin inhibits autophagy. Stimulation of the insulin receptor tyrosine kinase on the plasma membrane ultimately results in activation of class I PI3K (PI3K-I). PI-3,4,5-trisphosphate (PIP₃), the main lipid product of PI3K-I, goes on to activate the protein kinase AKT (also known as protein kinase B). Several recent reports have shown that AKT directly phosphorylates TSC2 on multiple residues [99–101]. Furthermore, the data suggest that phosphorylation by AKT can promote instability of the TSC1–TSC2 complex and degradation of TSC2 [99]. As such, increased AKT activity, e.g. after insulin treatment, would be predicted to decrease the overall GAP activity of TSC1–TSC2 and result in higher amounts of Rheb-GTP and TORC1 activity. In accordance, overexpression of TSC2 mutants that lack the AKT phosphorylation sites inhibited the ability of insulin to activate TORC1 and p70S6K [101]. In addition to providing a molecular basis for insulin in autophagy, this mechanism also appears to explain why exogenously added PIP₃ or overexpression of a constitutively active AKT inhibited autophagy in HT-29 cells [102, 103]. Conversely, the regulation of TSC1–TSC2 by AKT can also explain why overexpression of the PTEN phosphatase [which catabolizes PIP₃] stimulated autophagy in HT-29 cells.

Recently, additional evidence has been presented to suggest that signals from amino acids are also transmitted to TORC1 via the TSC complex and Rheb. It could be shown in 293 cells that amino acid starvation can decrease (although not completely disrupt) the binding of Rheb to TORC1 [104]. By decreasing Rheb interaction, amino acid starvation would be predicted to repress TORC1-function (since Rheb-GTP binding is required full kinase activity). Thus, the TSC–Rheb machinery provides a model that can account for amino acid dependency, inhibition of TORC1–p70S6K signaling and autophagy induction. It is still unclear how decreases in amino acid disrupt Rheb–TORC1. However, withdrawal of Leu alone from the cell medium was able to disrupt Rheb–TORC1 binding to a similar extent as withdrawal of all amino acids, further suggesting that this mechanism is directly relevant to autophagy regulation. Despite this attractive model, the regulation of autophagy function (e.g. protein degradation) by manipulating TSC or Rheb in mammalian cells remains to be directly demonstrated.

The necessity to confirm any putative mechanisms with direct studies of autophagy is highlighted by the reports regarding 5′-AMP-activated protein kinase (AMPK). AMPK acts as a sensitive sensor of cellular energy by responding to increases in the AMP/ATP ratio (for reviews, see Refs. [105, 106]). It is understood that small increases in the ADP/ATP ratio are amplified by adenylate kinase to produce a much larger increase in the AMP/ATP ratio, which goes on to activate AMPK via multiple mechanisms. For example, AMPK is readily phosphorylated and activated by glucose starvation or treatment with 2-deoxy glucose, an analog that inhibits glucose utilization [107]. AMPK was found to phosphorylate TSC2 on multiple residues and this phosphorylation enhanced the ability of the TSC1–TSC2 complex to inhibit TORC1 activity, presumably via a reduction of Rheb-GTP levels. According to the model described above, decreased TORC1 and p70S6K activity stemming from low cellular energy and activated AMPK would be predicted to result in stimulated autophagy. Indeed, increased autophagy has been reported in certain situations of cellular energy crisis and some data in PC12 cells indicated that treatment with 2-deoxyglucose resulted in higher clearance of huntingtin polyglutamine aggregates via autophagy [108, 109]. In contrast, Samari et al. could demonstrate that treatment of hepatocytes with AICAR, an adenosine analog that potently activates AMPK, inhibits autophagy [110]. Why are there these apparently conflicting observations? Samari et al. argue that the cell may favor lower levels of autophagy during times of ATP depletion since the early steps of autophagosome formation appear to be energy dependent [111, 112]. Further studies will be required to resolve the relationship between AMPK, TSC1–TSC2 activity, TORC1 and autophagy. However, due to our current limited understanding, it is apparent that the tendency to draw premature linear conclusions will have pitfalls.

2.3.6
Downstream of TOR

Downstream of TORC1, what is the role of p70S6K? Several studies have proposed a model in which p70S6K regulates protein translation by phosphorylating ribosomal protein S6 [113–115]. It was shown that translational efficiency of the 5′-terminal oligopyrimidine (TOP) class of mRNAs was increased by S6 phosphorylation. Since many 5′-TOP mRNAs encode ribosomal proteins, p70S6K has been proposed to positively regulate ribosome biogenesis. However, recent reports present an alternative model in which 5′-TOP mRNA translation is largely independent of S6 phosphorylation and more robustly regulated by PI3K-I [116, 117]. This latter model is intriguing since it suggests that p70S6K may have additional roles (such as the regulation of autophagy). In this context, the strong correlation between decreased p70S6K activity and increased autophagy in rat hepatocytes is highly intriguing, but evidence showing direct cause and effect has yet to be presented. Seemingly in contradiction, autophagy in the developing *Drosophila* fat body was completely abolished by a p70S6K null mutation [74]. As such, the regulation of autophagy in *Drosophila* and mammalian cells could be relying differently on p70S6K. Alternatively, lower p70S6K activities may trigger autophagy, but some residual p70S6K activity is required for autophagy to take place. It is still unclear which of these explanations is more accurate and so the role of p70S6K in autophagy is currently controversial. It is noteworthy that a TOR null mutation resulted in constitutively higher levels of autophagy in the fat body, suggesting that perhaps decreases in TOR activity are sufficient to stimulate autophagy and that decreases in p70S6K activity have essentially been molecular markers of TORC1 function.

What are the other potential mechanisms downstream of TORC1 that could regulate autophagy? The only data that have been described to suggest such a mechanism has been obtained from studying yeast. Using two-hybrid screening methods, the protein Atg13 was identified as a binding partner of Atg1, a protein kinase identified in the original screens for autophagy defective mutants [118]. In yeast (and other organisms, see below), Atg1 is absolutely essential for autophagy. Interestingly, Atg13 is phosphorylated, and both amino acid starvation and inhibition of TOR with rapamycin led to lower levels of Atg13 phosphorylation. Atg13 with lower amounts of phosphorylation appeared to have a higher affinity for Atg1. In addition, Atg13 binding was required for Atg1 to positively regulate autophagy. These findings suggest a model in which starvation decreases levels of Atg13 phosphorylation via a TOR-dependent pathway and thereby promote increased Atg1–Atg13 interaction and higher levels of Atg1 signaling to downstream effectors. Currently, the link between TOR and Atg13 phosphorylation is unknown; direct phosphorylation by TOR or activation of a phosphatase by starvation (or rapamycin) remain as possible scenarios. Studies have also proposed that the kinase activity of Atg1 is not essential to regulate autophagy [119]. In addition, further studies will be needed to determine the extent of conservation in this pathway for mammalian cells. The variable binding between Atg1 and Atg13 has been proposed to be a mechanism that allows Atg1 to switch from regulating autophagy to the cytoplasm-to-vacuole targeting (Cvt) pathway. Since the Cvt pathway does not exist outside of yeast and no Atg13 homolog is apparent in the human genome sequence, the identity of the mammalian Atg1–Atg13 module remains unknown. However, Atg1 is known to be required for autophagy in *Dictyostelium*, *C. elegans* and *Drosophila*, and it is likely that many insights into Atg1 function gleaned from these model organisms can be applied to mammalian systems [66, 73, 74].

In parallel with the nutrient sensor, another protein that has been widely appreciated as an autophagy regulator is PI3K. The earliest indication of an involvement was reported by Seglen and Gordon [120] who searched for methylated adenosine derivatives of amino acids that could inhibit autophagy. In this manner, their screen identified 3-methyladenine (3-MA) to have potent autophagy inhibitory actions. The molecular target of 3-MA was not recognized until a later investigation showed that wortmannin, LY290002 and 3-MA all robustly inhibited autophagy by targeting PI3K [121]. Molecular evidence for PI3K in autophagy was revealed with the discovery that in yeast, Atg6/beclin exists in two distinct complexes, both of which contain Vps34p (a type of PI3K) and the protein kinase Vps15p [122]. One these complexes, which specifically regulates Cvt trafficking, also contains Vps38p, while the other complex, which specifically regulates autophagy, also contains Atg14p. No mammalian Vps14 (or Vps38) homolog has been identified yet. However, beclin is bound to hVps34p in mammalian cells [123, 124], and hVps15p (also known as p150) binds hVps34p and regulates its activity [125]. As such, most of the original beclin–Vps34p–Vps15p complex described in yeast appears to be conserved in mammals. In mammalian cells, hVps34 is also known as class III PI3K (PI3K-III) since it specifically generates as its product PI3P, which elicits effects distinct from those of other phosphoinositides, e.g. the products generated by PI3K-I.

A widely used control in mammalian cells has been to inhibit autophagy using PI3K inhibitors such as wortmannin and 3-MA (see specific section below on experimental methods used to inhibit autophagy), which unfortunately target class I and class III enzymes indiscriminately. In addition, small interfering (si) RNA knockdown of Atg6 has also been fairly effective at inhibiting autophagy in a number of cell types. These findings suggest that PI3K-III activity positively regulates autophagy and this idea was supported by experiments in which addition of exogenous PI3P to HT-29 cells was sufficient to stimulate autophagy [126]. In addition, some data has been presented to show that amino acid starvation of C2C12 myotubes results in higher levels of beclin-associated PI3K-III activity [123].

Recently, findings from several studies have presented a modified model that links amino acid levels to regulation of PI3K-III and downstream effects on TORC1 [124, 127]. The novel theme of this model is the activation of PI3K-III by amino acid (or treatment with insulin), which implies that PI3K-III activity inhibits autophagy, contrary to our previous understanding. In addition, it was concluded that PI3K-III activity positively regulated TORC1 and p70S6K activity. For example, overexpression of PI3K-III or its binding partner p150 in 293 cells increased p70S6K activity to phosphorylate ribosomal protein S6. Conversely, siRNA towards PI3K-III could decrease the activation of TORC1 signaling. Furthermore, it could be shown that PI3P was mediating the effects on TOR since overexpression of a recombinant protein containing the FYVE domain (which binds to and effectively inactivates the PI3P) decreased the phosphorylation of p70S6K by TORC1. How PI3P regulates TORC1 will require further characterization. It was especially intriguing that amino acid starvation rapidly led to decreased PI3K-III activity and decreased levels of PI3P in a number of different mammalian cell types. These authors also noted that beclin-bound PI3K-III activity was inhibited after amino acid starvation, although the amount of beclin-associated PI3K-III was unaltered. As such, some additional protein modification or binding partner might be underlying how amino acid starvation inhibits beclin-associated PI3K-III. A potential mechanism has been suggested by another recent study that focused on the role of a beclin–Bcl-2 interaction [128]. This work indicated that Bcl-2 can bind beclin and thereby inhibit beclin's function to positively regulate autophagy. Surprisingly, the amount of Bcl-2 that was bound to beclin could be increased by incubating the cells in amino acid-rich conditions. As such, these findings on Bcl-2 provide yet another possible molecular mechanism to link amino acid starvation and autophagy. Taken together, it is possible that amino acid starvation leads to disrupted binding of Bcl-2 to a beclin–PI3K-III complex. In this model, the decreased binding of Bcl-2 could be transmitting an allosteric conformational change that results in lower PI3K-III activity and lower TORC1 function. Undoubtedly, this recent progress provides an abundance of new hypotheses that await experimental testing.

In sum, the discussion above shows the firm foundation in our understanding of autophagy regulation. For example, the implications of TOR, rapamycin and PI3K have been widely supported by many studies from different laboratories (although exceptional cases also exist). On top of this foundation, waves of recent work have expanded the model to include new parts such as Rheb, TSC, AMPK and bcl-2, which feed into the existing scheme. As for future directions, it will be interesting to determine how amino acid regulation of Rheb–TORC1 binding, bcl-2–beclin binding and PI3K-III are coordinated. It remains to be described how signals are transmitted from the putative plasma membrane Leu receptor into the cytoplasmic pathways. In addition, it makes biological sense that autophagy should also be regulated by intracellular amino acid levels and this argument has some experimental basis [129]. If so, how do intracellular amino acids do this? Downstream events are also largely unclear as only Atg1 is well appreciated as the TOR effector controlling autophagy. Of course, once we understand the mammalian TORC1–Atg1 connection, it will be interesting to determine further how this module regulates the known processes of autophagy initiation such as Atg5–Atg12 localization change, modification of LC3 and nucleation of autophagosome membranes.

2.4
Methods to Measure Autophagy

When autophagy was first described [1], electron microscopy was the only way to identify autophagosomes. Since then a number of biochemical techniques to measure autophagic induction and degradation have been developed, including lactate dehydrogenase (LDH) sequestration assays and protein degradation assays. A big step forward was made in the mid-1990s with the identification of the Atg genes in yeast by several groups [130–132]. One of the proteins identified then was Atg8. So far, Atg8 and its mammalian homologs are the only proteins which bind specifically to, and remain associated with, newly formed autophagosomes. Since then, Atg8 tagged with GFP has been used in numerous studies to study the induction of autophagy and to specifically identify autophagosomal structures in light microscopy experiments. In the following discussion, we describe the most commonly used methods of measuring and manipulating autophagy. We aim to outline the principles of the techniques, and the strengths and weaknesses, in order to provide a basis to design experiments that study autophagy.

2.4.1
Microscopic Methods

2.4.1.1 Electron Microscopy

Conventional electron microscopy is particularly useful to identify autophago-somes and to measure the extent of autophagic induction. It allows also an identification and distinction to be made between early autophagosomes (AVi) and later structures (see Fig. 2.3). Early, immature autophagosomes (AVis) can be clearly defined as double-membrane vesicles, varying in size between 200 and 2000 nm depending on the cell type, and containing undegraded cytosol as well as organelles including mitochondria, ER and small vesicles. Late, degrada-tive autophagosomes (AVd) are defined by their degraded electron-dense con-tent. In cases where autophagosomes cannot be defined by morphology alone, immunoelectron microscopy with antibodies against LC3 or GFP–LC3 allows clearer identification of autophagosomes.

In the past, electron microscopy has been used extensively to quantify the number of autophagosomes in a variety of cell types and tissues. Basically, the number or area within autophagosomes is quantified from an unbiased sam-pling of several grids and images, and then expressed normalized to the total cellular area. The big disadvantage of electron microscopy, however, is that it re-quires a lot of time as well as experience. However, the advantages are that au-tophagosomes can be unequivocally identified based on their morphology as well as by the presence of specific markers in the membrane.

2.4.1.2 Light Microscopy

GFP–LC3 and endogenous LC3

The identification of Atg8 in two yeast screens for autophagy proteins [131, 132] proved to be an important step forward in the development of autophagy assays. Atg8 is the only specific marker for autophagosomes found so far which remains bound after the closure of the sequestering isolation membrane. Mammalian cells contain several homologs, LC3 [63], GABARAP and GATE16 [133], all of which have been shown to bind to the membrane of autophagosomes. Of these three proteins, LC3 [134], has been investigated the most. Pro-LC3 gets cleaved after a highly conserved glycine residue in the C-terminus by the protease Atg4 [135] to give rise to the cytosolic form LC3-I. Upon induction of autophagy LC3-I is modified in a ubiquitin-like process, which involves Atg7 and Atg3 and the attach-ment of a lipid residue, presumably phosphatidylethanolamine to the C-terminal glycine residue [136, 137], in an Atg5–Atg12-dependent manner [10]. The resulting form LC3-II binds to isolation membranes (phagophores) and autophagosomes, and can be used as a specific marker for autophagy. The other Atg8 homologs, GA BARAP and GATE16, are less well characterized, but have also been shown to be processed by the same machinery prior to binding to autophagosomes [138]. In human cells there exist at least three distinct LC3 genes with different tissue expression patterns, termed LC3A, LC3B and LC3C [139].

Endogenous LC3 or LC3sp fused at its N-terminal to GFP, or GFP-derivatives, has proven to be a valuable tool to investigate autophagy on a light microscopy level. In full growth medium GFP–LC3 has a diffuse cytosolic distribution. When autophagy is induced, GFP–LC3 gets processed and binds to newly formed autophagosomes. In cell types with large autophagosomes, like embryonic stem cells or hepatocytes, autophagosomes can be clearly identified as ring-shaped structures, while in HeLa cells, autophagosomes are smaller and can be detected as spots only (also seen in HEK293 cells, Fig. 2.8A [59, 140]). The appearance of GFP–LC3 spots can be quantified and used to measure the extent of the autophagic induction after any treatment. Another advantage of GFP–LC3 is that it can be used in colocalization experiments with other markers. Using light microscopy, colocalization of autophagosomes with a structure of choice can be easily assessed, e.g. Rab7 [40, 41] or LysoTracker Red-positive structures [74]. Additional experiments have provided evidence about the colocalization of LC3 with protein aggregates, e.g. in Presenillin1$^{-/-}$ hippocampal neurons LC3 colocalizes with aggregates of telencephalin [141]. A protein called Alfy (autophagy-linked FYVE protein) has also been demonstrated to colocalize with GFP–LC3 positive structures [142] and it was proposed that Alfy plays a role in targeting cytosolic proteins to autophagosomes for their degradation. Additional information was provided in 2004, when Mizushima et al. reported the generation of a GFP–LC3 transgenic mouse and investigated the tissue-specific autophagy responses after nutrient starvation using light microscopy [25]. Extension of this approach in other transgenic or knockout mice has demonstrated the importance of this marker [27, 128].

Monodansylcadaverine (MDC)

In 1995, Biederbick et al. [143] initially identified the fluorescent molecule MDC as a specific marker for autophagosomes, after which it was used in a number of studies to identify autophagosomes and as a measurement of autophagic induction [144–147]. However, recent reports showed that MDC primarily stains acidic compartments and colocalizes to a high degree with LysoTracker Red- or CD63-stained organelles, but only partially with LC3-positive compartments [140]. Therefore, MDC may not label autophagosomes before they become acidic through fusion with compartments of the endocytic system. Results obtained with MDC therefore have to be interpreted with care as it may not be specific for autophagosomes and may only mark late autophagosomal structures. Using this dye, it might not be possible to distinguish between lysosomes and AVd (degradative AVs) at the light microscopy level.

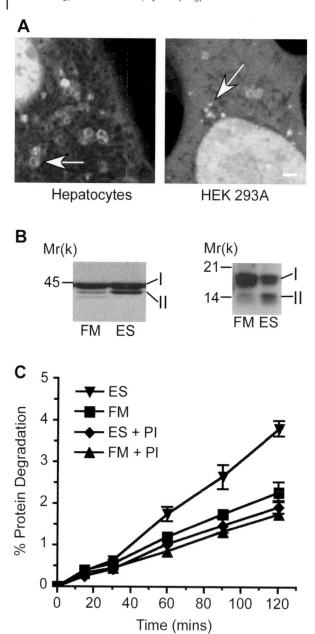

2.4.2
Biochemical Methods

2.4.2.1 **Formation and Induction**

LDH and [¹⁴C]sucrose sequestration

LDH and sucrose sequestration assays are particular useful methods to measure the percentage of cytosol sequestered into autophagosomes, and are therefore a direct measurement of the extent of autophagy. LDH is a cytosolic enzyme and its sequestration into autophagosomes in the presence of lysosomal protease inhibitors can be measured with a simple spectrophotometer-based assay [148]. [¹⁴C]sucrose-, as well [¹⁴C]lactose- and [¹⁴C]raffinose-based sequestration assays require electropermeabilization of the cells before autophagy can be induced. Both assays have been widely used [149–155]. Since both assays are based on separating cytosol from membranes, a critical factor is that the cells can be reproducibly disrupted to release cytosol, leaving organelles intact.

LC3 processing

As described above, cytosolic LC3-I is modified by a ubiquitin-like process to give rise to LC3-II, the autophagosome-bound form [138]. LC3-II has a higher mobility than LC3-I on denaturing sodium dodecylsulfate–polyacrylamide gel electrophoresis (SDS–PAGE). Induction of autophagy leads to an increase in LC3-II [138] and this can be detected using specific antibodies for LC3 in Western blots (see Fig. 2.8 B, right). One problem of this approach, which may underestimate the to-

Fig. 2.8 Methods to study autophagy. (A) Comparison of the size of autophagosomes in hepatocytes and HEK 293A cells by confocal microscopy. Hepatocytes were infected with adenoviruses encoding GFP–LC3 and HEK 293A cells were transfected with plasmid to express GFP–LC3. Both cell types were starved for 2 h in Earle's buffered saline solution (ES). Arrows point to autophagosomes. In hepatocytes, autophagosomes can be identified as ring-shaped structures up to 2 μm in diameter. In HEK 293A cells autophagosomes are smaller and can mostly be seen as distinct puncta only. Note that GFP–LC3 can also be found in the nucleus of cells, although the mechanism behind this is not known. Bar=2 μm. (B) Analysis of LC3 processing by Western blotting. Hepatocytes were either GFP–LC3 infected (left panel) or not (right panel) and then incubated in either full medium (FM) or starvation medium (ES) for 2 h before solubilization and Western blotting with either an anti-GFP (left) or anti-LC3 (right) antibody. The appearance of the membrane-bound form [GFP–LC3-II or LC3-II (II)] can be clearly seen to be increased in the samples of starved cells. (C) Long-lived protein degradation time course in rat hepatocytes incubated in FM or ES with or without the lysosomal protease inhibitors E64d and Pepstatin A (PI). Samples were taken from the medium at 0, 15, 30, 60, 90 and 120 min to measure the release of ¹⁴C-valine into the medium.

tal induction of autophagy, is that the LC3-II bound on the inside of an autophagosome can be degraded in AVds. This degradation can be effectively blocked by adding lysosomal protease inhibitors such as leupeptin or E64d, also known as EST, and pepstatin A to the experimental medium [59, 156]. Which combination of lysosomal protease inhibitors works best depends on the cell type and has to be experimentally determined. GFP–LC3-I and -II behave similarly to LC3-I and LC3-II and have also been used (see Fig. 2.8 B, left). Many researchers have reported that resolving and identifying LC3-I and LC3-II is difficult, and requires optimization, such as loading a high amount of protein. Furthermore, saponin extraction of cells before solubilization selectively enriches the relative amounts of LC3-II as all cytosolic proteins are removed, including LC3-I [59]. Another approach, which takes advantage of the LC3 lipidation and increase in hydrophobicity, has been used by Bampton et al. [140]. HeLa cells expressing GFP–LC3 were labeled with [^{3}H]ethanolamine and incubated in NH_4Cl to accumulate autophagosomes. After Triton X-114 phase separation, GFP–LC3-II was selectively enriched in the Triton X-114 detergent phase as detected by Western blotting and autoradiography of SDS–PAGE gels, while GFP–LC3-I was in the aqueous phase.

2.4.2.2 Fusion

Protein degradation

The biochemical methods mentioned above are very useful to investigate the extent of autophagic induction; however, they do not measure the flux of cytosolic content through the autophagosome enroute to its degradation in lysosomes. For this question, protein degradation assays are very useful and have been used in a number of studies [10, 40, 157–160] and see discussion above with Fig. 2.4. The assay is based on the release of radioactivity from [^{14}C]/[^{3}H]valine- or [^{14}C]/[^{3}H]leucine-labeled cells after the induction of autophagy. Trichloroacetic acid is used to precipitate whole proteins, while individual amino acids arising from proteins degraded via autophagy remain soluble. An additional point is that the long chase period decreases the pool of labeled short-lived proteins, which get degraded via the proteasome. Therefore the assay measures the degradation of long-lived proteins through autophagy. The ratio of the soluble released radioactivity over the sum of the total radioactivity (medium and cells) gives the percentage of protein degradation (see Fig. 2.8C). Since leucine on its own is a strong inhibitor of autophagy (see Section 2.3), valine might be the amino acid of choice [58, 161]. It is important to note that this method does not measure the extent of autophagic induction, e.g. the "number" of autophagosomes, but only the flux through the system. For example, microtubule-depolymerizing drugs such as vinblastine greatly reduce the amount of protein degradation by inhibiting the fusion of autophagosomes with compartments of the endosomal system. However, other assays using electron microscopy or LDH sequestration have shown that vinblastine treated cells contain a high number of autophagosomes [162].

2.4.2.3 Purification of Autophagosomes from Rat Liver

So far there is only one report [23] that describes a method to isolate highly purified autophagosomes. The method has been developed for isolated primary rat hepatocytes, which are starved in the presence of vinblastine to accumulate early autophagosomes and amphisomes. Hepatocytes are then disrupted and incubated with glycyl-phenylalanine-naphthylamide, a cathepsin C substrate, to selectively disrupt lysosomes. A post-nuclear supernatant is prepared and then subjected to two sequential gradients. The first stage involves a nycodenz step gradient that separates autophagosomes and ER from the rest of the organelles. The second stage involves a Percoll step gradient that generates highly purified autophagosomes. To remove the Percoll, autophagosomes are further purified on a final OptiPrep gradient. AVs prepared in this manner have been analyzed using electron microscopy and two-dimensional electrophoresis [163]. It is likely that purification of autophagosomes from other cell types than hepatocytes will require extensive adaptation of this method.

2.4.2.4 Inhibition of Autophagy

Autophagy can be inhibited at several steps by a number of pharmacological drugs and also more recently using siRNAs against specific autophagy proteins. As described above, commonly used pharmacological inhibitors of autophagosome formation are PI3K inhibitors, such as wortmannin, LY290002 [121] and 3-MA [120]. Other drugs, such as okadaic acid and AICAR, an AMPK activator [110], inhibit autophagy by less well understood mechanisms. Furthermore, the amino acid leucine has been shown to inhibit autophagy when added to starvation medium and to induce autophagy when removed from the growth medium [58, 65, 164]. Drugs that act at a later stage are the microtubule-depolymerizing drugs vinblastine and nocodazole. These inhibit the fusion of autophagosomes with lysosomes [162, 165] and thereby inhibit protein degradation, although both drugs also affect the formation of AVs [59].

More recently, Yu et al. [79] used siRNAs to knockdown Atg7 and Atg6 from L929 and U937 cells to inhibit autophagy, and found that zVAD induced cell death was inhibited. Jackson et al. [166] proposed that autophagosome membranes are the site of viral RNA synthesis, and used siRNAs to deplete Atg12 and LC3. This reduced the yield of intracellular polio virus by 3- and 4-fold, respectively. Boya et al. [167], on the other hand, showed that when autophagy was inhibited with siRNAs against Atg5, Atg6, Atg10 and Atg12, the cells died through apoptosis when nutrients were depleted. These are just some of the recent results that highlight the usefulness of the siRNA approach to knockdown Atg genes to address a number of questions.

Table 2.1 Common questions and approaches to answer

Question	Method
Identification of autophagosomes	Conventional transmission electron microscopy, immunoelectron microscopy, light microscopy with GFP–LC3 or antibodies against LC3
Quantitation of early and late autophagosomes	Electron microscopy, colocalization of GFP–LC3 or LC3 with LysoTracker Red
Induction of autophagy	LDH and sucrose sequestration assays, LC3 processing
Effect of different treatments on the extent of autophagic induction	GFP–LC3 vesicle formation, LC3 processing, quantitative electron microscopy

2.4.2.5 Common Questions

Common Questions and approaches to answer them are listed in Table 2.1.

2.4.3
Summary and Outlook

A major advance was made in the autophagy field with the identification of the Atg genes in the mid-1990s. Experimental methods based on the use of LC3 have made possible a quick and accurate analysis of autophagy, and do not require the time and expertise one would need, for example, for electron microscopy-based assays. This, and also the fact that technologies such as siRNA or the generation of transgenic mice are now widely available, should facilitate the development of new assays. In the future the generation of more autophagy knockout mice combined with crossbreeding with other transgenic mice will provide more information to investigate the role of autophagy, especially in cancer, neurological disorders (characterized by protein aggregates), and in infection and immunity. An under-represented area so far has been the development of high-throughput screens; however, the availability of new cell-based assays should now allow the screening of genome wide libraries in a relatively short time. Additionally it would be interesting to see *in vitro* assays to reconstitute the various steps (sequestration, formation and fusion) in autophagy developed to investigate the molecular requirements of autophagosome formation, fusion and degradation in mammalian cells.

References

1 Clark, S. L., Jr. **1957**. *J Cell Biol 3*, 349–362.
2 de Duve, C., Wattiaux, R. **1966**. *Annu Rev Physiol 28*, 435–492.
3 Ashford, T. P., Porter, K. R. **1962**. *J Cell Biol 12*, 198–202.
4 Seglen, P. O. **1987**. Regulation of autophagic protein degradation in isolated liver cells. In: *Lysosomes: Their Role in Protein Breakdown.* Glaumann, H., Ballard, F. (Eds.). Academic Press, London.
5 Novikoff, A. B., Essner, E., Quintana, N. **1964**. *Fedn Proc 23*, 1010–1022.
6 Arstila, A. U., Trump, B. F. **1968**. *Am J Pathol 53*, 687–733.
7 Fengsrud, M., Sneve, M. L., Overbye, A., Seglen, P. O. **2004**. Structural aspects of mammalian autophagy. In: *Autophagy.* Klionsky, D. J. (Ed). Landes Bioscience, Georgetown, TX.
8 Kim, J., Huang, W. P., Stromhaug, P. E., Klionsky, D. J. **2002**. *J Biol Chem 277*, 763–773.
9 Suzuki, K., Kirisako, T., Kamada, Y., Mizushima, N., Noda, T., Ohsumi, Y. **2001**. *EMBO J 20*, 5971–5981.
10 Mizushima, N., Yamamoto, A., Hatano, M., Kobayashi, Y., Kabeya, Y., Suzuki, K., Tokuhisa, T., Ohsumi, Y., Yoshimori, T. **2001**. *J Cell Biol 152*, 657–668
11 Reggiori, F., Klionsky, D. **2005**. *Curr Opin Cell Biol 17*, 1–8
12 Kopitz, J., Kisen, G. O., Gordon, P. B., Bohley, P., Seglen, P. O. **1990**. *J Cell Biol 111*, 941–953.
13 Kovacs, A. L., Seglen, P. O. **1981**. *Biochim Biophys Acta 676*, 213–220.
14 Seglen, P. O., Gordon, P. B., Holen, I. **1990**. *Semin Cell Biol 1*, 441–448
15 Reunanen, H., Hirsimaki, P., Punnonen, E. L. **1988**. *Comp Biochem Physiol A 90*, 321–327
16 Reunanen, H., Punnonen, E. L., Hirsimaki, P. **1985**. *Histochemistry 83*, 513–517
17 Punnonen, E. L., Pihakaski, K., Mattila, K., Lounatmaa, K., Hirsimaki, P. **1989**. *Cell Tissue Res 258*, 269–276.
18 Fengsrud, M., Erichsen, E. S., Berg, T. O., Raiborg, C., Seglen, P. O. **2000**. *Eur J Cell Biol 79*, 871–882.
19 Dunn, W. A., Jr. **1990**. *J Cell Biol 110*, 1923–1933.
20 Locke, M., Sykes, A. K. **1975**. *Tissue Cell 7*, 143–158
21 Frank, A. L., Christensen, A. K. **1968**. *J Cell Biol 36*, 1–13.
22 Yamamoto, A., Masaki, R., Fukui, Y., Tashiro, Y. **1990**. *J Histochem Cytochem 38*, 1571–1581.
23 Stromhaug, P. E., Berg, T. O., Fengsrud, M., Seglen, P. O. **1998**. *Biochem J 335*, 217–224.
24 Purhonen, P., Pursiainen, K., Reunanen, H. **1997**. *Eur J Cell Biol 74*, 63–67
25 Mizushima, N., Yamamoto, A., Matsui, M., Yoshimori, T., Ohsumi, Y. **2004**. *Mol Biol Cell 15*, 1101–1111.
26 Yue, Z., Jin, S., Yang, C., Levine, A. J., Heintz, N. **2003**. *Proc Natl Acad Sci USA 100*, 15077–15082.
27 Kuma, A., Hatano, M., Matsui, M., Yamamoto, A., Nakaya, H., Yoshimori, T., Ohsumi, Y., Tokuhisa, T., Mizushima, N. **2004**. *Nature 432*, 1032–1036.
28 Komatsu, M., Waguri, S., Ueno, T., Iwata, J., Murata, S., Tanida, I., Ezaki, J., Mizushima, N., Ohsumi, Y., Uchiyama, Y., Kominami, E., Tanaka, K., Chiba, T. **2005**. *J Cell Biol 169*, 425–434.
29 Eskelinen, E. L. **2005**. *Autophagy 1*, 1–10.
30 Hoyvik, H., Gordon, P. B., Berg, T. O., Stromhaug, P. E., Seglen, P. O. **1991**. *J Cell Biol 113*, 1305–1312.
31 Gruenberg, J., Stenmark, H. **2004**. *Nat Rev Mol Cell Biol 5*, 317–323.
32 Touret, N., Paroutis, P., Grinstein, S. **2005**. *J Leukoc Biol 77*, 878–885.
33 Tooze, S. A., Martens, G. J. M., Huttner, W. B. **2001**. *Trends Cell Biol 11*, 116–122.
34 Aniento, F., Emans, N., Griffiths, G., Gruenberg, J. **1993**. *J Cell Biol 123*, 1373–1387
35 Munro, S. **2002**. *Curr Opin Cell Biol 14*, 506–514.
36 Rodman, J., Wandinger-Ness, A. **2000**. *J Cell Sci 113 Pt 2*, 183–192.
37 Darsow, T., Rieder, S. E., Emr, S. D. **1997**. *J Cell Biol 138*, 517–529.
38 Ishihara, N., Hamasaki, M., Yokota, S., Suzuki, K., Kamada, Y., Kihara, A., Yoshimori, T., Noda, T., Ohsumi, Y. **2001**. *Mol Biol Cell 12*, 3690–3702.

39 Kirisako, T., Baba, M., Ishihara, N., Miyazawa, K., Ohsumi, M., Yoshimori, T., Noda, T., Ohsumi, Y. **1999**. *J Cell Biol 147*, 435–446.

40 Gutierrez, M. G., Munafo, D. B., Beron, W., Colombo, M. I. **2004**. *J Cell Sci 117*, 2687–2697

41 Jager, S., Bucci, C., Tanida, I., Ueno, T., Kominami, E., Saftig, P., Eskelinen, E.-L. **2004**. *J Cell Sci 117*, 4837–4848

42 Eskelinen, E. L. **2004**. Macroautophagy in mammalian cells. In: *Lysosomes*. Saftig, P. (Ed.). Landes Bioscience/Eurekah, Georgetown, TX.

43 Dunn, W. A., Jr. **1990**. *J Cell Biol 110*, 1935–1945.

44 De Duve, C. **1959**. Lysosomes, a new group of cytoplasmic particles. In: Hayashi, T. (Ed.). *Subcellular Particles*. Roland Press, New York.

45 Eskelinen, E.-L., Tanaka, Y., Saftig, P. **2003**. *Trends Cell Biol 13*, 137–145.

46 Bonifacino, J. S., Traub, L. M. **2003**. *Annu Rev Biochem 72*, 395–447

47 Mullins, C., Bonifacino, J. S. **2001**. *BioEssays 23*, 333–343.

48 Mullock, B. M., Bright, N. A., Fearon, C. W., Gray, S. R., Luzio, J. P. **1998**. *J Cell Biol 140*, 591–601.

49 Abeliovich, H., Darsow, T., Emr, S. D. **1999**. *EMBO J 18*, 6005–6016.

50 Efe, J. A., Botelho, R. J., Emr, S. D. **2005**. *Curr Opin Cell Biol 17*, 402–408

51 Gagnon, E., Duclos, S., Rondeau, C., Chevet, E., Cameron, P. H., Steele-Mortimer, O., Paiement, J., Bergeron, J. J., Desjardins, M. **2002**. *Cell 110*, 119–131.

52 McNew, J. A., Parlati, F., Fukuda, R., Johnston, R. J., Paz, K., Paumet, F., Sollner, T. H. **2000**. *Nature 407*, 153–159.

53 Gagnon, E., Bergeron, J. J., Desjardins, M. **2005**. *J Leukoc Biol 77*, 843–845.

54 Woodside, K. H., Mortimore, G. E. **1972**. *J Biol Chem 247*, 6474–6481.

55 Schworer, C., Shiffer, K., Mortimore, G. **1981**. *J Biol Chem 256*, 7652–7658

56 Mortimore, G. E., Schworer, C. M. **1977**. *Nature 270*, 174–176.

57 Blommaart, E. F. C., Luiken, J. J. F. P., Blommaart, P. J. E., van Woerkom, G. M., Meijer, A. J. **1995**. *J Biol Chem 270*, 2320–2326.

58 Kanazawa, T., Taneike, I., Akaishi, R., Yoshizawa, F., Furuya, N., Fujimura, S., Kadowaki, M. **2004**. *J Biol Chem 279*, 8452–8459.

59 Köchl, R., Hu, X., Chan, E., Tooze, S. A. **2006**. *Traffic 7*, 129–145.

60 Seglen, P. O., Gordon, P. B., Poli, A. **1980**. *Biochim Biophys Acta 630*, 103–118.

61 Fulks, R. M., Li, J. B., Goldberg, A. L. **1975**. *J Biol Chem 250*, 290–298.

62 Chua, B., Siehl, D. L., Morgan, H. E. **1979**. *J Biol Chem 254*, 8358–8362.

63 Kabeya, Y., Mizushima, N., Ueno, T., Yamamoto, A., Kirisako, T., Noda, T., Kominami, E., Ohsumi, Y., Yoshimori, T. **2000**. *EMBO J 19*, 5720–5728

64 Ogier-Denis, E., Pattingre, S., El Benna, J., Codogno, P. **2000**. *J Biol Chem 275*, 39090–39095.

65 Mordier, S., Deval, C., Bechet, D., Tassa, A., Ferrara, M. **2000**. *J Biol Chem 275*, 29900–29906.

66 Otto, G. P., Wu, M. Y., Clarke, M., Lu, H., Anderson, O. R., Hilbi, H., Shuman, H. A., Kessin, R. H. **2004**. *Mol Microbiol 51*, 63–72.

67 Rusten, T. E., Lindmo, K., Juhasz, G., Sass, M., Seglen, P. O., Brech, A., Stenmark, H. **2004**. *Dev Cell 7*, 179–192.

68 Poso, A., Wert, J., Jr., Mortimore, G. **1982**. *J Biol Chem 257*, 12114–12120.

69 Miotto, G., Venerando, R., Khurana, K. K., Siliprandi, N., Mortimore, G. E. **1992**. *J Biol Chem 267*, 22066–22072.

70 Miotto, G., Venerando, R., Marin, O., Siliprandi, N., Mortimore, G. E. **1994**. *J Biol Chem 269*, 25348–25353.

71 Pfeifer, U. **1978**. *J Cell Biol 78*, 152–167

72 Neely, A. N., Cox, J. R., Fortney, J. A., Schworer, C. M., Mortimore, G. E. **1977**. *J Biol Chem 252*, 6948–6954.

73 Melendez, A., Talloczy, Z., Seaman, M., Eskelinen, E.-L., Hall, D. H., Levine, B. **2003**. *Science 301*, 1387–1391.

74 Scott, R. C., Schuldiner, O., Neufeld, T. P. **2004**. *Dev Cell 7*, 167–178.

75 Schworer, C. M., Mortimore, G. E. **1979**. *Proc Nat Acad Sci USA 76*, 3169–3173

76 Daido, S., Kanzawa, T., Yamamoto, A., Takeuchi, H., Kondo, Y., Kondo, S. **2004**. *Cancer Res 64*, 4286–4293.

77 Scarlatti, F., Bauvy, C., Ventruti, A., Sala, G., Cluzeaud, F., Vandewalle, A., Ghido-

ni, R., Codogno, P. 2004. *J Biol Chem* 279, 18384–18391.

78 Yu, L., Alva, A., Su, H., Dutt, P., Freundt, E., Welsh, S., Baehrecke, E. H., Lenardo, M. J. 2004. *Science 304,* 1500–1502.

79 Shimizu, S., Kanaseki, T., Mizushima, N., Mizuta, T., Arakawa-Kobayashi, S., Thompson, C. B., Tsujimoto, Y. 2004. *Nat Cell Biol 6,* 1221–1228

80 Noda, T., Ohsumi, Y. 1998. *J Biol Chem* 273, 3963–3966.

81 Gutierrez, M. G., Master, S. S., Singh, S. B., Taylor, G. A., Colombo, M. I., Deretic, V. 2004. *Cell 119,* 753–766.

82 Ravikumar, B., Vacher, C., Berger, Z., Davies, J. E., Luo, S., Oroz, L. G., Scaravilli, F., Easton, D. F., Duden, R., O'Kane, C. J., Rubinsztein, D. C. 2004. *Nat Genet 36,* 585–595.

83 Jacinto, E., Hall, M. N. 2003. *Nat Rev Mol Cell Biol 4,* 117–126.

84 Fingar, D. C., Blenis, J. 2004. *Oncogene 23,* 3151–3171.

85 Lynch, C. J., Fox, H. L., Vary, T. C., Jefferson, L. S., Kimball, S. R. 2000. *J Cell Biochem 77,* 234–251.

86 Kim, D. H., Sarbassov dos, D., Ali, S. M., King, J. E., Latek, R. R., Erdjument-Bromage, H., Tempst, P., Sabatini, D. M. 2002. *Cell 110,* 163–175.

87 Kim, D. H., Sarbassov, D. D., Ali, S. M., Latek, R. R., Guntur, K. V., Erdjument-Bromage, H., Tempst, P., Sabatini, D. M. 2003. *Mol Cell 11,* 895–904.

88 Hara, K., Maruki, Y., Long, X., Yoshino, K., Oshiro, N., Hidayat, S., Tokunaga, C., Avruch, J., Yonezawa, K. 2002. *Cell 110,* 177–189.

89 Loewith, R., Jacinto, E., Wullschleger, S., Lorberg, A., Crespo, J. L., Bonenfant, D., Oppliger, W., Jenoe, P., Hall, M. N. 2002. *Mol Cell 10,* 457–468.

90 Sarbassov, D. D., Ali, S. M., Kim, D. H., Guertin, D. A., Latek, R. R., Erdjument-Bromage, H., Tempst, P., Sabatini, D. M. 2004. *Curr Biol 14,* 1296–1302.

91 Martin, K. A., Blenis, J. 2002. *Adv Cancer Res 86,* 1–39.

92 Isotani, S., Hara, K., Tokunaga, C., Inoue, H., Avruch, J., Yonezawa, K. 1999. *J Biol Chem 274,* 34493–34498.

93 Burnett, P. E., Barrow, R. K., Cohen, N. A., Snyder, S. H., Sabatini, D. M. 1998. *Proc Natl Acad Sci USA 95,* 1432–1437

94 Biondi, R. M., Kieloch, A., Currie, R. A., Deak, M., Alessi, D. R. 2001. *EMBO J 20,* 4380–4390.

95 Pearson, R. B., Dennis, P. B., Han, J. W., Williamson, N. A., Kozma, S. C., Wettenhall, R. E., Thomas, G. 1995. *EMBO J 14,* 5279–5287.

96 Kwiatkowski, D. J. 2003. *Ann Hum Genet 67,* 87–96.

97 Long, X., Lin, Y., Ortiz-Vega, S., Yonezawa, K., Avruch, J. 2005. *Curr Biol 15,* 702–713.

98 Gao, X., Zhang, Y., Arrazola, P., Hino, O., Kobayashi, T., Yeung, R. S., Ru, B., Pan, D. 2002. *Nat Cell Biol 4,* 699–704.

99 Inoki, K., Li, Y., Zhu, T., Wu, J., Guan, K. L. 2002. *Nat Cell Biol 4,* 648–657

100 Potter, C. J., Pedraza, L. G., Xu, T. 2002. *Nat Cell Biol 4,* 658–665.

101 Manning, B. D., Tee, A. R., Logsdon, M. N., Blenis, J., Cantley, L. C. 2002. *Mol Cell 10,* 151–162.

102 Petiot, A., Ogier-Denis, E., Blommaart, E. F. C., Meijer, A. J., Codogno, P. 2000. *J Biol Chem 275,* 992–998

103 Arico, S., Petiot, A., Bauvy, C., Dubbelhuis, P. F., Meijer, A. J., Codogno, P., Ogier-Denis, E. 2001. *J Biol Chem 276,* 35243–35246.

104 Long, X., Ortiz-Vega, S., Lin, Y., Avruch, J. 2005. *J Biol Chem 280,* 23433–23436.

105 Hardie, D. G., Carling, D., Carlson, M. 1998. *Annu Rev Biochem 67,* 821–855.

106 Hardie, D. G. 2005. *Curr Opin Cell Biol 17,* 167–173.

107 Inoki, K., Zhu, T., Guan, K. L. 2003. *Cell 115,* 577–590.

108 Ravikumar, B., Stewart, A., Kita, H., Kato, K., Duden, R., Rubinsztein, D. C. 2003. *Hum Mol Genet 12,* 985–994.

109 Lum, J. J., Bauer, D. E., Kong, M., Harris, M. H., Li, C., Lindsten, T., Thompson, C. B. 2005. *Cell 120,* 237–248

110 Samari, H. R., Seglen, P. O. 1998. *J Biol Chem 273,* 23758–23763.

111 Plomp, P. J., Gordon, P. B., Meijer, A. J., Hoyvik, H., Seglen, P. O. 1989. *J Biol Chem 264,* 6699–6704.

112 Plomp, P.J., Wolvetang, E.J., Groen, A.K., Meijer, A.J., Gordon, P.B., Seglen, P.O. **1987**. *Eur J Biochem 164*, 197–203.

113 Jefferies, H.B., Fumagalli, S., Dennis, P.B., Reinhard, C., Pearson, R.B., Thomas, G. **1997**. *EMBO J 16*, 3693–3704.

114 Jefferies, H.B., Reinhard, C., Kozma, S.C., Thomas, G. **1994**. *Proc Natl Acad Sci USA 91*, 4441–4445.

115 Terada, N., Patel, H.R., Takase, K., Kohno, K., Nairn, A.C., Gelfand, E.W. **1994**. *Proc Natl Acad Sci USA 91*, 11477–11481.

116 Tang, H., Hornstein, E., Stolovich, M., Levy, G., Livingstone, M., Templeton, D., Avruch, J., Meyuhas, O. **2001**. *Mol Cell Biol 21*, 8671–8683.

117 Stolovich, M., Tang, H., Hornstein, E., Levy, G., Cohen, R., Bae, S.S., Birnbaum, M.J., Meyuhas, O. **2002**. *Mol Cell Biol 22*, 8101–8113.

118 Kamada, Y., Funakoshi, T., Shintani, T., Nagano, K., Ohsumi, M., Ohsumi, Y. **2000**. *J Cell Biol 150*, 1507–1513.

119 Abeliovich, H., Zhang, C., Dunn, W.A., Jr., Shokat, K.M., Klionsky, D.J. **2003**. *Mol Biol Cell 14*, 477–490.

120 Seglen, P.O., Gordon, P.B. **1982**. *Proc Natl Acad Sci USA 79*, 1889–1892.

121 Blommaart, E.F., Krause, U., Schellens, J.P., Vreeling-Sindelarova, H., Meijer, A.J. **1997**. *Eur J Biochem 243*, 240–246.

122 Kihara, A., Noda, T., Ishihara, N., Ohsumi, Y. **2001**. *J Cell Biol 152*, 519–530.

123 Tassa, A., Roux, M.P., Attaix, D., Bechet, D.M. **2003**. *Biochem J 376*, 577–586.

124 Byfield, M.P., Murray, J.T., Backer, J.M. **2005**. *J Biol Chem 280*, 33076–33082.

125 Volinia, S., Dhand, R., Vanhaesebroeck, B., MacDougall, L.K., Stein, R., Zvelebil, M.J., Domin, J., Panaretou, C., Waterfield, M.D. **1995**. *EMBO J 14*, 3339–3348

126 Petiot, A., Ogier-Denis, E., Blommaart, E.F., Meijer, A.J., Codogno, P. **2000**. *J Biol Chem 275*, 992–998

127 Nobukuni, T., Joaquin, M., Roccio, M., Dann, S.G., Kim, S.Y., Gulati, P., Byfield, M.P., Backer, J.M., Natt, F., Bos, J.L., Zwartkruis, F.J., Thomas, G. **2005**. *Proc Natl Acad Sci USA 102*, 14238–14243.

128 Pattingre, S., Tassa, A., Qu, X., Garuti, R., Liang, X.H., Mizushima, N., Packer, M., Schneider, M.D., Levine, B. **2005**. *Cell 122*, 927–939.

129 Beugnet, A., Tee, A.R., Taylor, P.M., Proud, C.G. **2003**. *Biochem J 372*, 555–566.

130 Harding, T.M., Morano, K.A., Scott, S.V., Klionsky, D.J. **1995**. *J Cell Biol 131*, 591–602.

131 Thumm, M., Egner, R., Koch, B., Schlumpberger, M., Straub, M., Veenhuis, M., Wolf, D.H. **1994**. *FEBS Lett 349*, 275–280.

132 Tsukada, M., Ohsumi, Y. **1993**. *FEBS Lett 333*, 169–174.

133 Tanida, I., Komatsu, M., Ueno, T., Kominami, E. **2003**. *Biochem Biophys Res Commun 300*, 637–644.

134 Mann, S.S., Hammarback, J.A. **1994**. *J Biol Chem 269*, 11492–11497

135 Hemelaar, J., Lelyveld, V.S., Kessler, B.M., Ploegh, H.L. **2003**. *J Biol Chem 278*, 51841–51850.

136 Ichimura, Y., Kirisako, T., Takao, T., Satomi, Y., Shimonishi, Y., Ishihara, N., Mizushima, N., Tanida, I., Kominami, E., Ohsumi, M., Noda, T., Ohsumi, Y. **2000**. *Nature 408*, 488–492.

137 Tanida, I., Ueno, T., Kominami, E. **2004**. *Int J Biochem Cell Biol 36*, 2503–2518

138 Kabeya, Y., Mizushima, N., Yamamoto, A., Oshitani-Okamoto, S., Ohsumi, Y., Yoshimori, T. **2004**. *J Cell Sci 117*, 2805–2812.

139 He, H., Dang, Y., Dai, F., Guo, Z., Wu, J., She, X., Pei, Y., Chen, Y., Ling, W., Wu, C., Zhao, S., Liu, J.O., Yu, L. **2003**. *J Biol Chem 278*, 29278–29287

140 Bampton, E.T.W., Goemans, C.G., D., N., Mizushima, N., Tolkovsky, A.M. **2005**. *Autophagy 1*, 23–36.

141 Esselens, C., Oorschot, V., Baert, V., Raemaekers, T., Spittaels, K., Serneels, L., Zheng, H., Saftig, P., De Strooper, B., Klumperman, J., Annaert, W. **2004**. *J Cell Biol 166*, 1041–1054.

142 Simonsen, A., Birkeland, H.C.G., Gillooly, D.J., Mizushima, N., Kuma, A., Yoshimori, T., Slagsvold, T., Brech, A.,

Stenmark, H. **2004**. *J Cell Sci 117*, 4239–4251.

143 Biederbick, A., Kern, H. F., Elasser, H. P. **1995**. *Eur J Cell Biol 66*, 3–14.

144 Munafo, D. B., Colombo, M. I. **2001**. *J Cell Sci 114*, 3619–3629.

145 Bera, A., Singh, S., Nagaraj, R., Vaidya, T. **2003**. *Mol Biochem Parasitol 127*, 23–35.

146 Amer, A. O., Swanson, M. S. **2005**. *Cell Microbiol 7*, 765–778

147 Contento, A. L., Xiong, Y., Bassham, D. C. **2005**. *Plant J 42*, 598–608

148 Seglen, P. O., Bohley, P. **1992**. *Experientia 48*, 158–172.

149 Mousavi, S. A., Brech, A., Berg, T., Kjeken, R. **2003**. *Biochem J 372*, 861–869.

150 Berg, T. O., Fengsrud, M., Stromhaug, P. E., Berg, T., Seglen, P. O. **1998**. *J Biol Chem 273*, 21883–21892.

151 Gordon, P. B., Seglen, P. O. **1982**. *Exp Cell Res 142*, 1–14.

152 Gordon, P. B., Kisen, G. O., Kovacs, A. L., Seglen, P. O. **1989**. *Biochem Soc Symp 55*, 129–143.

153 Mousavi, S. A., Kjeken, R., Berg, T. O., Seglen, P. O., Berg, T., Brech, A. **2001**. *Biochim Biophys Acta Biomembr 1510*, 243–257

154 Seglen, P. O., Gordon, P. B., Tolleshaug, H., Hoyvik, H. **1986**. *Exp Cell Res 162*, 273–277

155 Seglen, P. O., Gordon, P. B., Hoyvik, H. **1986**. *Biomed Biochim Acta 45*, 1647–1656.

156 Asanuma, K., Tanida, I., Shirato, I., Ueno, T., Takahara, H., Nishitani, T., Kominami, E., Tomino, Y. **2003**. *FASEB J 17*, 1165–1167

157 Furuta, S., Hidaka, E., Ogata, A., Yokota, S., Kamata, T. **2004**. *Oncogene 23*, 38983904.

158 Tanaka, Y., Guhde, G., Suter, A., Eskelinen, E. L., Hartmann, D., Lullmann-Rauch, R., Janssen, P. M., Blanz, J., von Figura, K., Saftig, P. **2000**. *Nature 406*, 902–906.

159 Liang, X. H., Jackson, S., Seaman, M., Brown, K., Kempkes, B., Hibshoosh, H., Levine, B. **1999**. *Nature 402*, 672–676.

160 Gronostajski, R. M., Pardee, A. B. **1984**. *J Cell Physiol 119*, 127–132.

161 Mortimore, G. E., Pösö, A. R., Kadowaki, M., Wert, J. J. J. **1987**. *J Biol Chem 262*, 16322–16327

162 Fengsrud, M., Roos, N., Berg, T., Liou, W., Slot, J. W., Seglen, P. O. **1995**. *Exp Cell Res 221*, 504–519.

163 Fengsrud, M., Raiborg, C., Berg, T. O., Stromhaug, P. E., Ueno, T., Erichsen, E. S., Seglen, P. O. **2000**. *Biochem J 352*, 773–781.

164 Mortimore, G. E., Poso, A. R. **1987**. *Annu Rev Nutr 7*, 539–564.

165 Alpin, A., Jasionowski, T., Tuttle, D. L., Lenk, S. E., Dunn, W. A. J. **1992**. *J Cell Physiol 152*, 458–466.

166 Jackson, W. T., Giddings, T. H., Jr., Taylor, M. P., Mulinyawe, S., Rabinovitch, M., Kopito, R. R., Kirkegaard, K. **2005**. *PLoS Biol 3*, e156.

167 Boya, P., Gonzalez-Polo, R.-A., Casares, N., Perfettini, J.-L., Dessen, P., Larochette, N., Metivier, D., Meley, D., Souquere, S., Yoshimori, T., Pierron, G., Codogno, P., Kroemer, G. **2005**. *Mol Cell Biol 25*, 1025–1040.

3
Transgenic Models of Autophagy

Noboru Mizushima

3.1
Molecular Mechanism of Mammalian Autophagy

Although macroautophagy was discovered in the 1950s, it has not been understood at the molecular level until recently. A breakthrough came from yeast genetic studies, which have identified more than 20 genes required for autophagy, most of which function in autophagosome formation (*atg* genes) [1, 2]. Importantly, most of them are conserved in higher eukaryotes, including mammals and plants. These Atg proteins are classified into several functional groups.

3.1.1
Atg12 Conjugation System

Atg12, a ubiquitin-like protein, is covalently attached to Atg5 through a ubiquitination-like reaction [3–5]. Atg7 and Atg10 function as Atg12-activating (E1) and Atg12-conjugating (E2) enzymes, respectively [6–13]. The resulting Atg12–Atg5 conjugate behaves as if it is a single molecule. Mammalian and yeast Atg12–Atg5 conjugate interact with Atg16L and Atg16, respectively, and form large protein complexes [14–16]. Mammalian Atg16L is much larger than yeast Atg16 and possesses a C-terminal domain containing seven WD repeats that is from yeast Apg16. Therefore, it is speculated that Atg16L could have some additional function, although it has not been demonstrated.

Most Apg12–Apg5 · Apg16L complexes resides in the cytosol and a very small fraction localizes on the outer side of the isolation membrane throughout its elongation process [17] (Fig. 3.1). The complex dissociates from the membrane upon completion of autophagosome formation. Atg12–Atg5 is required for elongation of the isolation membranes [17].

Autophagy in Immunity and Infection. Edited by Vojo Deretic
Copyright © 2006 WILEY-VCH Verlag GmbH & Co. KGaA, Weinheim
ISBN: 3-527-31450-4

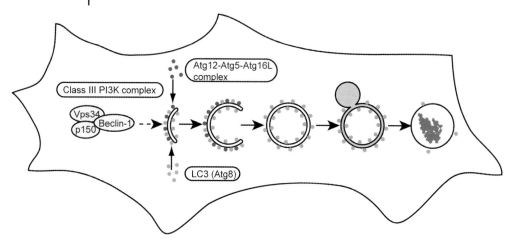

Fig. 3.1 Model of autophagosome formation in mammalian cells. The Atg12–Atg5 conjugate and Atg16L localize to the isolation membrane throughout its elongation process. LC3 is recruited to the membrane in an Atg5-dependent manner. Atg12–Atg5 and Atg16L dissociate from the membrane upon completion of autophagosome formation, while LC3 remains on the autophagosome membrane. Atg5 and its modification by Atg12 are required for elongation of the isolation membrane. The class III PI3K complex containing beclin 1 likely functions upstream of the elongation step.

3.1.2
Atg8/Microtubule-associated Protein 1 Light Chain 3 (LC3) Conjugation System

Yeast Atg8 is another ubiquitin-like protein [18]. It has at least three mammalian homologs: LC3 [19], the Golgi-associated ATPase enhancer of 16 kDa (GATE-16) [20] and γ-aminobutyric acid (GABA)$_A$ receptor-associated protein (GABARAP) [21]. The Atg8 family proteins are conjugated to PE [22, 23]. Atg7 is shared with the Atg12 system as the activating enzyme and Atg3 functions as an E2 enzyme specific for Atg8 proteins [10, 24]. Among these Atg8 homologs, LC3 has been analyzed most extensively in the autophagy pathway. In contrast to Atg12–Atg5, LC3 localizes on the membrane of complete spherical autophagosomes as well as on the isolation membranes (Fig. 3.1) [25]. The membrane association of LC3 depends on the function of Atg12–Atg5 and phosphatidylethanolamine (PE) conjugation [17, 26, 27]. LC3 is also detected on the membrane of autolysosomes, but in amounts less than that of autophagosomes [25, 28]. LC3–PE on the outer membrane of autophagosome is deconjugated by Atg4 homologs (also called autophagins) and free LC3 is released again into the cytoplasm [22, 23, 29–31]. LC3–PE on the inner membrane is degraded by lysosomal proteinases [32]. The role of LC3 is unknown, but yeast studies indicated that it is also required for autophagosome generation [18].

The function and localization of other two Atg8 homologs have been proposed, but are still controversial. GABARAP was suggested to be involved in GABA$_A$ receptor clustering [33] or transport [34]. GATE-16 has been suggested to be an intra-Golgi transport modulator that interacts with *N*-ethylmaleimide-sensitive factor (NSF) and Golgi v-SNARE (soluble *N*-ethylmalemide-sensitive fusion protein (NSF) attachment receptor) GOS-28 [20]. However, transfection experiments showed that both GABARAP and GATE-16 are able to reside on autophagosomal membranes [22]. It is not know whether these three homologs are functionally redundant or have some specific roles.

3.1.3
Class III Phosphatidylinositol-3-kinase (PI3K) Complex

Yeast Atg6/Vps30 and its mammalian homolog beclin 1 are the components of class III PI3K complex [35, 36]. This complex is thought to function upstream of autophagosome formation. In yeast, autophagosomes seem to be generated from a structure near the vacuole, termed the pre-autophagosomal structure (PAS) [26, 37]. Formation of PAS requires a PI3K complex made up of Vps15, Atg6, Atg14 and Vps34 [36]. Apart from this type of class III complex, there is another kind of PI3K complex in yeast consisting of Vps15, Atg6, Vps38 and Vps34, which functions in the vacuolar protein sorting pathway, not for autophagy. Thus, Atg6 is not a protein specific for autophagy. Recently, it was shown that *Caenorhabditis elegans* beclin is also essential for both autophagy and endocytosis [38].

Mammalian Atg6 was originally identified as a Bcl-2-interacting protein, beclin 1 [39]. Later it was suggested that beclin 1 is a negative regulator of tumorigenesis and also required for autophagy in mammalian cells [40]. Primary localization of beclin 1 is the *trans*-Golgi Network (TGN) [35]. As the transport of lysosomal hydrolases from TGN to endosomes depends on PI3K activity in mammalian cells [41, 42], TGN-associated beclin 1 would function for this pathway. However, it was also shown that beclin 1 may not be essential for the transport of lysosomal enzymes in mammalian cells [43].

3.1.4
Atg1 Kinase Complex

Atg1 is a protein kinase, which forms a complex with Atg13 and Atg17 in yeast [44–46]. The kinase activity of Apg1 is upregulated by starvation signaling, which is mediated by Atg13. Under normal conditions, Atg13 is hyperphosphorylated in a target of rapamycin (TOR)-dependent manner, but it is immediately dephosphorylated by starvation treatment. Thus, Atg1 complex could be one of the signal transducers. However, yeast genetic analyses have suggested that the Atg1 complex functions at a later step of autophagosome formation, not at the PAS formation step [26]. This complex seems to have a direct role in membrane dynamism or autophagosome size regulation, not only in signal transduction [46].

Homologs of Atg1 have been identified in *C. elegans* (UNC51) [47] and mammals (ULK1 and ULK2) [48, 49]. ULK1 interacts with GATE-16 and GABARAP [50], suggesting that Atg8 homologs are functionally related to Atg1 protein complex. Although the function of these Atg1 homologs in autophagy is not clear, they were reported to have an additional function in axon outgrowth of neurons [51, 52]. This finding is consistent with the "UNCordinate" phenotype of the worm mutant of Atg1 [53].

3.1.5
Other Factors

Atg9 is a transmembrane protein and is required for PAS formation in yeast [26, 54]. Retrieval of Atg9 from the PAS requires Atg18 and Atg2 [55]. Indeed, Atg9 interacts with Atg2 [56] and Atg18 [55]. Mammalian homologs of Atg9 [57] and Atg18 [WD-repeat protein interacting with phosphoinosides (WIPI)] [58, 59] have been identified, but their role in autophagy remains unclear.

3.2
Autophagy Indicator Mice: Green Fluorescent Protein (GFP)–LC3 Transgenic Mice

To date, electron microscopy has been the only morphological method to monitor autophagy. Unfortunately, this is a method requiring many skills and much time, and sometimes it is difficult to distinguish autophagic vacuoles (AVs) from other structures just by morphology. Recent studies on the molecular mechanism of autophagy provided several marker proteins for autophagosomes [60]. For example, Atg12–Atg5 and Atg16L are specific markers for the isolation membrane, and LC3 can be used as a general marker for autophagic membranes (Fig. 3.1). These localizations are easily examined by generating chimeric proteins fused with GFP or its derivatives. Autophagosomes can be recognized as ring-shaped structures by fluorescence microscopy if their diameters are larger than 1 m in cultured cells. The examination of GFP–LC3 localization is a very simple method, which requires only a high-resolution fluorescence microscope. In addition, real-time observation in living cells is feasible.

This method has been applied to *in vivo* analysis, by generating GFP–LC3 transgenic mice [61]. In this mouse model, GFP–LC3 is overexpressed in almost all tissues under the control of the constitutive CAG promoter. Using this transgenic mouse model, the occurrence of autophagy in mouse tissues can be directly monitored by simply creating cryosections and subsequent fluorescence microscopic analysis. In general, autophagy is induced in almost all tissues during starvation. Active autophagy was observed in skeletal muscle, liver, heart, exocrine glands such as pancreatic acinar cells and seminal gland cells, and podocytes in kidney after 24-h food withdrawal (Fig. 3.2) [61]. This suggests that the major role of autophagy is degradation of self proteins as a starvation response.

In some tissues, autophagy even occurs actively without starvation treatments [61]. Among these, thymic epithelial cells show the highest basal levels under nutrient-rich conditions. Younger mice and late-stage embryos also show very active autophagy in thymic epithelial cells (our unpublished observation). In

a)

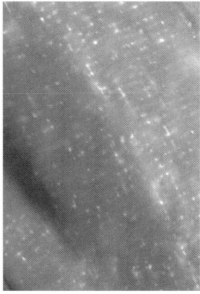

b)

Fig. 3.2 *In vivo* analysis of autophagy using GFP–LC3 mice. Gastrocnemius muscle samples were prepared from GFP–LC3 transgenic mice before (A) or after 24 h of starvation (B) and fixed with 4% paraformaldehyde. Cryosections were analyzed by fluorescence microscopy. GFP–LC3 dots represent autophagosomes. Bar = 10 μm.

general, cytoplasmic antigens such as viral proteins are processed by the proteasome and presented via major histocompatibility (MHC) class I to CD8$^+$ T cells. On the other hand, exogenous antigens are taken up by antigen-presenting cells via the endocytic pathway and processed in the lysosome or MHC Class II compartment. There, antigen peptides become associated with class II MHC molecules, and are then delivered to the plasma membrane and presented to specific CD4$^+$ T cells. However, many reports have demonstrated that endogenous proteins, particularly those that are long lived, are also presented on class II MHC [62–65]. Although how cytoplasmic proteins are loaded on class II MHC has not been completely understood, autophagy is now thought to be one of the pathways [66–69]. Since thymic epithelial cells are not thought to have phagocytic activity, it is reasonable to hypothesize that they provide self-antigens from their own cytoplasm. In this scenario, autophagy in thymic epithelial cells might be involved in T cell development and central tolerance.

It is interesting that autophagy was not induced in the brain even after 48-h food withdrawal. However, this phenomenon is only observed *in vivo*. In primary cultured neural cells of GFP–LC3 mice, numerous GFP–LC3 dots can be induced when they are cultured in starvation media (our unpublished observation). Thus, neural cells have the ability to induce autophagy. The brain may not be starved by such simple food withdrawal treatment. These observations revealed that, although the primary role of autophagy seems to be starvation adaptation, autophagy is differently induced among tissues, suggesting that the regulation of autophagy is not uniform, but rather is organ dependent.

GFP–LC3 dot formation at the embryonic and perinatal stages was also analyzed. Autophagy seems to be suppressed throughout the embryonic period, except some in tissues such as thymic epithelial cells. However, the formation of the GFP–LC3 dots was extensively induced in various tissues soon after a natural delivery [70]. In particular, the heart muscle, diaphragm, alveolar cells and skin displayed massive autophagy. The autophagic activity reached a maximum level 3–6 h after birth, although the neonatal mice began suckling before that time. The number of GFP–LC3 dots gradually decreased to basal levels by day 1 or 2. These results suggest that massive autophagy is transiently induced in normal neonatal mice under physiological conditions, probably in response to the nutrient limitations imposed by the sudden termination of the transplacental supply (as discussed below).

Although this transgenic mouse model is very useful for *in vivo* studies, there are a few possible limitations. First, GFP–LC3 localization only represents autophagosome formation. Since autolysosomes have less membrane-bound LC3 than autophagosomes, the appearance of GFP–LC3 dots does not guarantee autophagic "degradation". On the contrary, too fast fusion of autophagosomes with lysosomes may result in a lower number of GFP–LC3 dots, which would underestimate the autophagic activity.

Another problem could be the difficulty of strict quantification of the number of GFP–LC3 structures using software. Uneven cytosolic background signals makes thresholding difficult or unable to extract weak dot signals. To better extract the dot signals, the "Top Hat" algorithm of MetaMorph Series Version 6 (Molecular Devices, Sunnyvale, CA) is useful. Small dot peaks can be extracted from the surrounding relatively lower background signals irrespective of the absolute signal intensity.

It is also very important to distinguish the true GFP–LC3 dot signals from autofluorescent signals. Some cells such as neurons show autofluorescent dot structures like lipofuscin. Such artifacts may be avoided by the following two methods. First, it is particularly important to compare samples expressing GFP–LC3 with nontransgenic control samples. Second, specific GFP–LC3 signals should not be detected using other fluorescence filter sets such as rhodamine, Cy5 and UV. True GFP–LC3 signals should be detected specifically by the GFP or FITC filter set.

3.3
Mouse Models Deficient for Autophagy-related Genes

As discussed above, the most fundamental and evolutionarily conserved functions of autophagy would be the starvation adaptation. Autophagy-defective yeast mutants lose viability during starvation [71]. However, there has been no direct evidence showing that autophagy is also important in mammals in response to starvation. Moreover, it has remained unclear whether autophagy has some critical roles other than starvation adaptation. However, recent mouse genetic approaches using the gene-targeting technique have revealed the fundamental role of autophagy in mammals.

3.3.1
Atg5-deficient Mice

Atg5 is an acceptor molecule for the ubiquitin-like molecule, Atg12 [3, 4]. Atg5 and its proper modification with Atg12 are required for the elongation of the autophagic isolation membrane [17]. $atg5^{-/-}$ mice are born at the expected Mendelian frequency and they appear almost normal at birth, although the body weight of $atg5^{-/-}$ mice is slightly lower than that of wild-type mice [70]. Electron microscopic analysis confirmed that autolysosomes were not present in tissues from homozygous mutants. These data suggest that $atg5^{-/-}$ mice survive fetal development almost normally, although many studies have suggested possible roles for autophagy in development and cell death.

Despite the minimal abnormalities present at birth, most of $atg5^{-/-}$ neonates died within 1 day of delivery [70]. Heterozygous mice did not display any abnormal phenotypes. The majority of the $atg5^{-/-}$ neonates were found with no milk in their stomachs, suggesting that they may have had a suckling defect. However, the early death of the homozygous mice was not simply due to suckling failure because the survival time of $atg5^{-/-}$ mice was much shorter than wild-type mice when compared under nonsuckling conditions after Caesarean delivery. The survival time of $atg5^{-/-}$ neonates could be delayed by forced milk ingestion, suggesting that a major problem of $atg5^{-/-}$ neonates was a lack of nutrients. Soon after the Caesarian delivery, plasma amino acid concentrations in the $atg5^{-/-}$ neonates were not different from those of wild-type littermates. However, at 10 h after the Caesarian delivery, the amino acid concentration of $atg5^{-/-}$ mice was significantly lower than that of wild-type mice. In particular, the plasma concentration of essential amino acids and branched-chain amino acids showed large differences. A similar pattern was observed for amino acid concentrations in various tissues such as liver, heart and brain. In addition, tissue energy levels estimated by activity of AMP-activated protein kinase were also low in 10-h fasting $atg5^{-/-}$ mice. Therefore, early neonatal $atg5^{-/-}$ mice suffer from systemic amino acid and energy insufficiency, suggesting that neonates could use the amino acids produced by autophagy for energy homeostasis (Fig. 3.3).

This is the first demonstration of the importance of autophagy in mammals as the starvation response. To overcome life-threatening problems, autophagy must be activated to maintain an adequate amino acid pool. Amino acids produced by autophagy can be used as an energy source, for gluconeogenesis in the liver or for the new protein synthesis required for the proper starvation response. Thus far, a number of developmental defects have been reported in autophagy mutants in several species. In *Saccharomyces cerevisiae*, autophagy mutants are defective in spore formation [71], while autophagy mutants of *Dictyos-*

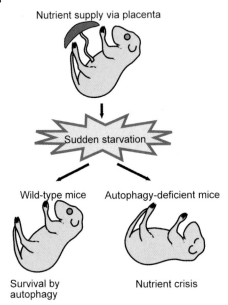

Nutrient supply via placenta

Sudden starvation

Wild-type mice   Autophagy-deficient mice

Survival by   Nutrient crisis
autophagy

Fig. 3.3 Schematic representation of the phenotype of $atg5^{-/-}$ and $atg7^{-/-}$ mice.

telium discoideum are defective in normal multicellular developmental processes such as aggregation formation and fruiting body formation [72]. Premature death from the third larval to pupal stages was reported in *Drosophila melanogaster* mutants [73, 74]. Finally, dauer formation is abnormal in *C. elegans* autophagy mutants [75]. Considering that these developmental defects are related to nutrient starvation [76], the insufficiency of amino acids could explain many of these phenotypes that result from the loss of autophagy.

Recently, the involvement of autophagy in cell death has been suggested in a number of physiological and pathological processes. One of them is a type of cell death during development, so-called autophagic degeneration or type 2 cell death, which is distinct from typical apoptosis [77, 78]. Numerous AVs are accumulated in the cytoplasm with relatively intact nuclei. However, the role of autophagy in this type of cell death is still controversial [79–81]. Some studies demonstrated that autophagy is causative of this type of cell death [82, 83], while other groups reported that autophagy is a survival mechanism against cell death [84, 85]. Finally, a recent study suggests that autophagy is a secondary phenomenon, which does not affect cell death/survival [86]. The fact that Atg5 knockout mice are born without an apparent increase or decrease in cell number suggests that autophagy is not critical for cell death, at least during mammalian embryogenesis [70]. Therefore, the physiological role of autophagy in cell death should be carefully reconsidered.

3.3.2
Atg7-deficient Mice

Atg7 is a ubiquitin-activating enzyme (E1)-like protein that catalyzes both Atg12 and Atg8 family proteins [11]. The phenotype of Atg7 knockout mice was very similar to that of Atg5 knockout mice [87]. They are born at Mendelian frequency and their body weight is slightly lower than that of wild-type mice. The survival times of $atg7^{-/-}$ neonates under starvation conditions are significantly shorter and they exhibit reduced amino acid concentrations in plasma. This phenotype confirmed the crucial role of autophagy in survival of the starvation period.

To determine the role of autophagy in adult tissues, liver-specific $atg7^{-/-}$ mice were also generated [87]. Mice homozygous for the $atg7$flox allele ($atg7^{f/f}$) were crossed with Mx1-Cre transgenic mice. In the resulting mice, the expression of the Cre recombinase can be induced by injection of interferon-γ or polyinosinic acid–polycytidylic acid (pIpC) so that the exon 14 of the $atg7$ gene between the two flox sequences is excised. Effective recombination was confirmed in the liver and spleen, and detailed analysis of the liver was reported.

In wild-type mice, liver proteins in the cytosol and organelles such as mitochondria decrease to about 70%, while the decrease is not significant in $atg7^{f/f}$:Mx1 mice, suggesting that autophagy accounts for the majority of starvation-induced protein degradation in the liver. Furthermore, $atg7^{f/f}$:Mx1 mice show hepatomegaly as early as 20 days after pIpC injection. It indicates that autophagy constitutively occurs at low levels irrespective of nutrient conditions. Some abnormal organelles such as deformed mitochondria and endoplasmic reticulum also accumulate in hepatocytes. The most striking finding is that many ubiquitin-positive aggregates are generated in hepatocytes [87]. The mechanism underlying the accumulation of ubiquitin-positive aggregates is unknown. Inhibition of continuous turnover of soluble proteins might lead to aggregate formation. Alternatively, aggregates may naturally be degraded by autophagy. Finally, $atg7^{f/f}$:Mx1 mice at 90 days after pIpC injection showed severe hepatomegaly with disorganized hepatic lobules, cell swelling and cell death. Elevation of serum alanine aminotransferase and aspartate aminotransferase is also observed at this stage. Taken together, these results suggest baseline autophagy is critical for intracellular clearance and homeostasis of hepatocytes.

However, autophagy-defective yeast cells [71], embryonic stem cells [17] and embryonic fibroblasts [70] show no apparent abnormalities under growing conditions. In rapidly dividing cells, abnormal proteins may be quickly diluted even if they are not degraded. Therefore, the intracellular clearance by autophagy should be particularly important in post-mitotic cells.

3.3.3
Beclin 1-deficient Mice

The genomic localization of *beclin 1* is within a tumor susceptibility locus in 17q21, which monoallelically deleted in 40–75% of sporadic human breast and ovarian cancers [40]. A breast cancer cell line, MCF-7, expresses very low levels of beclin 1. MCF-7 cells show low autophagic activity, which can be restored by forced expression of beclin 1. Beclin 1 expression also inhibits tumorigenesis when MCF-7 cells are implanted into nude mice [40]. These findings suggest that beclin 1 is required for both autophagy and protection from cancer.

Two groups independently generated beclin 1 knockout mice [88, 89]. The homozygous mutant mice show early embryonic lethality, probably around embryonic day 7.5 [89]. Electron microscopic analysis of *beclin 1$^{-/-}$* embryonic stem cells revealed that these cells are defective in autophagy, confirming that beclin 1 is required for autophagy [89].

Although beclin 1 heterozygous mice develop normally and are fertile, they develop spontaneous tumors such as lung cancer, hepatocellular carcinoma and lymphoma [88, 89]. By crossing with the GFP–LC3 mice, *beclin 1$^{+/-}$* mice were shown to be partially defective in autophagy in various tissues including tissues associated with increased spontaneous tumorigenesis [88].

As mentioned above, beclin 1/Atg6 is a component of the class III PI3K complex, and at least in yeast and worms it has multiple functions not restricted to autophagy [36, 38]. In addition, beclin 1 has another binding partner, Bcl-2/CED-9 [38, 39], which may provide beclin 1 with additional functions in cell death. Finally, the phenotype of *beclin 1$^{-/-}$* mice is much more severe than that of *atg5$^{-/-}$* and *atg7$^{-/-}$* mice, which appear almost normal at birth [70, 87, 89]. Therefore, beclin 1 likely has additional roles beyond its function in autophagy. It remains to be clarified which phenotypes of *beclin 1$^{-/-}$* and *beclin 1$^{+/-}$* mice are caused by autophagy defects.

3.4
Concluding Remarks

Recently it has been rapidly revealed that autophagy has pleiotropic roles in protein metabolism during starvation, quality control inside cells, degradation of intracellular bacteria, antigen presentation of cytosolic proteins, anti-aging, tumorigenesis and cell death. The genetically manipulated animals described above are now extensively used for autophagy studies and should provide further new insights into this mysterious cellular function.

References

1 D.J. Klionsky, J.M. Cregg, W.A. Dunn, Jr., S.D. Emr, Y. Sakai, I.V. Sandoval, A. Sibirny, S. Subramani, M. Thumm, M. Veenhuis, Y. Ohsumi. *Dev Cell* **2003**, *5*, 539–545.

2 D.J. Klionsky. *J Cell Sci* **2005**, *118*, 7–18.

3 N. Mizushima, T. Noda, T. Yoshimori, Y. Tanaka, T. Ishii, M.D. George, D.J. Klionsky, M. Ohsumi, Y. Ohsumi. *Nature* **1998**, *395*, 395–398.

4 N. Mizushima, H. Sugita, T. Yoshimori, Y. Ohsumi. *J Biol Chem* **1998**, *273*, 33889–33892.

5 Y. Ohsumi. *Nat Rev Mol Cell Biol* **2001**, *2*, 211–216.

6 T. Shintani, N. Mizushima, Y. Ogawa, A. Matsuura, T. Noda, Y. Ohsumi. *EMBO J* **1999**, *18*, 5234–5241.

7 I. Tanida, N. Mizushima, M. Kiyooka, M. Ohsumi, T. Ueno, Y. Ohsumi, E. Kominami. *Mol Biol Cell* **1999**, *10*, 1367–1379.

8 J. Kim, V. M. Dalton, K. P. Eggerton, S. V. Scott, D. J. Klionsky. *Mol Biol Cell* **1999**, *10*, 1337–1351.

9 W. Yuan, P. E. Stromhaug, W. A. Dunn Jr. *Mol Biol Cell* **1999**, *10*, 1353–1366.

10 I. Tanida, E. Tanida-Miyake, M. Komatsu, T. Ueno, E. Kominami. *J Biol Chem* **2002**, *277*, 13739–13744.

11 I. Tanida, E. Tanida-Miyake, T. Ueno, E. Kominami. *J Biol Chem* **2001**, *276*, 1701–1706.

12 N. Mizushima, T. Yoshimori, Y. Ohsumi. *FEBS Lett* **2002**, *532*, 450–454.

13 T. Nemoto, I. Tanida, E. Tanida-Miyake, N. Minematsu-Ikeguchi, M. Yokota, M. Ohsumi, T. Ueno, E. Kominam. *J Biol Chem* **2003**, *278*, 39517–39526.

14 N. Mizushima, T. Noda, Y. Ohsumi. *EMBO J* **1999**, *18*, 3888–3896.

15 A. Kuma, N. Mizushima, N. Ishihara, Y. Ohsumi. *J Biol Chem* **2002**, *277*, 18619–18625.

16 N. Mizushima, A. Kuma, Y. Kobayashi, A. Yamamoto, M. Matsubae, T. Takao, T. Natsume, Y. Ohsumi, T. Yoshimori. *J Cell Sci* **2003**, *116*, 1679–1688.

17 N. Mizushima, A. Yamamoto, M. Hatano, Y. Kobayashi, Y. Kabeya, K. Suzuki, T. Tokuhisa, Y. Ohsumi, T. Yoshimori. *J Cell Biol* **2001**, *152*, 657–667.

18 T. Kirisako, M. Baba, N. Ishihara, K. Miyazawa, M. Ohsumi, T. Yoshimori, T. Noda, Y. Ohsumi. *J Cell Biol* **1999**, *147*, 435–446.

19 S. S. Mann, J. A. Hammarback. *J Biol Chem* **1994**, *269*, 11492–11497.

20 Y. Sagiv, A. Legesse-Miller, A. Porat, Z. Elazar. *EMBO J* **2000**, *19*, 1494–1504.

21 H. Wang, F. K. Bedford, N. J. Brandon, S. J. Moss, R. W. Olsen. *Nature* **1999**, *397*, 69–72.

22 Y. Kabeya, N. Mizushima, A. Yamamoto, S. Oshitani-Okamoto, Y. Ohsumi, T. Yoshimori. *J Cell Sci* **2004**, *117*, 2805–2812.

23 T. Kirisako, Y. Ichimura, H. Okada, Y. Kabeya, N. Mizushima, T. Yoshimori, M.

Ohsumi, T. Takao, T. Noda, Y. Ohsumi. *J Cell Biol* **2000**, *151*, 263–275.

24 Y. Ichimura, T. Kirisako, T. Takao, Y. Satomi, Y. Shimonishi, N. Ishihara, N. Mizushima, I. Tanida, E. Kominami, M. Ohsumi, T. Noda, Y. Ohsumi. *Nature* **2000**, *408*, 488–492.

25 Y. Kabeya, N. Mizushima, T. Ueno, A. Yamamoto, T. Kirisako, T. Noda, E. Kominami, Y. Ohsumi, T. Yoshimori. *EMBO J* **2000**, *19*, 5720–5728.

26 K. Suzuki, T. Kirisako, Y. Kamada, N. Mizushima, T. Noda, Y. Ohsumi. *EMBO J* **2001**, *20*, 5971–5981.

27 J. Kim, W.-P. Huang, D. J. Klionsky. *J Cell Biol* **2001**, *152*, 51–64.

28 E. T. W. Bampton, C. G. Goemans, D. Niranjan, N. Mizushima, A. M. Tolkovsky. *Autophagy* **2005**, *1*, 23–36.

29 G. Marino, J. A. Uria, X. S. Puente, V. Quesada, J. Bordallo, C. Lopez-Otin. *J Biol Chem* **2003**, *278*, 3671–3678.

30 J. Hemelaar, V. S. Lelyveld, B. M. Kessler, H. L. Ploegh. *J Biol Chem* **2003**, *278*, 51841–51850.

31 I. Tanida, Y. S. Sou, J. Ezaki, N. Minematsu-Ikeguchi, T. Ueno, E. Kominami. *J Biol Chem* **2004**, *279*, 36268–36276.

32 I. Tanida, N. Minematsu-Ikeguchi, T. Ueno, E. Kominami. *Autophagy* **2005**, *1*, 84–91.

33 L. Chen, H. Wang, S. Vicini, W. Olsen. *Proc Natl Acad Sci USA* **2000**, *97*, 11557–11562.

34 M. Kneussel, S. Haverkamp, J. C. Fuhrmann, H. Wang, H. Wassle, R. W. Olsen, H. Betz. *Proc Natl Acad Sci USA* **2000**, *97*, 8594–8599.

35 A. Kihara, Y. Kabeya, Y. Ohsumi, T. Yoshimori. *EMBO Rep* **2001**, *2*, 330–335.

36 A. Kihara, T. Noda, N. Ishihara, Y. Ohsumi. *J Cell Biol* **2001**, *152*, 519–530.

37 T. Noda, K. Suzuki, Y. Ohsumi. *Trends Cell Biol* **2002**, *12*, 231–235.

38 K. Takacs-Vellai, T. Vellai, A. Puoti, M. Passannante, C. Wicky, A. Streit, A. L. Kovacs, F. Muller. *Curr Biol* **2005**, *15*, 1513–1517.

39 X. H. Liang, L. Kleeman, H. H. Jaing, G. Gordon, J. E. Goldman, G. Berry, B. Herman, B. Levine. *J Virol* **1998**, *72*, 8586–8596.

40 X. H. Liang, S. Jackson, M. Seaman, K. Brown, B. Kempkes, H. Hibshoosh, B. Levine. *Nature* **1999**, *402*, 672–676.

41 W. J. Brown, D. B. DeWald, S. D. Emr, H. Plutner, W. E. Balch. *J Cell Biol* **1995**, *130*, 781–796.

42 H. W. Davidson. *J Cell Biol* **1995**, *130*, 797–805.

43 N. Furuya, J. Yu, M. Byfield, B. Levine. *Autophagy* **2005**, *1*, 46–52.

44 Y. Kamada, T. Funakoshi, T. Shintani, K. Nagano, M. Ohsumi, Y. Ohsumi. *J Cell Biol* **2000**, 1507–1513.

45 Y. Kabeya, Y. Kamada, M. Baba, H. Takikawa, M. Sasaki, Y. Ohsumi. *Mol Biol Cell* **2005**, *16*, 2544–2553.

46 H. Cheong, T. Yorimitsu, F. Reggiori, J. E. Legakis, C. W. Wang, D. J. Klionsky. *Mol Biol Cell* **2005**, *16*, 3438–3453.

47 A. Matsuura, M. Tsukada, Y. Wada, Y. Ohsumi. *Gene* **1997**, *192*, 245–250.

48 J. Yan, H. Kuroyanagi, A. Kuroiwa, Y. Matsuda, H. Tokumitsu, T. Tomoda, T. Shirasawa, M. Muramatsu. *Biochem Biophys Res Commun* **1998**, *246*, 222–227.

49 J. Yan, H. Kuroyanagi, T. Tomemori, N. Okazaki, K. Asato, Y. Matsuda, Y. Suzuki, Y. Ohshima, S. Mitani, Y. Masuho, T. Shirasawa, M. Muramatsu. *Oncogene* **1999**, *18*, 5850–5859.

50 N. Okazaki, J. Yan, S. Yuasa, T. Ueno, E. Kominami, Y. Masuho, H. Koga, M. Muramatsu. *Brain Res Mol Brain Res* **2000**, *85*, 1–12.

51 T. Tomoda, R. S. Bhatt, H. Kuroyanagi, T. Shirasawa, M. E. Hatten. *Neuron* **1999**, *24*, 833–846.

52 T. Tomoda, J. H. Kim, C. Zhan, M. E. Hatten. *Genes Dev* **2004**, *18*, 541–558.

53 K. Ogura, C. Wicky, L. Magnenat, H. Tobler, I. Mori, F. Muller, Y. Ohshima. *Genes Dev* **1994**, *8*, 2389–2400.

54 T. Noda, J. Kim, W. P. Huang, M. Baba, C. Tokunaga, Y. Ohsumi, D. J. Klionsky. *J Cell Biol* **2000**, *148*, 465–480.

55 F. Reggiori, K. A. Tucker, P. E. Stromhaug, D. J. Klionsky. *Dev Cell* **2004**, *6*, 79–90.

56 C. W. Wang, J. Kim, W. P. Huang, H. Abeliovich, P. E. Stromhaug, W. A. Dunn, Jr., D. J. Klionsky. *J Biol Chem* **2001**, *276*, 30442–30451.

57 T. Yamada, A. R. Carson, I. Caniggia, K. Umebayashi, T. Yoshimori, K. Nakabayashi, S. W. Scherer. *J Biol Chem* **2005**, *280*, 18283–18290.

58 T. R. Jeffries, S. K. Dove, R. H. Michell, P. J. Parker. *Mol Biol Cell* **2004**, *15*, 2652–2663.

59 T. Proikas-Cezanne, S. Waddell, A. Gaugel, T. Frickey, A. Lupas, A. Nordheim. *Oncogene* **2004**, *23*, 9314–9325.

60 N. Mizushima. *Int J Biochem Cell Biol* **2004**, *36*, 2491–2502.

61 N. Mizushima, A. Yamamoto, M. Matsui, T. Yoshimori, Y. Ohsumi. *Mol Biol Cell* **2004**, *15*, 1101–1111.

62 R. M. Chicz, R. G. Urban, J. C. Gorga, D. A. Vignali, W. S. Lane, J. L. Strominger. *J Exp Med* **1993**, *178*, 27–47.

63 R. Lechler, G. Aichinger, L. Lightstone. *Immunol Rev* **1996**, *151*, 51–79.

64 H. Rammensee, J. Bachmann, N. P. Emmerich, O. A. Bachor, S. Stevanovic. *Immunogenetics* **1999**, *50*, 213–219.

65 A. R. Dongre, S. Kovats, P. deRoos, A. L. McCormack, T. Nakagawa, V. Paharkova-Vatchkova, J. Eng, H. Caldwell, J. R. Yates, 3rd, A. Y. Rudensky. *Eur J Immunol* **2001**, *31*, 1485–1494.

66 M. I. Brazil, S. Weiss, B. Stockinger. *Eur J Immunol* **1997**, *27*, 1506–1514.

67 F. Nimmerjahn, S. Milosevic, U. Behrends, E. M. Jaffee, D. M. Pardoll, G. W. Bornkamm, J. Mautner. *Eur J Immunol* **2003**, *33*, 1250–1259.

68 C. Paludan, D. Schmid, M. Landthaler, M. Vockerodt, D. Kube, T. Tuschl, C. Munz. *Science* **2004**, *307*, 593–596.

69 D. Dorfel, S. Appel, F. Grunebach, M. M. Weck, M. R. Muller, A. Heine, P. Brossart. *Blood* **2005**, *105*, 3199–3205.

70 A. Kuma, M. Hatano, M. Matsui, A. Yamamoto, H. Nakaya, T. Yoshimori, Y. Ohsumi, T. Tokuhisa, N. Mizushima. *Nature* **2004**, *432*, 1032–1036.

71 M. Tsukada, Y. Ohsumi. *FEBS Lett* **1993**, *333*, 169–174.

72 G. P. Otto, M. Y. Wu, N. Kazgan, O. R. Anderson, R. H. Kessin. *J Biol Chem* **2003**, *278*, 17636–17645.

73 G. Juhasz, G. Csikos, R. Sinka, M. Erdelyi, M. Sass. *FEBS Lett* **2003**, *543*, 154–158.

74 R. C. Scott, O. Schuldiner, T. P. Neufeld. *Dev Cell* **2004**, *7*, 167–178.

75 A. Melendez, Z. Talloczy, M. Seaman, E. L. Eskelinen, D. H. Hall, B. Levine. *Science* **2003**, *301*, 1387–1391.

76 B. Levine, D. J. Klionsky. *Dev Cell* **2004**, *6*, 463–477.

77 P. G. Clarke. *Anat Embryol (Berl)* **1990**, *181*, 195–213.

78 J. U. Schweichel, H. J. Merker. *Teratology* **1973**, *7*, 253–266.

79 E. H. Baehrecke. *Cell Death Differ* **2003**, *10*, 940–945.

80 D. Gozuacik, A. Kimchi. *Oncogene* **2004**, *23*, 2891–2906.

81 J. Debnath, E. H. Baehrecke, G. Kroemer. *Autophagy* **2005**, *1*, 66–74.

82 L. Yu, A. Alva, H. Su, P. Dutt, E. Freundt, S. Welsh, E. H. Baehrecke, M. J. Lenardo. *Science* **2004**, *304*, 1500–1502.

83 S. Shimizu, T. Kanaseki, N. Mizushima, T. Mizuta, S. Arakawa-Kobayashi, C. B. Thompson, Y. Tsujimoto. *Nat Cell Biol* **2004**, *6*, 1221–1228.

84 J. J. Lum, D. E. Bauer, M. Kong, M. H. Harris, C. Li, T. Lindsten, C. B. Thompson. *Cell* **2005**, *120*, 237–248.

85 P. Boya, R. A. Gonzalez-Polo, N. Casares, J. L. Perfettini, P. Dessen, N. Larochette, D. Metivier, D. Meley, S. Souquere, T. Yoshimori, G. Pierron, P. Codogno, G. Kroemer. *Mol Cell Biol* **2005**, *25*, 1025–1040.

86 A. Degterev, Z. Huang, M. Boyce, Y. Li, P. Jagtap, N. Mizushima, G. D. Cuny, T. J. Mitchison, M. A. Moskowitz, J. Yuan. *Nat Chem Biol* **2005**, *1*, 112–119.

87 M. Komatsu, S. Waguri, T. Ueno, J. Iwata, S. Murata, I. Tanida, J. Ezaki, N. Mizushima, Y. Ohsumi, Y. Uchiyama, E. Kominami, K. Tanaka, T. Chiba. *J Cell Biol* **2005**, *169*, 425–434.

88 X. Qu, J. Yu, G. Bhagat, N. Furuya, H. Hibshoosh, A. Troxel, J. Rosen, E. L. Eskelinen, N. Mizushima, Y. Ohsumi, G. Cattoretti, B. Levine. *J Clin Invest* **2003**, *112*, 1809–1820.

89 Z. Yue, S. Jin, C. Yang, A. J. Levine, N. Heintz. *Proc Natl Acad Sci U S A* **2003**, *100*, 15077–15082.

4
Autophagy in Disease and Aging

Marta Martinez-Vicente, Susmita Kaushik and Ana Maria Cuervo

4.1
Introduction

The identification of previously unknown molecular components involved in autophagy (Chapter 1) and of novel regulators for this process (Chapter 2) in recent years has allowed the development of new methods to track autophagy, in both culture cells and in whole organisms (Chapter 3). In addition, some of the new components identified have become suitable targets for manipulations aimed at activating or blocking autophagy. Being able to measure and compare levels of autophagy under different conditions and to analyze the consequences of altering the normal course of this intracellular process has provided new insights on the physiological relevance of autophagy (reviewed in Refs. [1–3]). New roles have been added to the classically recognized role of autophagy in catabolism and maintenance of cellular homeostasis. The demonstrated contribution of autophagy, in its different forms, to development, cell differentiation, intrinsic and extrinsic immunity, host-to-pathogen response, adaptation to stress, and cell death now places this intracellular process at a completely new level regarding its relevance for proper cellular functioning. Both the new tracking methods for autophagy and the elucidation of its diverse cellular functions have allowed us to link autophagy with a growing number of human pathologies. For some of them, failure or malfunctioning of the autophagic process seems to be the basis of their pathogenesis. In others, the observed changes in autophagy are likely consequences of the progressive course of the disease or a defensive response of the cell to the pathological condition. This possible dual role of autophagy, as a survival mechanism or as cell death effector, is changing the way in which autophagy is analyzed in different disorders. Identifying changes in autophagic activity is no longer enough for most pathologies, but it is critical to determine when those changes occur, since their cellular consequences could be completely different depending on when they occur in the course of the disease.

In this chapter, we discuss the most recent findings linking autophagy to pathological conditions. The different length of each of the sections reflects the

Autophagy in Immunity and Infection. Edited by Vojo Deretic
Copyright © 2006 WILEY-VCH Verlag GmbH & Co. KGaA, Weinheim
ISBN: 3-527-31450-4

situation of our current understanding. For some disorders, such as neurode-generative disorders or cancer, the field has moved faster and mechanistic expla-nations for the contribution of autophagy to the disease are already available. Other disorders are still in a more descriptive stage, but they should soon bene-fit from the findings of the better-studied pathologies. We will not cover the connections of autophagy with immunological and infectious-related disorders, since they are addressed in detail in other chapters of this book.

Of the three different forms of autophagy identified in mammalian cells, i.e. macroautophagy, microautophagy and chaperone-mediated autophagy (CMA), connections to particular disorders have already been established for macroau-tophagy and CMA. The lack of reliable methods for tracking microautophagy has precluded establishing a direct relation of this form of autophagy with hu-man diseases. Interestingly, in a growing number of disorders, changes in au-tophagy do not seem restricted to one type of autophagy or another. We de-scribe, when known, the primary system affected, but a new lesson derived from all these studies has been the existence of crosstalk and compensatory mechanisms among the different proteolytic systems in the cell.

In the last part of this chapter, we summarize the changes that the lysoso-mal/autophagic system undergoes with age. Although many of the pathologies covered here are considered age-related disorders, a decline in autophagic activ-ity also occurs during normal physiological aging. Even though not a pathology, it is still pertinent to discuss it here, since these age-related changes in autopha-gy could explain, at least in part, why these pathologies aggravate with age.

4.2
Autophagy in Neurodegenerative Disorders

Accumulation of autophagic vacuoles (AVs) and protein aggregates, in the form of inclusion bodies, is a common feature of protein conformational disorders. Among these disorders, those directly affecting the central nervous system, i.e. neurodegenerative disorders, have received particular attention in recent years. Here, we first comment on the relation between autophagy and protein aggrega-tion in a general context. We then compile the information currently available on the role of autophagy in the pathogenesis of three major degenerative disor-ders, i.e. Parkinson's disease, Alzheimer's disease and Huntington's disease, highlighting both analogies and peculiarities for each of these disorders. In ad-dition, we comment on two other forms of neurodegeneration, i.e. prion disor-ders and Niemann–Pick Type C, for which connections to autophagy are still in their beginnings, but that, as discussed below, have the potential to critically contribute to our understanding of their pathogenesis and future development of therapeutic approaches.

4.2.1
Protein Misfolding and Aggregation

Protein aggregates or aggresomes are insoluble stable oligomeric complexes of misfolded and unfolded proteins that, when accumulate intracellularly, are called inclusion bodies [4]. Protein aggregation is precipitated by the presence of mutations and/or post-transcriptional modifications in the proteins or under particular intracellular conditions such as oxidative stress (reviewed in Ref. [5]).

Cells avoid accumulation of aggregate proteins by different sequential mechanisms (illustrated in Fig. 4.1). As the first "front of action", intracellular chaperones interact with the unfolded or partially folded protein intermediates to pre-

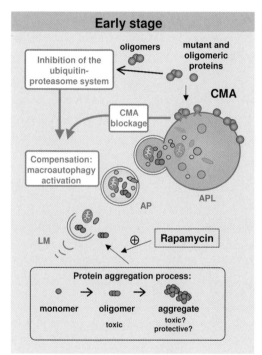
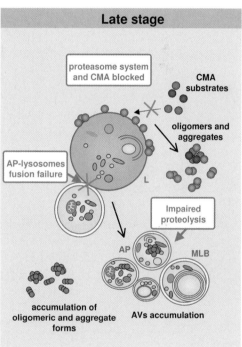

Fig. 4.1 Autophagy in protein conformational disorders. Protein conformational disorders result from abnormal conformational changes in particular proteins (due to mutations or post-translational modifications) that make them prone to aggregation. In the early stages of the disorder, the abnormal proteins often block the activity of proteolytic systems normally responsible for the degradation of soluble proteins (proteasome and CMA by the lysosome), resulting in compensatory activation of macroautophagy to eliminate the oligomeric toxic forms. As the diseases progress (late stage), a macroautophagic failure often occurs, probably due to problems in the clearance of the autophagocytosed materials, leading to the accumulation of AVs with partially degraded contents and eventually to cell death. Abbreviations: L=lysosome; LM=limiting membrane; AP=autophagosome; APL=autophagolysosome; MLB=multilamelar bodies. (This figure also appears with the color plates).

vent exposure of hydrophobic residues and to promote folding of nascent proteins or refolding of misfolded ones [4]. If proper folding is not attained, the ubiquitin–proteasome system [6, 7] and, in some particular instances, CMA [8] can eliminate most of the soluble misfolded proteins. However, as oligomerization of the toxic intermediates progresses, insoluble aggregates, unsuitable for proteasome or CMA degradation, form and accumulate intracellularly [4, 9]. The presence of trapped ubiquitinated proteins, chaperones and proteases has been reported in almost all types of aggregate inclusion bodies, supporting that the inclusion body nucleates from proteins that could not be refolded or degraded before their aggregation [4].

Once this stage is reached, only activation of macroautophagy has proven able to efficiently remove aggresomes [6, 10–12]. Activation of macroautophagy has been shown to decrease levels of intracellular aggregates and protect against neurodegeneration [10, 13, 14]. Accordingly, macroautophagy inhibitors reduce aggregate clearance in several neurodegenerative disorders [10, 15]. The mechanisms involved in the autophagic degradation of intracellular aggregates are currently being elucidated. Impaired activity of the proteasome–ubiquitin system, likely because of the direct action of the misfolded proteins on this system, induces autophagy [7]. Essential macroautophagy components and lysosomes are recruited, via a microtubule-dependent mechanism, to the perinuclear area of the cells, where inclusion bodies normally accumulate. This mechanism has been proposed to increase the efficiency and selectivity of the autophagic degradation of the aggregates [16]. Autophagic vacuoles then engulf the aggresomes and degrade them by fusion with lysosomes. Often, the number of aggregates increases as the disease advances, probably reflecting both decreased susceptibility of the aggregate protein to the lysosomal enzymes (as irreversible covalent modifications and crosslinking occur) and a primary impairment of the autophagic process (as macroautophagy effectors and cytoskeletal components are sequestered in the inclusion bodies). It is, at this stage, when general cell dysfunction occurs and the affected cells often end up being eliminated by, what has been called, programmed cell death type II or autophagy [2, 3, 17].

The recently described degradation of protein aggregates via macroautophagy has introduced two main changes in the way that aggregates are viewed now in the context of protein conformational disorders. First, it has revealed that protein aggregates might not be the pathogenic basis of these disorders, but, instead, that protein aggregation could have a cytoprotective function (since aggregates are in general less toxic than the oligomeric soluble forms of the altered proteins) [18–20]. Second, the susceptibility of aggresomes to macroautophagy degradation opens the possibility of therapeutic approaches oriented both to promote aggregation of the toxic products and to facilitate their removal by activating macroautophagy. Although the effectiveness of these types of approaches in late stages of the disease, when the autophagic machinery is already compromised, is doubtful, it is now generally accepted that activation of macroautophagy in the early stages of most protein conformational disorders could be beneficial.

4.2.2
Parkinson's Disease

Parkinson's disease is an age-related neurodegenerative disorder characterized by the loss of dopaminergic neurons in the substantia nigra and in the striatum. Intracellular protein inclusions, called Lewy bodies (LB), are found in different Parkinson's disease brain regions and, although currently controversial, have been considered for a long time the histological signature of this disorder [21–23].

The main component of the LB is α-synuclein, a soluble cytoplasmic protein localized in the presynaptic terminal of most neuronal types. In pathologic conditions, α-synuclein undergoes conformational changes that lead to the formation of insoluble α-synuclein fibrils often found in the core of the LB. Although α-synuclein fibrils were thought to be the cause of the neuronal death, recent evidence instead supports that the soluble oligomeric or proto-fibrillar structures, which are formed previous to insoluble fibrils, are the pathogenic/toxic forms. The fact that sequestration of these toxic forms into fibrils in inclusion bodies may protect cells from neurotoxicity confers LB a cytoprotective role in the Parkinson's disease pathology [24, 25]. The mechanisms by which these α-synuclein protofibrils may cause dysfunction and neuronal death are still unknown. It has been proposed that the ability of protofibrils to bind and permeabilize vesicles may lead to inappropriate redistribution of vesicular components and to organelle content leakage [26].

In familial forms of Parkinson's disease, the conformational changes in α-synuclein result from point mutations in its amino acid sequence. Two of those mutations (A30P and A53T) have shown increased tendency to form toxic oligomeric intermediates [27, 28]. Although mutations in α-synuclein only occur in familial forms of Parkinson's disease, α-synuclein is still the main component of LB in all forms of Parkinson's disease. Analysis of the LB has revealed several post-transcriptional modifications in wild-type α-synuclein, including nitration, phosphorylation, oxidation, glycosylation, ubiquitination and formation of dopamine adducts, which could all be involved in its fibrillation [29–34]. Of particular interest are the α-synuclein dopamine adducts, which might explain why dopaminergic neurons are most susceptible to neurodegeneration in Parkinson's disease. It is possible that, under oxidizing conditions, the cytosolic dopamine could form an adduct with α-synuclein, which slows down conversion of protofibrils to fibrils. Hence, dopamine would become an endogenous protofibril stabilizer, causing the accumulation of the cytotoxic soluble forms of α-synuclein [26, 35]. These findings indicate that cytosolic dopamine in dopaminergic neurons could promote the accumulation of toxic protofibrils.

Apart from the formation of dopamine adducts, phosphorylation [33] and methionine oxidation [36] also favor the maintenance of α-synuclein in a soluble toxic form, by inhibiting fibrillation. Furthermore, these modified α-synucleins could also inhibit the fibrillation of the unmodified α-synuclein, as has already been shown for the oxidized forms [34]. The relationships between the presence of these post-transcriptional modifications, protein fibrillization and neurodegeneration in Parkinson's disease are still unclear.

Both supporting and opposing evidence of degradation of α-synuclein via the ubiquitin–proteasome pathway has been presented [25, 37, 38]. Only recently, it has been shown that α-synuclein can indeed be degraded by both proteasome and autophagic pathways, depending on its conformational state [6]. Wild-type α-synuclein can be degraded by the proteasome [6], but it is also a substrate for CMA [8]. α-Synuclein contains a CMA-targeting motif in its amino acid sequence that, when recognized by the cytosolic hsc70 chaperone, delivers the protein to the lysosomal membrane receptor [the lysosome-associated membrane protein type 2A (LAMP-2A)]. Once in the membrane, the protein is translocated into the lysosomal lumen for its degradation [8]. The percentage of α-synuclein degraded by each of these two pathways is currently unknown, but it may vary depending on the cellular conditions. Mutant synucleins cannot be degraded by either of these two mechanisms [8, 38], but are efficiently removed via macroautophagy [6, 19]. Impaired degradation of mutant α-synucleins by the ubiquitin–proteasome system is a combination of both the inability of this protease to degrade proteins, once they organize into complex fibrillar structures or aggregates, and also the direct inhibitory effect of the mutant proteins on the proteasome [7, 39]. Mutant α-synucleins are poorly degraded by CMA because, despite binding to the lysosomal membrane receptor with high affinity, their translocation across the lysosomal membrane is severely impaired [8]. Experimental overexpression of mutant α-synuclein activates macroautophagy in culture cells, facilitating the removal of the abnormal protein [39]. How macroautophagy is activated under these conditions remains elusive. However, it could be a consequence of the effect of mutant α-synucleins on other proteolytic systems. Synuclein mutants have been shown to directly interfere with the activity of both the ubiquitin/proteasome system [39] and of CMA [8], and blockage of these two systems has been experimentally shown to lead to macroautophagy activation [7, 188].

Although rapamycin, a stimulator of macroautophagy, increases clearance of mutant α-synuclein in culture cells [6], the particular forms of the protein removed by this process are still unclear. It has been shown that the oligomeric forms of mutant α-synuclein are susceptible to degradation by macroautophagy, whereas the insoluble fibrillar inclusion bodies are not [25]. It is thus possible that only "semi-aggregates", but not long-time formed aggregates, could be removed by activation of macroautophagy.

Based on these findings, activation of macroautophagy in the early stages of the disease could help to remove the α-synuclein oligomeric intermediates before these toxic forms accumulate in the cytosol. However, as the disease progresses, the increasing number of aggregates and their decreased susceptibility for degradation by lysosomes may lead to a general cell dysfunction and cellular death. Further studies should be oriented to identify "the point of no return" after which activation of macroautophagy is no longer beneficial for the affected neurons. In addition, most of the studies have focused so far on the degradation of mutant α-synuclein. Mutations in the protein are only found in a small number of patients. More information is necessary about the effect of the other synuclein modifications in its intracellular turnover. In that respect, we have recently found that distinct post-translational modifications of α-synuclein have different effects on its degradation via CMA [166].

4.2.3
Alzheimer's Disease

Alzheimer's disease is a progressive neurodegenerative disorder that involves gradual neuronal loss. Alzheimer's disease is characterized by the presence of intraneuronal formation of neurofibrillary tangles (NFT), composed mainly of aggregated tau (a microtubule-associated protein) and also by the presence of extracellular senile plaques, containing mainly β-amyloid (Aβ) deposits [fibrillar aggregates of Aβ, a peptide resulting from the proteolytic cleavage of the amyloid precursor protein (APP)] [40, 41]. Alzheimer's disease is a sporadic disorder in more than 80% of patients. Hereditary forms of Alzheimer's disease are mainly due to mutations in the *presenilin-1* gene (PS1), coding for one of the proteins of the γ-secretase, the proteolytic complex responsible for APP cleavage.

Abnormalities in the endosomal–lysosomal pathway during Alzheimer's disease progression have been well documented (reviewed in Ref. [42]). In the early stages of the disease, there is a general upregulation of the endocytic–lysosomal system (lysosomal proliferation and increase of lysosomal hydrolases), evident even before substantial Aβ is deposited in the brain [43].

As the disease progresses, the neuronal lysosomal system becomes less efficient and lysosomal enzymatic activities decrease, eventually leading to impaired clearance of AVs and to their accumulation in the Alzheimer's disease neurons [44]. The persistence of AVs in the Alzheimer's disease brain is particularly important in the pathogenesis of the disease, since they become new sources for Aβ production. It has recently been shown that both APP and components of the β- and γ-secretase complexes can be detected in AVs. Under normal conditions, the half-life of AVs, from their formation until their elimination via lysosomal fusion, is very short (of the order of minutes), and, thus, the coexistence of enzymes and substrate in the same compartment does not have any cellular consequence. However, in the disease, the persistence of AVs in the cellular cytosol for longer periods of time, due to their impaired lysosomal clearance, results in vacuolar cleavage of APP by the secretases, and release of β- and γ-secretase APP cleavage products from AVs which become new Aβ-generating compartments [45]. This Aβ overproduction not only contributes significantly to Aβ deposition in Alzheimer's disease neurons, but also further impairs AV clearance, thus perpetuating this novel pathogenic mechanism [45].

In the last stages of the disease, the persistence of undigested material inside AVs (autophagosomes and/or autophagolysosomes) could lead to oxidation and nonspecific crosslinking of the content, membrane instability, and leakage of lysosomal enzymes in the cytosol. Whether or not these series of events are the trigger for programmed cell death of the affected neurons needs future investigation.

4.2.4
Huntington's Disease

Huntington's disease is the best known of the group of inherited neurodegenerative disorders caused by polyglutamine tracts expansion [46]. This autosomal dominant disorder is characterized by the formation of intraneuronal inclusion bodies enriched with mutant huntingtin (Htt), the protein altered in this disorder [47]. When the number of glutamine repeats (poly-Q) associated to the N-terminus of Htt exceeds 37, the protein becomes prone to aggregation into insoluble inclusion bodies [48]. The presence of an increased number of glutamine repeats in Htt correlates with gain in toxicity and aggregation. As in other protein conformational disorders, the toxic effect initially proposed for the inclusion bodies [49] is currently under revision, since recent studies have suggested a possible protective function for the protein aggregates [18, 50].

The function of Htt and the molecular mechanisms that contribute to the pathogenesis of Huntington's disease are still being elucidated. Recent studies suggest that neuronal degeneration in Huntington's disease results from the combined effects of gain of function of the poly-Q-expanded form of Htt, along with the loss of function of wild-type Htt (reviewed in Ref. [51]). Htt is ubiquitously expressed and present in many subcellular compartments [52, 53]. Htt knockout mice are nonviable, but different mouse models, expressing the full-length or truncated forms of the human protein, are now available [54]. These truncated forms of Htt are particularly interesting, since Htt can be cleaved by caspases, calpains and aspartic proteases, resulting in fragments that are believed to be the toxic species found in the aggregates [49, 53]. Htt interacts with numerous proteins [53, 55], suggesting that it is, indeed, a multifunctional protein involved in processes as diverse as gene transcription, signaling and intracellular trafficking [52]. Based on the large intracellular pool of Htt-interacting proteins, it has been proposed that mutant Htt may recruit these proteins during the aggregation process and drag them into the inclusion bodies. The loss of those particular proteins would have different cellular consequences depending on their normal function. Thus, the presence of transcription factors such as p53, CREB-binding protein (CBP), specificity protein 1 (SP1) and TATA-binding protein (TBP) in the Htt aggregates could explain the transcriptional deregulation observed in Huntington's disease models [56, 57]. Intraneuronal poly-Q aggregates are often ubiquitinated and contain chaperones (hsc40 and hsp70) and proteasome subunits, supporting failure of the protein refolding and degradation machinery [58, 59].

Very little information is currently available about the mechanisms responsible for the degradation of wild-type full-length Htt. Although Htt contains in its amino acid sequence three CMA-targeting motifs, the unmodified form of the protein does not seem to be degraded through this autophagic pathway (Martinez-Vicente, unpublished results). Soluble mutant Htts and some of the toxic fragments could be the substrates for the proteasome–ubiquitin system [4, 7]. In contrast, macroautophagy has been revealed as the main pathway for the clearance of Htt aggregates formed during the progression of the disease [10, 15]. The first observations linking Huntington's disease and macroautophagy reported that overexpression of mutant Htt in culture cells resulted in stimulation of lysosomal activity and increased formation of AVs [47]. The exact mechanisms leading to macroautophagy activation are still unknown, but it has been proposed that the observed sequestration of mammalian target of rapamycin (mTOR; the major negative regulator of macroautophagy) [60] in the poly-Q aggregates would inhibit its kinase activity and, consequently, induce macroautophagy [13, 14]. Stimulation of macroautophagy by treatment with rapamycin (a mTOR inhibitor) protects against neurodegeneration in fly models of Huntington's disease, and decreases aggregate formation and cell death in cell culture models of Huntington's disease [13]. In contrast, treatment with two macroautophagy inhibitors, 3-methyladenine (3-MA) and bafilomycin A1, reduces the capacity of cells to eliminate poly-Q-expanded Htt [10, 15]. As in other protein conformational disorders, these findings support a protective role for cellular aggregates by means of activation of macroautophagy. Procedures aiming at activating macroautophagy have become the basis for new therapeutic approaches in these aggregate-prone pathologies. Although proven effective experimentally, a major concern was the possible negative consequences that massive degradation of other intracellular components by macroautophagy could have. However, recent work has revealed a certain degree of selectivity in the removal of poly-Q aggregates via macroautophagy [12, 16]. Thus, to preferentially favor the degradation of the aggregate proteins, both aggresomes and the molecular machinery involved in macroautophagy are relocated to the perinuclear region of the cell in a microtubule-dependent manner [12, 16]. Although macroautophagy does not facilitate the removal of nuclear aggregates, also common in this disease, the fact that the removal of only cytosolic aggregates is enough to improve cell function supports its therapeutic potential [12]. Finally, as described for the other neurodegenerative disorders, it is likely that, in late stages of the disease, activation of macroautophagy may no longer help, but instead it could contribute to neuronal death. This would explain the elevated rates of cellular death found when macroautophagy was activated by the combined action of mutant Htt and a source of oxyradical stress (dopamine) [61].

4.2.5
Prion Diseases

Human transmissible spongiform encephalopathies (TSEs; also known as prion diseases) are a group of fatal neurodegenerative disorders of genetic, infectious or sporadic origin, characterized by spongiform degeneration of the grey matter, neuronal loss, astrocytic gliosis and neuronal accumulation of the pathological prion protein isoform (PrP^{Sc}) [62, 63]. The most important characteristic that differentiates prion diseases from the rest of the neurodegenerative disorders is that the pathological isoform PrP^{Sc} is itself infectious [64]. This isoform arises from the conversion of PrP^c, the nonpathogenic form of PrP (mostly a-sheets), into the pathogenic isoform PrP^{Sc} (mostly β-sheets) [65, 66]. The mechanisms involved in this conversion, the subsequent intracellular accumulation of the pathogenic isoform and the resulting neurotoxicity have yet to be completely elucidated.

Although less explored than the neurodegenerative disorders described in the previous sections, several pieces of evidence have connected prion disorders with autophagy. Despite initial reports claiming resistance of PrP to protease cleavage, it has recently been shown that PrP^{Sc} can be degraded by lysosomal enzymes, such as cathepsin L and B, thus supporting the participation of the lysosomal system in PrP^{Sc} degradation [67]. In addition, accumulation of AVs has been described in neurons from patients and in cells with experimentally induced prion disease [68, 69]. In fact, this intracellular accumulation of AVs, along with other organelle deformities [mitochondrial dilatation, enlargement of Golgi and endoplasmic reticulum (ER)] and the absence of typical apoptotic features in the dying neurons, has led to the conclusion that neuronal loss in these disorders occurs via autophagic cell death [68, 70, 71]. Whether this accumulation of AVs occurs only in the advanced stages of the disease, while early activation of autophagy could have a protective effect in the progression of prion diseases, as shown for other neurodegenerative disorders, still needs to be elucidated.

4.2.6
Niemann–Pick Type C

Niemann–Pick Type C is a neurodegenerative lysosomal storage disorder characterized by the accumulation of cholesterol and other lipids [sphingomyelin, lyso-bis-phosphoatidic acid (LBPA), glycosphingolipids and phospholipids] in endocytic-lysosomal organelles, collectively referred to as lysosome-like storage organelles [72]. Niemann–Pick Type C patients present pathogenic symptoms in early childhood, in the form of progressive neurological degeneration and hepatosplenomegaly (enlarged liver and spleen), which leads to death in their early teens. Mutations in the *npc1* gene are found in approximately 95% of the patients and the in *npc2* gene in the rest (for review, see Ref. [73]).

Although the exact biological role of Npc1 remains unknown, this integral membrane protein has been proposed to participate in lipid trafficking, as it binds cholesterol [74] and contains a "sterol-sensing domain" [75]. Normal Npc1 is present in late endosomes and lysosomes [76–78], but the mutant protein is partially mislocalized toward the cell periphery [79]. Npc2, also a late endosomal and lysosomal cholesterol-biding protein, when mutated accumulates in cholesterol-storing endocytic organelles [79]. Loss-of-function mutations, in either Niemann–Pick Type C1 or Niemann–Pick Type C2, lead to lipid storage in endocytic compartments and delayed transport of lipids from the endocytic pathway to other intracellular destinations. Although, based on the lipids accumulated in the lysosome-like storage organelles, Niemann–Pick Type C is considered primarily a glycosphingolipid storage disorder [80], secondary accumulation of cholesterol in these compartments is possibly responsible for the most severe pathology. Impaired transport of cholesterol from the cell body to the axons in neurons seems to be one of the main reasons for the neurological dysfunction in Niemann–Pick Type C disease [81, 82].

The relation between the lipid accumulation and the neuronal death in Niemann–Pick Type C still remains unknown. It was suggested that the elevated accumulation of lipids could be toxic for neurons; however, recent studies in Purkinje neurons from Niemann–Pick Type C mice propose that neurodegeneration is a consequence of the activation of the autophagy cell death pathway in the affected neurons [83]. In contrast to the other neurodegenerative disorders described in this chapter, in which initial activation of macroautophagy could serve a protective role, the contribution of autophagy to the pathogenesis of Niemann–Pick Type C should be looked at from a completely different perspective. In this case, there is a primary defect that modifies the properties of the endocytic/lysosomal compartment and, likely, it affects autophagocytic vesicles too. Although the consequences of changes in lipid composition in autophagy are still poorly understood, several phospholipids [phosphatidylethanolamine and phosphatidylinositol-3,4,5-triphosphate (PIP_3)] are essential for nucleation and elongation of the limiting membrane of the endosome [84], and alterations in cholesterol trafficking have been found in a mouse model with a primary defect in autophagy (reviewed in more detail in Section 4.4.1) [85]. Consequently, malfunctioning of the lysosomal/autophagic system (i.e. impaired autophagosome/lysosome fusion, leakage from autophagic compartments, insufficient delivery of lysosomal enzymes) could lead to the observed accumulation of AVs in the affected neurons and the inability to orchestrate a proper autophagic response, rather than the proposed exacerbation of this process, could be the cause of cellular death.

4.3
Autophagy and Cancer

Although numerous studies have reported altered macroautophagy in different forms of cancer over the past three decades, whether autophagy inhibits tumorigenesis, is mechanistically prosurvival or has a dual role remains largely unanswered.

4.3.1
Proteolysis of long-lived Proteins and Effect of Nutrient Deprivation in Cancer Cells

Total rates of protein degradation decrease in most types of cancer cells, as compared to their normal counterparts [86, 87]. The lysosomal system may play an important role in oncogenesis biology since the proteolysis of long-lived proteins, the preferential substrates of the lysosomal system, is predominantly altered in cancer cells. Indeed, very early studies already reported reduced autophagic activity in cells from carcinogen-treated rats [88]. In most types of tumor-derived cells (colon, breast, skin, liver, brain and lung), this decreased activity reverses under conditions of nutrient deprivation to levels similar to those observed in untransformed cells [89–92]. Also supporting changes in the lysosomal/autophagic system, recent morphological surveys for autophagic structures in different primary human tumors have revealed the presence of AVs in seven of 12 types of tumors investigated, including ganglioneuroma, infiltrating ductal carcinoma of the breast, adenocarcinoma of the lung, pancreatic adenocarcinoma and pancreatic islet cell tumor [93]. This increase in the number of AVs could, however, be the morphological expression of either increased rates of macroautophagy or decreased clearance of AVs by lysosomes. In fact, in contrast to the commonly reported decrease in protein degradation in most cancers, undifferentiated colon cancer cells, displaying a similar morphological vacuolar phenotype, maintain high rates of autophagic proteolysis. The different nature of the analyzed cells supports that upregulation or downregulation of macroautophagy may be directly related to factors such as type of cancer, tumoral stage and the degree of cellular differentiation. It has been proposed that, during early stages of carcinogenesis, the decrease in macroautophagy would, in theory, provide an anabolic advantage to the preneoplastic cells, allowing them to selectively outgrow neighboring populations of "normal" cells (as illustrated in Fig. 4.2). On the other hand, upregulation of macroautophagy in later stages of cancer development (reviewed in Ref. [94]) would be particularly beneficial for cells residing in the central regions of solid tumors. In these hypoxic and low-nutrient regions of tumors, macroautophagy would serve as a prosurvival mechanism, providing cells with basic nutrients. The whole picture increases in complexity when considering that larger tumors have microenvironmental niches with gradients of essential metabolites like oxygen, glucose and other nutrients, thereby resulting in a heterogeneous population of cells with varying autophagy levels (reviewed in Refs. [3, 94]).

4.3.2
Autophagic Cell Death in Response to Anticancer Treatment

Adding to the role of autophagy in the normal progression of the oncogenic process, several studies have reported activation of macroautophagy in cancer cells in response to anti-cancer therapy. Gamma-irradiation induces macroautophagy in cancer cells of epithelial origin, such as colon, breast and prostate cancer [95, 96]. Tamoxifen, an estrogen antagonist commonly used to treat patients with advanced breast cancer, also induces macroautophagy in cultured breast cancer cells [97]. Similarly, rapamycin induces macroautophagy and suppresses proliferation in malignant glioma cells [98]. Other examples of cancer chemotherapeutic agents that activate macroautophagy include temozolomide, a DNA alkylating agent [99], and natural products such as, arsenic trioxide (in malignant glioma cells) [100, 101], resveratrol (in ovarian cancer cells) [102] and soybean B-group triterpenoid saponins (in colon cancer cells) [103].

Most of these studies point toward macroautophagy being activated as a mechanism of cell death, since blockage of macroautophagy prevents cell death. However, in other studies, inhibition of the autophagic activation accelerates cell death. In these particular cases, the autophagic response to therapeutic drugs

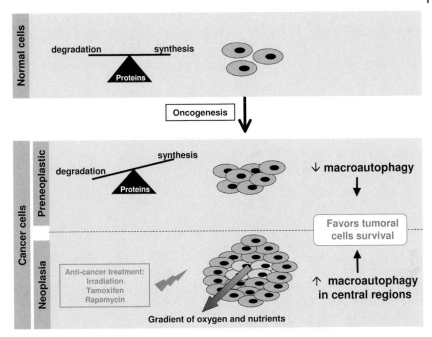

Fig. 4.2 Dual role of macroautophagy in cancer. In normal cells, there is a balance between protein synthesis and degradation, which is deranged in cancer cells. However, in early stages of carcinogenesis (preneoplastic cells), low levels of macroautophagy would serve as a prosurvival mechanism for the cells. In contrast, in late stages (neoplasia) upregulation of macroautophagy in the hypoxic, low-nutrient center regions of tumors would be advantageous for cell survival. Possible therapeutic attempts try to activate macroautophagy by means of γ-irradiation, tamoxifen or rapamycin (red callout). However, these treatments at later stages of tumorigenesis, during which macroautophagy activity is required for survival can, in fact, promote tumor cell proliferation rather than cell death. (This figure also appears with the color plates).

may be an adaptive response to the drug itself [94]. Therefore, it would be important to preclude such drug resistance when developing future therapeutics.

Although it is still controversial whether macroautophagy is really an effector of cellular death or rather the manifestation of a last survival attempt from the cells, recent reports support that activation of macroautophagy beyond normal levels results in cellular death, in a way directly depending on macroautophagy components [104]. It is thus likely that, in response to the cellular damage induced by these agents, autophagy is activated with two possible outcomes: if autophagy is maintained at levels enough to eliminate damaged components, this would help cell survival; however, if the activation of autophagy in response to the treatment reaches certain levels, this would make it incompatible with cell survival (Fig. 4.2).

As discussed below, the particular molecular components that determine the degree of autophagic activation, and the switch between cell survival and cell death are still unknown for the most part, but recent reports have started to elucidate the role of some molecules in the crosstalk between autophagy and apoptosis [104]. The following subsections focus only in macroautophagy, since there is currently no information on possible changes in CMA or microautophagy in oncogenic processes.

4.3.3
Molecular Mechanisms

Of the different novel players for the autophagic process described in recent years (reviewed in Chapters 1 and 2), some of them have been revealed to be of particular interest in oncogenesis.

4.3.3.1 Beclin 1

One of the first genetic links between tumor suppression and autophagic pathways came through the identification of decreased levels of beclin 1, an autophagy effector, in several human breast epithelial carcinoma cell lines and tissues [105]. Evidence for tumor development *per se*, due to the deficiency of beclin 1, was put forth when monoallelic ablation of beclin 1 in mice resulted in increased incidence of spontaneous malignancies, increased cellular proliferation and reduced macroautophagy *in vivo* [106, 107], thus supporting beclin 1's role as a haploinsufficient tumor suppressor.

Beclin 1 binds to class III phosphotidylinositol-3-kinase (PI3K; involved in the elongation of the limiting membrane of the AV, see Chapter 1) and it is also part of the class III PI3K complex in the *trans*-Golgi network (TGN). Consequently, beclin 1 could be involved in sorting autophagosomal components [108] and putatively in trafficking of lysosomal enzymes from the TGN to the lysosomes [109]. Beclin 1 also binds to the antiapoptotic protein, Bcl-2 [110], and this interaction has recently been shown to play a role in the crosstalk between macroautophagy and apoptosis [104]. Binding of Bcl-2 to beclin 1 prevents its association with class III PI3K complex, thereby inhibiting macroautophagy activation. Disruption of the interaction between the antiapoptotic factor, Bcl-2, and beclin 1 results in the activation of macroautophagy, and, at least in some cell types, cellular death dependent on autophagy [104]. Therefore, added to the antiapoptotic effect of Bcl-2, this novel role as inhibitor of macroautophagy may also contribute to cancer development.

4.3.3.2 PI3K–Akt–mTOR Pathway

The other major step in macroautophagy for which connections with oncogenic processes have been reported is the regulatory axis centered around mTOR, a highly conserved member of the PI3-related kinase family of serine/threonine protein kinases (Chapter 2). *In vivo*, mTOR exists in at least two distinct complexes with different functions: a rapamycin-sensitive complex [which interacts with rap-

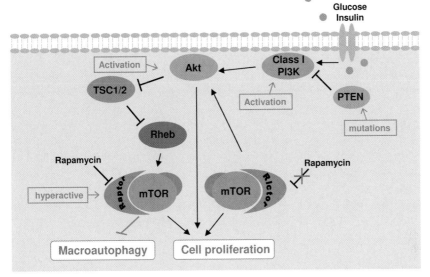

Fig. 4.3 A model for the PI3K–Akt–mTOR pathway. Akt, a serine/threonine kinase, downstream of class I PI3K regulates cell proliferation and cell survival, and also activates mTOR indirectly, in response to extrinsic stimuli such as nutrients and growth factors. PTEN, a lipid phosphatase, dephosphorylates and inactivates class I PI3K. mTOR forms complexes with two cytosolic proteins: raptor (rapamycin-sensitive) and rictor (rapamycin-insensitive). mTOR inhibits macroautophagy and upregulates protein-synthesis. In various cancers, mutations in PTEN, amplifications of both PI3K and Akt, and hyperactivation of the mTOR have been observed, resulting in the blockage of macroautophagy, thus providing an anabolic advantage (shown in red callouts) to the cancer cells. Abbreviations: mTOR = mammalian target of rapamycin; PTEN = phosphatase and tensin homolog deleted on chromosome 10; TSC = tuberous sclerosis complex. (This figure also appears with the color plates).

tor (regulatory-associated protein of mTOR)] and a rapamycin-insensitive complex [which interacts with rictor (rapamycin-insensitive companion of mTOR)] [111, 112] (Fig. 4.3). mTOR inhibits macroautophagy under basal conditions by mediating the hyperphosphorylation of a macroautophagy effector (Atg13) and also by inhibiting transcription/translation of different proteins, including some macroautophagy players (Chapter 2). The raptor–mTOR pathway is hyperactive in some cancers and, therefore, a possible target for therapeutic intervention [113]. Activated mTOR would result in macroautophagy inhibition and upregulation of protein synthesis, providing the anabolic advantage to the cancer cells [114].

Several other regulators upstream of mTOR have also been related to oncogenic processes. Akt is a serine-threonine kinase of the cAMP-dependent, cGMP-dependent protein kinase C (AGC) family functioning downstream of the class I PI3K (Fig. 4.3) [115]. Numerous studies have reported Akt alterations in human cancer and in experimental models of tumorigenesis. Akt regulates cell proliferation and survival, cell size, and mediates responses to nutrient availability, intermediary metabolism, angiogenesis and tissue invasion, which are all attributes of cancer [116]. Akt indirectly activates mTOR, thereby resulting in the downregulation of macroautophagy and, at least in theory, facilitating cancer progression. Different oncoproteins and tumor suppressors intersect in the Akt pathway. The counter regulatory enzyme phosphatase and tensin homolog deleted on chromosome 10 (PTEN), a tumor-suppressor lipid phosphatase, dephosphorylates and inhibits class I PI3K, which in turn inhibits Akt activity and allows macroautophagy to occur (Fig. 4.3) [117]. Interestingly, mutations in PTEN, and amplifications of both PI3K and Akt, are prevalent in various cancers, including, breast, prostate and endometrial cancers [118], providing evidence that macroautophagy is blocked in these cells, apart from beclin 1 mutations. The extrinsic stimuli for the class I PI3K–Akt–mTOR pathway are nutrients, such as amino acids or glucose, and growth factors, such as insulin or insulin-like growth factor, which upregulate mTOR, resulting in changes in transcription/translation as well as in macroautophagy repression. The typically low levels of nutrients and growth factors in poor vascularized regions of solid tumors (Fig. 4.2) could explain the downregulation of this signaling axis and, consequently, the activation of macroautophagy as a prosurvival mechanism in those tumor regions (reviewed in Refs. [119, 120]).

Other signaling components known to play a role in programmed cell death, such as the death-associated protein kinase (DAPK) or the death-associated related protein kinase 1 (DRP1), have also recently been shown to regulate macroautophagy [121]. However, the molecular players involved in this regulation and their possible involvement in oncogenesis still remains unknown.

4.3.4
Possible Therapeutic Attempts

In recent years, autophagy has revealed itself as an important modulator of tumorigenesis and, therefore, future therapeutic agents aimed at augmenting autophagic cell death may prove fruitful. Different cancer types have defective macroautophagy, thus becoming resistant to autophagic cell death. Restoration of normal levels of beclin 1 or PTEN, which in turn would activate macroautophagy, presents as a possible therapeutic intervention in such cells. In this regard, adenovirus-mediated transfer of PTEN has been shown to inhibit growth of prostate cancer cells [122], and stable transfection of beclin 1 promoted macroautophagy and decreased tumorigenesis in breast cancer cells [105]. In contrast, cells with intact autophagic machinery may proceed to autophagic cell death when treated with drugs like rapamycin, which releases the mTOR-mediated repression on macroautophagy. Preclinical and early clinical trails have shown that mTOR inhibitors (rapamycin and its derivatives) prevent tumor growth in a variety of tumors, including breast cancer, renal cell carcinoma, malignant glioma, cervical cancer and uterine cancers [113]. Furthermore, combinatorial therapies, with conventional antioncogenic cocktails and macroautophagy activators, have shown promising results [123]. Remaining challenges to increase the specificity of these treatments would require us first to identify the associated signaling events that occur downstream to mTOR in order to dissociate its effect on macroautophagy from that on transcription/translation. Finally, cancer cells which exhibit resistance to different antioncogenic treatments, via induction of macroautophagy, may respond better if treated with macroautophagy inhibitors, such as bafilomycin A1, 3-MA and nocodazole [95, 99].

4.4
Myopathies

Although classically the lysosomal system has been considered to play a very modest role in skeletal muscle protein turnover, recent work supports that, like many other tissues, muscles upregulate macroautophagy upon nutrient deprivation [124–126]. In addition, an abnormal increase in the number of AVs has been reported in various muscle pathologies (myopathies), supporting a role for autophagy, or at least macroautophagy, in muscle pathophysiology (reviewed in Ref. [127]).

Cytosolic membranous tubules derived from the sarcoplasmic reticulum have been observed in various myopathies. However, it was only after extensive immunocytochemical and histological studies that, in some myopathies, these membranous tubules were found positive for macroautophagy-related proteins and were, thus, identified as AVs (reviewed in Ref. [127]). Myopathies with accumulated AVs have been sub-classified as autophagic vacuolar myopathies (AVMs). Here, we briefly review the most common AVMs [Danon disease and X-linked myopathy with excessive autophagy (XMEA)], and those vacuolar myopathies with a primary defect unrelated to autophagy, but which also display altered autophagic activity as part of their phenotype (rimmed vacuolar myopathies). Myopathies directly resulting from a lysosomal enzymatic defect, such as those observed in many lysosomal storage disorders, are beyond the scope of this chapter. Because not all the myopathies affect cardiac muscle, and because there are changes in autophagy in heart that obey different mechanisms than for skeletal muscle, cardiomyopathies are reviewed separately in the last section.

4.4.1
Danon Disease

Danon disease was first described in 1981 in patients with hypertrophic cardiomyopathy, proximal muscle weakness and mental retardation as the major symptoms [128]. The histological comparison of skeletal and cardiac muscles between knockout mice for LAMP-2 and those from patients suffering from Danon disease, resulted in the identification of a mutation in LAMP-2 as the primary defect in this disease [129, 130]. All of these mutations result in the complete absence, or severely compromised levels, of the LAMP-2 protein [129, 131]. Because patients with mutations in only one particular splicing variant of this gene, LAMP-2B, present the whole muscle phenotype, it is now believed that this is the isoform directly involved in macroautophagy. This could also explain why in the complete LAMP-2 knockout mice, AVs accumulate in nonmuscular tissues, in contrast to the restricted muscular phenotype observed in Danon's disease patients [130]. Autophagic vacuoles, containing cytosolic components or sarcoplasmic membranes, have been proposed to accumulate due to defective fusion of AVs and lysosomes [130]. As in the other vacuolar myopathies, the accumulation of these AVs in skeletal muscles results in muscle weakness, contraction failure and eventually in muscle degeneration. A recent attempt to study morphological features of major AVMs reported that in Danon disease some autolysosomes are surrounded by membranes with sarcolemmal proteins, acetylcholinesterase activity and basal lamina [132]. This unique intracytosolic membranous structure has not been observed in other AVMs.

4.4.2
XMEA

XMEA is characteristic of skeletal muscle alone, with no involvement of other muscle types or other tissues, and it is associated with normal levels of LAMP-2. Clinically, it presents as an early onset disease with progressive muscular weakness. Histopathological investigations have revealed vacuolation of myofibers [133]. Interestingly, not only do these vacuoles contain lysosomal enzymes and cellular debris, but their membranes stain positive for dystrophin and laminin (sarcolemmal proteins), suggesting sarcolemmal origin of these vacuoles. XMEA has been linked to chromosome Xq28, but the primary defect still remains to be identified. Despite the lack of information on the primary defect, it is still considered an AVM, because of the autophagic nature of the accumulated vesicles.

4.4.3
Rimmed Vacuolar Myopathies

The rimmed vacuolar myopathies are autosomal recessive inherited disorders and the most common forms of vacuolar myopathies (reviewed in Ref. [127]). The most prominent pathologic finding in these myopathies is the presence of rimmed vacuoles in muscle fibers. In contrast to the AVMs, these are secondary lysosomal myopathies, because all of the identified causative genes encode extralysosomal proteins. In these instances, mutations in different genes result in the accumulation of potentially harmful misfolded protein aggregates and the vacuoles are the expression of the activation of macroautophagy to eliminate these altered proteins from muscle fibers [134].

One of the better characterized rimmed vacuolar myopathies is the chloroquine myopathy, caused by chronic intoxication with the antimalarial drug chloroquine and its derivative hydroxychloroquine, an anti-inflammatory drug. Both drugs raise the lysosomal pH-inactivating acid hydrolases [135]. Localization of macroautophagy markers on the vacuoles provided direct molecular evidence for the role of macroautophagy in this disease [136].

4.4.4
Other Myopathies

There are reports of accumulation of vacuoles with sarcolemmal features in some other myopathies. There have been two cases of infantile AVM [137]. This disorder has infantile onset and is genetically distinct from Danon disease. A different type, the adult-onset AVM, is characterized by slow progression, cardiomyopathy and multisystem involvement [138]. Similar to other AVMs, there is an accumulation of AVs with sarcolemmal features containing cytosolic debris and electron-dense material in the muscle fibers. Further analysis has revealed the presence of vacuoles in cardiac muscles and liver, thus distinguishing this disorder from the two other forms of AVM described above. There is also another type of infantile-onset myopathy, linked with defective autophagy, X-linked myotubular myopathy, which is characterized by hypotonia and muscle weakness. This myopathy is caused by a mutation in the myotubularin gene, which encodes for a ubiquitously expressed phosphatase with an established function in vesicular trafficking and in autophagy [139].

It is important to note that, as pointed out for other pathologies, increased numbers of AVs in muscle fibers do not necessarily correlate with enhanced protein degradation. For example, in Danon disease, despite a massive increase in AVs in skeletal and cardiac muscle, rates of protein degradation are characteristically low, thus revealing defective lysosomal degradation [129, 130]. As the primary defects and mutations have been identified for only a few of the myopathies, extensive immunohistopathological and biochemical analyses are being carried out to better understand the molecular mechanism behind these muscular disorders. This would not only be beneficial for the design of therapeutic strategies against these myopathies in the future, but would also help us better understand autophagy in muscles, an area still poorly studied.

4.4.5
Cardiomyopathies and Myocardial Cell Death

Only a small percentage of vacuolar myopathies extend beyond the skeletal muscle, affecting cardiac fibers also. Danon disease is a classic example of accumulation of AVs in cardiac muscle, thus responsible for the cardiomyopathy observed in these patients [130].

Apart from this inherited disease, defective autophagy has also been implicated as the major mechanism of myocardial cell death [140, 141]. In dilated cardiomyopathy and in hearts with severe aortic stenosis, degenerated cardiomyocytes accumulate AVs. Based on these electron microscopic and immunohistochemical observations, it has been proposed that autophagy could be directly responsible for the observed progressive destruction of cardiomyocytes [140]. There has been growing evidence that autophagy, along with apoptosis and oncosis (acute ischemic cell death), are indeed responsible for myocyte degeneration, the major phenomena in heart failure. Work on explanted hearts from patients with idiopathic dilated cardiomyopathy revealed that these three pathways act in concert, at varying degrees, to mediate loss of myocytes [142]. The cells from these hearts not only depicted extensive accumulation of AVs but, also a decrease in lysosomal degradation (because of reduced cathepsin D levels). Death, by one or all of the above-mentioned pathways, results in loss of contractile muscle mass, chiefly irreplaceable in these terminally differentiated cells.

Sporadic examples of cardiomyopathies containing prominent AVs in myocytes, but with minimal consequences in cardiac function [143], leave open the possibility of autophagy being activated in some circumstances as an adaptation, rather than as a cell death mechanism. In fact, in a recent study, increases in macroautophagy-related proteins and in levels of the chaperones involved in CMA were found in chronically ischemic myocardium [144]. Interestingly, the induction of autophagy was evidenced only after two or three episodes of coronary stenosis, suggesting that autophagy could act as a prosurvival mechanism by limiting harmful effects of ischemia.

4.5
Liver Diseases

Despite liver being the organ in which most of the initial characterization of mammalian autophagy was carried on and for which more information about the regulation of autophagy has been compiled through the years, there are still relatively few liver pathologies directly linked to alterations in autophagy. The best studied has been α_1-antitrypsin (AAT) deficiency, first described in 1963 as an autosomal-recessive metabolic disorder, with chronic hepatitis and early-onset emphysema. AAT is a member of the serine proteinase inhibitor superfamily, synthesized in the liver, and then distributed into the interstitial and alveolar lining fluids. In the lung, it functions as a protective screen for the alveolar walls by inactivating neutrophil-derived proteases, released during normal phagocytosis of microorganisms. The causative mechanism of the disease is a conformational defect in the structure of AAT, which results in its aggregation in hepatocytes. This aggregation does not only interfere with the secretory pathway, but also leads to hepatitis and cirrhosis [145]. The very low circulating and alveolar levels of AAT in these patients makes them very sensitive to "insults" like cigarette smoke, infections, dust and fumes.

The mutant form of the protein (AATZ, the most common point mutation) [146] aggregates in the rough ER [147]. Accumulating evidence supports that not only the proteasome [148, 149], but also macroautophagy [150, 151] are part of the quality control mechanisms involved in the elimination of these protein aggregates in hepatocytes. The induction of autophagy under these conditions is thus a mechanism to protect hepatocytes from the toxic effects of the aggregated mutant protein. Evidence from cell culture and mouse models of AAT deficiency, and from livers of AAT-deficient patients, demonstrate that mitochondrial damage also plays a role in the mechanism of liver cell injury [145]. The autophagic response may, in part, remove these damaged organelles, thus preventing the activation of the apoptotic program [145]. Augmentation of the autophagic response therefore presents as a possible therapeutic option.

The important role of macroautophagy in liver physiology has been recently confirmed in a mouse model deficient for a critical macroautophagy effector (Atg7) [152]. The liver-specific Atg7 conditional knockout mouse presents severe hepatomegaly and hepatocyte swelling, attributable to poor protein and organelle turnover. Concentric membranous structures (derived from the ER, encircling mitochondria and lipid droplets), peroxisomes and deformed mitochondria, as well as ubiquitin aggregates, all accumulate in the cytosol of Atg7-defective livers. One of the most surprising findings, regarding the role of autophagy in liver, arising from the studies in the Atg7 knockout mice, is that protein degradation was impaired even under normal nutritional conditions, suggesting that macroautophagy may not only be activated in response to starvation, but as a constitutive mechanism for organelle turnover. However, an indirect effect of macroautophagy blockage on the systems that normally take care of basal degradation is still possible. The presence of ubiquitinated protein aggregates in the cytosol of Atg7 knockout mice livers reinforces the role of macroautophagy in the removal of abnormal and aggregate proteins as discussed in Section 4.2.1. Consequently, this mouse model could provide valuable insights into liver pathology, particularly related to protein conformational disorders.

4.6
Diabetes Mellitus

The inhibitory effect of insulin on macroautophagy was reported way back in the 1970s, when it was shown that intraperitoneal injection of insulin in rats dramatically decreased the fractional volume of AVs in the cytosol of hepatocytes [153, 154]. Later, this was confirmed by measuring protein degradation, by pulse–chase experiments, in perfused liver from nonfasted rats [155], hepatocyte monolayers from fed animals [156] and in rat kidney proximal tubular cells [157].

A link between autophagy and insulin-dependent diabetes mellitus (type I diabetes) was established after examining the effect of streptozotocin (STZ) (a drug that decreases insulin levels by ablating pancreatic B cells) in fed and fasted rats [158]. The STZ-treated animals exhibited elevated rates of protein degradation, increased fractional volume of AVs and higher levels of lysosomal hydrolases, both in liver and in kidney distal tubular cells [158, 159]. All these effects were reversed upon insulin administration. It was only recently that the mechanism of inhibition of autophagy via the mTOR pathway by insulin was elucidated [160] (as discussed in more detail in Chapter 2).

Interestingly, a novel connection between macroautophagy and diabetes has recently been reported in relation to the neuropathy observed in patients with type 2 diabetes [161]. Serum from these patients induced autophagosome formation in specific types of neuronal cells in culture, suggesting that induction of autophagy by a serum factor, most likely autoantibody(ies), could be the cause of the loss of function, and eventually, neuronal degeneration in type 2 diabetes.

Macroautophagy has also been associated with familial neurohypophyseal diabetes insipidus (FNDI), in this case as a prosurvival mechanism. Although classified as diabetes, because of the polyuria and polydipsia observed in these patients, the activation of macroautophagy in this pathology relates more to the described role of autophagy in protein conformational disorders (Section 4.2.1). In this autosomal dominant disorder, a mutant form of vasopressin (antidiuretic hormone) aggregates in the neuronal cell body. These protein aggregates are targets for lysosomal degradation by macroautophagy [162, 163], as in the other neurodegenerative disorders.

One of the secondary manifestations in diabetes mellitus, i.e. renal hypertrophy, could also result from autophagy malfunctioning. The decreased proteolysis and, hence, protein accumulation observed in renal hypertrophy is partly due to decreased CMA activity, which has been proposed as a major regulator of kidney growth [164]. Renal cortex cells from acutely diabetic rats display decreased lysosomal proteolysis and low levels of key CMA proteins (LAMP-2A and hsc70) in lysosomes [165]. Although future studies are required to elucidate the mechanisms by which decreased CMA favors kidney growth, it has been proposed that reduced degradation of particular CMA substrates, such as glycolytic enzymes (glyceraldehyde-3-phosphate dehydrogenase, aldolase, phosphoglucomutase) and specific transcription factors (Pax2) would create the appropriate cellular environment to promote cellular growth [165].

4.7
Aging

Many of the diseases discussed in previous sections can be considered age-related disorders, since, in many instances, their symptomatic manifestations do not appear until late in life. However, autophagy also undergoes major changes during physiological aging (reviewed in Refs. [166, 167]).

4.7.1
Changes in Protein Degradation with Age

A decrease in rates of protein degradation with age has been reported in almost every cellular and animal model analyzed ([168, 169]; and also reviewed in Refs. [166, 167, 170]). The inability of old cells to maintain normal protein and organelle turnover is probably the main reason for the accumulation of altered and damaged intracellular components in old organisms and tissues [171, 172]. The ubiquitin–proteasome system undergoes major changes during aging, but the decreased activity of this proteolytic complex seems the consequence, rather than the cause, of the accumulation of altered intracellular components in old cells (reviewed in Ref. [170]). Oxidatively modified proteins, products of lipid peroxidation and different types of nonspecific intracellular crosslinking have a direct inhibitory effect on the proteolytic activity of the proteasome [5].

The lysosomal system undergoes striking morphological changes in old organisms. In fact, expansion of the lysosomal vacuolar compartment and accumulation of undigested products inside lysosomes in the form of the autofluorescent pigment known as lipofuscin are often used as typical biomarkers of aging (reviewed in Ref. [174]). These morphological changes directly reflect lysosomal dysfunction [175] and correlate well with the temporal pattern of changes in intracellular levels of oxidized or modified proteins (reviewed in Refs. [167, 176]).

4.7.2
Age-related Changes in Autophagy

The activity of both macroautophagy and chaperone-mediated autophagy decreases with age. Although the mechanisms leading to their functional failure are, in principle, quite different, the recently discovered crosstalk between these two autophagic pathways (Massey, submitted), leaves open the possibility that their age-related changes are also interdependent.

Decreased macroautophagic proteolysis with age has been extensively reported both in culture cells and in livers of old rodents [177, 178]. Both morphometric and metabolic studies have revealed that while protein and organelle turnover in the presence of high concentrations of amino acids (normal nutritional conditions) does not significantly change with age, the ability of cells to upregulate protein degradation via macroautophagy in response to nutrient deprivation is impaired in aging [178]. The hormonal regulation of macroautophagy is differently affected by age – the stimulatory effect of glucagon is no longer observed in old animals, while the inhibitory effect of insulin is well preserved until late in life [177]. Three main steps have been proposed to be altered in the macroautophagic process with age: formation of autophagosomes, their clearance by lysosomal proteases and/or regulation of the autophagic process (Fig. 4.4) (reviewed in Refs. [166, 167]). The reasons for reduced formation of AVs with age remain unknown. A decrease in total levels of macroautophagy-related proteins or post-transcriptional modifications in these proteins could be behind the lower rate of AVs formed upon activation of this process. However, it is still possible that the defective formation of autophagosomes is a secondary consequence of the deregulation of macroautophagy signaling (see below). Lower autophagosome clearance has been proposed to result from a decrease in the proteolytic activity of lysosomes with age and/or of impaired ability of lysosomes to fuse with autophagosomes [174]. The consequent accumulation of undigested products in lysosomes, mostly in the form of lipofuscin, perpetuates the failure of AV clearance. The effects of age-related oxidative stress on the insulin receptor signaling pathway seem to play a critical role in decreased macroautophagy in old organisms. The inability of glucagon to upregulate macroautophagy in old rodents could result from the higher levels of basal activity of the insulin receptor during fasting in old individuals. Although normally weak, this basal insulin-independent activity increases under oxidative conditions (Fig. 4.4).

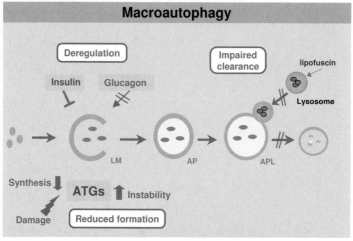

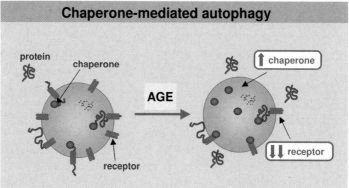

Fig. 4.4 Changes in autophagy during aging. The activity of both macroautophagy and CMA decreases with age. (Top) Impaired macroautophagic activity results from the combined effect of reduced formation of autophagic vesicles, impaired clearance and deregulation of the hormonal control. (Bottom) Levels of LAMP-2A, a receptor for CMA substrates at the lysosomal membrane, decrease with age. The decrease in levels of the receptor is initially compensated for by an increase in the levels of the chaperone that assist in substrate uptake. However, at advanced ages, the levels of the receptor decrease to a point for which compensation is no longer possible and failure of CMA becomes evident. Abbreviations: ATG = autophagy-related proteins; LM = limiting membrane; AP = autophagosome; APL = autophagolysosome. (This figure also appears with the color plates).

The strong linkage between oxidative stress and aging supports current efforts to decrease basal insulin receptor signaling with antioxidant compounds [179]. Recent genetic evidence supports the important role of insulin repression of autophagy in aging and longevity. Mutations in different proteins of the insulin receptor signaling cascade known to increase life span in the nematode *Caenorhabditis ele-*

gans fail to do that when autophagy is blocked, suggesting that activation of macro-autophagy is essential for life span extension [180].

Decreased CMA activity has been reported in senescent fibroblasts in culture and in different tissues of old rodents [181, 182]. In contrast to macroautophagy, the pool of lysosomes active for CMA, containing high levels of the luminal chaperone hsc70, do not undergo major morphological changes with age, neither do they accumulate undigested products in their lumen. Rather than changes in the proteolytic lysosomal enzymes, the defect in CMA function seems to be mostly due to a decrease with age in the levels of LAMP-2A, the CMA lysosomal receptor [182]. Levels of LAMP-2A decline before alterations in CMA activity are manifested. In fact, low levels of the lysosomal receptor are initially compensated through an increase in the number of lysosomes involved in this autophagic pathway. However, at advanced ages, levels of the receptor decrease to such an extent that the CMA defect becomes evident [182]. The reasons why levels of the receptor decrease with age are currently the subject of investigation. Altered turnover of the receptor itself and/or changes in its intracellular distribution could be behind the low LAMP-2A levels.

4.7.3
Consequences of the Failure of Autophagy in Aging

Because of the critical role of autophagy in the repair and turnover of intracellular components and as an essential mechanism for the adaptation to stressors (Chapter 1), the consequences of its functional decline with age are easy to infer.

Poor renewal of intracellular organelles with age has detrimental consequences, in particular in long-lived post-mitotic cells such as neurons, cardiac myocytes and skeletal muscle fibers. The persistence of poorly functioning organelles often contributes to generate further cellular stress. In particular, accumulation of senescent mitochondria, reported in all types of old tissues, largely contributes in large extent to increased oxidative stress. Although mitochondria are often targets of reactive oxygen species, once altered, they also become a source of these damaging species (reviewed in Ref. [183]). Slow turnover of intracellular proteins with age seems to be behind the increase in levels of oxidized, damaged and aggregated proteins in old tissues. Failure of CMA with age, in particular, has also been proposed to contribute to this accumulation, based on its role in the removal of damaged proteins after exposure to certain toxin compounds and during mild-oxidative stress [184]. Malfunctioning of CMA could also indirectly affect the activity of other proteolytic systems as specific subunits of the proteasome are normally degraded by CMA [185].

The inability to orchestrate a proper autophagy response to different types of intra- and extracellular stressors could play a critical role in the poor ability of old organisms to adapt to stress conditions. In addition, age-related changes in autophagy have also been linked to the loss of particular cell functions specific to some cells types. A well-documented example is the contribution of the decay of autophagic function with age in T cells to immunosenescence [186].

4.7.4
Slowing Down Aging?

The use of activators of autophagy to enhance this process in aging, and to thus favor the removal of damaged components, may not be as straight forward as described before for pathologies such as conformational disorders. The main difference is that, in many of those diseases, the autophagic/lysosomal system is intact and, consequently, it should be possible to enhance its activity. In contrast, there are primary alterations in the lysosomal system with age, which result in loss of function. As a result, the current interest is in identifying the age-related defects in this process to be able to either prevent them from happening or repair them once they have occurred.

Caloric restriction, the only intervention proven to slow down aging, preserves normal maroautophagic function until late in life [178]. Rather than increasing total rates of macroautophagy, caloric restriction preserves the response of this process to regulatory plasma nutrients and hormones [177]. Chronic administration of antilipolytic drugs also increases protein degradation by decreasing the availability of using lipids as a metabolic source of energy (reviewed in Ref. [187]). Caloric restricted animals show higher rates of CMA than normally fed littermates (Massey, unpublished), but the mechanisms behind this activation remain to be elucidated.

As mentioned before, attempts to downregulate basal insulin receptor signaling with cysteine have already shown significant beneficial effects on several muscle parameters that are typically affected by aging [188].

4.8
Concluding Remarks and Pending Questions

The current advances in our understanding of the molecular basis and regulation of autophagy are directly responsible for the exponentially growing number of reports linking autophagic alterations to particular diseases. As the field evolves, two very different pictures are becoming clear regarding the possible role of autophagy in disease. On the one hand, there are disorders in which failure of the autophagic system is the basis of, or at least contributes to, the pathological manifestations (i.e. AVMs, familial forms of Parkinson's disease, Niemann–Pick Type C). In these cases, therapeutic approaches should be aimed at correcting the primary defect and to restoring normal autophagic activity. In the second group of disorders, activation of autophagy is a cellular adaptive mechanism to the pathology (i.e. most protein conformational disorders, some cardiomyopathies and in neurodegeneration). In these disorders, the role that autophagy plays changes during the course of the pathology. Activation of autophagy is protective in the early stages of the disease, since it contributes to the removal of the injured/damaged components. However, as the disease progresses, failure of the autophagic process is common, contributing now to cellular dysfunction and, in many instances, cell death. This biphasic function of autophagy should be kept in mind while developing new therapeutic approaches.

There are still a large number of unanswered questions and many unexplored areas. Most of the current advances and therapeutics refer to macroautophagy. Although CMA malfunctioning has already been connected to some diseases, its participation in other diseases is unknown. Information about microautophagy and its role in pathology is still lacking. Even in the case of macroautophagy, where the field is moving faster, there is a growing need for functional markers. Despite the tremendous help that the introduction of macroautophagy-related proteins as markers has provided, most of them still give a static picture of this process. The fact that an increase in the number of AVs could be expression of both increased or decreased macroautophagy, makes a big claim for functional markers, applicable not only to culture systems, but also to whole organisms. Finally, because of the described crosstalk among the different autophagic pathways, and also with other proteolytic systems, it is becoming a priority to elucidate the mechanisms that regulate these interconnections to be able to utilize them for compensatory purposes.

References

1 D. J. Klionsky, Autophagy. *Curr Biol* **2005**, *15*, R282–283.

2 T. Shintani, D. J. Klionsky, Autophagy in health and disease: a double-edged sword. *Science* **2004**, *306*, 990–995.

3 A. Cuervo, Autophagy: in sickness and in health. *Trends Cell Biol* **2004**, *14*, 70–77.

4 R. R. Kopito, Aggresomes, inclusion bodies and protein aggregation. *Trends Cell Biol* **2000**, *10*, 524–530.

5 T. Grune, T. Jung, K. Merker, K. J. Davies, Decreased proteolysis caused by protein aggregates, inclusion bodies, plaques, lipofuscin, ceroid, and "aggresomes" during oxidative stress, aging, and disease. *Int J Biochem Cell Biol* **2004**, *36*, 2519–2530.

6 J. L. Webb, B. Ravikumar, J. Atkins, J. N. Skepper, D. C. Rubinsztein, Alpha-Synuclein is degraded by both autophagy and the proteasome. *J Biol Chem* **2003**, *278*, 25009–25013.

7 E. J. Bennett, N. F. Bence, R. Jayakumar, R. R. Kopito, Global impairment of the ubiquitin–proteasome system by nuclear or cytoplasmic protein aggregates precedes inclusion body formation. *Mol Cell* **2005**, *4*, 351–365.

8 A. M. Cuervo, L. Stefanis, R. Fredenburg, P. T. Lansbury, D. Sulzer, Impaired degradation of mutant alpha-synuclein by chaperone-mediated autophagy. *Science* **2004**, *305*, 1292–1295.

9 A. Michalik, C. Van Broeckhoven, Pathogenesis of polyglutamine disorders: aggregation revisited. *Hum Mol Genet* **2003**, *12 (Spec 2)*, R173–186.

10 B. Ravikumar, R. Duden, D. C. Rubinsztein, Aggregate-prone proteins with polyglutamine and polyalanine expansions are degraded by autophagy. *Hum Mol Genet* **2002**, *11*, 1107–1117.

11 H. J. Rideout, I. Lang-Rollin, L. Stefanis, Involvement of macroautophagy in the dissolution of neuronal inclusions. *Int J Biochem Cell Biol* **2004**, *36*, 2551–2562.

12 A. Iwata, J. C. Christianson, M. Bucci, L. M. Ellerby, N. Nukina, L. S. Forno, R. R. Kopito, Increased susceptibility of cytoplasmic over nuclear polyglutamine aggregates to autophagic degradation. *Proc Natl Acad Sci USA* **2005**, *102*, 13135–13140.

13 B. Ravikumar, C. Vacher, Z. Berger, J. E. Davies, S. Luo, L. G. Oroz, F. Scaravilli, D. F. Easton, R. Duden, C. J. O'Kane, D. C. Rubinsztein, Inhibition of mTOR

induces autophagy and reduces toxicity of polyglutamine expansions in fly and mouse models of Huntington disease. *Nat Genet* **2004**, *36*, 585–595.

14 B. Ravikumar, A. Stewart, H. Kita, K. Kato, R. Duden, D. C. Rubinsztein, Raised intracellular glucose concentrations reduce aggregation and cell death caused by mutant huntingtin exon 1 by decreasing mTOR phosphorylation and inducing autophagy. *Hum Mol Genet* **2003**, *12*, 985–994.

15 Z. H. Qin, Y. Wang, K. B. Kegel, A. Kazantsev, B. L. Apostol, L. M. Thompson, J. Yoder, N. Aronin, M. DiFiglia, Autophagy regulates the processing of amino terminal huntingtin fragments. *Hum Mol Genet* **2003**, *12*, 3231–3244.

16 A. Iwata, B. E. Riley, J. A. Johnston, R. R. Kopito, HDAC6 and microtubules are required for autophagic degradation of aggregated huntingtin. *J Biol Chem* **2005**, *280*, 40282–40292.

17 L. Stefanis, Caspase-dependent and -independent neuronal death: two distinct pathways to neuronal injury. *Neuroscientist* **2005**, *11*, 50–62.

18 J. P. Taylor, F. Tanaka, J. Robitschek, C. M. Sandoval, A. Taye, S. Markovic-Plese, K. H. Fischbeck, Aggresomes protect cells by enhancing the degradation of toxic polyglutamine-containing protein. *Hum Mol Genet* **2003**, *12*, 749–757.

19 M. Tanaka, Y. M. Kim, G. Lee, E. Junn, T. Iwatsubo, M. M. Mouradian, Aggresomes formed by alpha-synuclein and synphilin-1 are cytoprotective. *J Biol Chem* **2004**, *279*, 4625–4631.

20 C. W. Olanow, D. P. Perl, G. N. DeMartino, K. S. McNaught, Lewy-body formation is an aggresome-related process: a hypothesis. *Lancet Neurol* **2004**, *3*, 496–503.

21 R. L. Nussbaum, M. H. Polymeropoulos, Genetics of Parkinson's disease. *Hum Mol Genet* **1997**, *6*, 1687–1691.

22 M. C. Irizarry, W. Growdon, T. Gomez-Isla, K. Newell, J. M. George, D. F. Clayton, B. T. Hyman, Nigral and cortical Lewy bodies and dystrophic nigral neurites in Parkinson's disease and cortical Lewy body disease contain alpha-synuclein immunoreactivity. *J Neuropathol Exp Neurol* **1998**, *57*, 334–337.

23 P. J. Kahle, C. Haass, H. A. Kretzschmar, M. Neumann, Structure/function of alpha-synuclein in health and disease: rational development of animal models for Parkinson's and related diseases. *J Neurochem* **2002**, *82*, 449–457.

24 M. S. Goldberg, P. T. Lansbury, Jr., Is there a cause-and-effect relationship between alpha-synuclein fibrillization and Parkinson's disease? *Nat Cell Biol* **2000**, *2*, E115–119.

25 H. J. Lee, F. Khoshaghideh, S. Patel, S. J. Lee, Clearance of alpha-synuclein oligomeric intermediates via the lysosomal degradation pathway. *J Neurosci* **2004**, *24*, 1888–1896.

26 J. C. Rochet, T. F. Outeiro, K. A. Conway, T. T. Ding, M. J. Volles, H. A. Lashuel, R. M. Bieganski, S. L. Lindquist, P. T. Lansbury, Interactions among alpha-synuclein, dopamine, and biomembranes: some clues for understanding neurodegeneration in Parkinson's disease. *J Mol Neurosci* **2004**, *23*, 23–34.

27 K. A. Conway, S. J. Lee, J. C. Rochet, T. T. Ding, R. E. Williamson, P. T. Lansbury, Jr., Acceleration of oligomerization, not fibrillization, is a shared property of both alpha-synuclein mutations linked to early-onset Parkinson's disease: implications for pathogenesis and therapy. *Proc Natl Acad Sci USA* **2000**, *97*, 571–576.

28 C. W. Bertoncini, C. O. Fernandez, C. Griesinger, T. M. Jovin, M. Zweckstetter, Familial mutants of alpha-synuclein with increased neurotoxicity have a destabilized conformation. *J Biol Chem* **2005**, *280*, 30649–30652.

29 B. I. Giasson, J. E. Duda, I. V. Murray, Q. Chen, J. M. Souza, H. I. Hurtig, H. Ischiropoulos, J. Q. Trojanowski, V. M. Lee, Oxidative damage linked to neurodegeneration by selective alpha-synuclein nitration in synucleinopathy lesions. *Science* **2000**, *290*, 985–989.

30 H. Ischiropoulos, J. S. Beckman, Oxidative stress and nitration in neurodegeneration: cause, effect, or association? *J Clin Invest* **2003**, *111*, 163–169.

31 E. H. Norris, B. I. Giasson, H. Ischiropoulos, V. M. Lee, Effects of oxidative and nitrative challenges on alpha-synuclein fibrillogenesis involve distinct

mechanisms of protein modifications. *J Biol Chem* **2003**, *278*, 27230–27240.

32 R. Hodara, E. H. Norris, B. I. Giasson, A. J. Mishizen-Eberz, D. R. Lynch, V. M. Lee, H. Ischiropoulos, Functional consequences of alpha-synuclein tyrosine nitration: diminished binding to lipid vesicles and increased fibril formation. *J Biol Chem* **2004**, *279*, 47746–47753.

33 L. Chen, M. B. Feany, Alpha-synuclein phosphorylation controls neurotoxicity and inclusion formation in a *Drosophila* model of Parkinson disease. *Nat Neurosci* **2005**, *8*, 657–663.

34 C. B. Glaser, G. Yamin, V. N. Uversky, A. L. Fink, Methionine oxidation, alpha-synuclein and Parkinson's disease. *Biochim Biophys Acta* **2005**, *1703*, 157–169.

35 K. A. Conway, J. C. Rochet, R. M. Bieganski, P. T. Lansbury, Jr., Kinetic stabilization of the alpha-synuclein protofibril by a dopamine-alpha-synuclein adduct. *Science* **2001**, *294*, 1346–1349.

36 V. N. Uversky, G. Yamin, P. O. Souillac, J. Goers, C. B. Glaser, A. L. Fink, Methionine oxidation inhibits fibrillation of human alpha-synuclein *in vitro*. *FEBS Lett* **2002**, *517*, 239–244.

37 H. J. Rideout, K. E. Larsen, D. Sulzer, L. Stefanis, Proteasomal inhibition leads to formation of ubiquitin/alpha-synuclein-immunoreactive inclusions in PC12 cells. *J Neurochem* **2001**, *78*, 899–908.

38 K. Ancolio, C. Alves da Costa, K. Ueda, F. Checler, Alpha-synuclein and the Parkinson's disease-related mutant Ala53Thr-alpha-synuclein do not undergo proteasomal degradation in HEK293 and neuronal cells. *Neurosci Lett* **2000**, *285*, 79–82.

39 L. Stefanis, K. E. Larsen, H. J. Rideout, D. Sulzer, L. A. Greene, Expression of A53T mutant but not wild-type alpha-synuclein in PC12 cells induces alterations of the ubiquitin-dependent degradation system, loss of dopamine release, and autophagic cell death. *J Neurosci* **2001**, *21*, 9549–9560.

40 K. Beyreuther, P. Pollwein, G. Multhaup, U. Monning, G. Konig, T. Dyrks, W. Schubert, C. L. Masters, Regulation and expression of the Alzheimer's beta/A4 amyloid protein precursor in health, dis-ease, and Down's syndrome. *Ann NY Acad Sci* **1993**, *695*, 91–102.

41 M. Goedert, Tau protein and the neurofibrillary pathology of Alzheimer's disease. *Trends Neurosci* **1993**, *16*, 460–5.

42 R. A. Nixon, Endosome function and dysfunction in Alzheimer's disease and other neurodegenerative diseases. *Neurobiol Aging* **2005**, *26*, 373–382.

43 A. M. Cataldo, C. M. Peterhoff, J. C. Troncoso, T. Gomez-Isla, B. T. Hyman, R. A. Nixon, Endocytic pathway abnormalities precede amyloid beta deposition in sporadic Alzheimer's disease and Down syndrome: differential effects of APOE genotype and presenilin mutations. *Am J Pathol* **2000**, *157*, 277–286.

44 R. A. Nixon, J. Wegiel, A. Kumar, W. H. Yu, C. Peterhoff, A. Cataldo, A. M. Cuervo, Extensive involvement of autophagy in Alzheimer disease: an immuno-electron microscopy study. *J Neuropathol Exp Neurol* **2005**, *64*, 113–122.

45 W. H. Yu, A. Kumar, C. Peterhoff, L. Shapiro Kulnane, Y. Uchiyama, B. T. Lamb, A. M. Cuervo, R. A. Nixon, Autophagic vacuoles are enriched in amyloid precursor protein-secretase activities: implications for beta-amyloid peptide over-production and localization in Alzheimer's disease. *Int J Biochem Cell Biol* **2004**, *36*, 2531–2540.

46 S. Davies, D. Ramsden, Huntington's disease. *Mol Pathol* **2001**, *54*, 409–413.

47 K. B. Kegel, M. Kim, E. Sapp, C. McIntyre, J. G. Castano, N. Aronin, M. DiFiglia, Huntingtin expression stimulates endosomal-lysosomal activity, endosome tubulation, and autophagy. *J Neurosci* **2000**, *20*, 7268–7278.

48 C. M. Ambrose, M. P. Duyao, G. Barnes, G. P. Bates, C. S. Lin, J. Srinidhi, S. Baxendale, H. Hummerich, H. Lehrach, M. Altherr, et al., Structure and expression of the Huntington's disease gene: evidence against simple inactivation due to an expanded CAG repeat. *Somat Cell Mol Genet* **1994**, *20*, 27–38.

49 L. W. Ho, J. Carmichael, J. Swartz, A. Wyttenbach, J. Rankin, D. C. Rubinsztein, The molecular biology of Huntington's disease. *Psychol Med* **2001**, *31*, 3–14.

50 P. J. Muchowski, K. Ning, C. D'Souza-Schorey, S. Fields, Requirement of an intact microtubule cytoskeleton for aggregation and inclusion body formation by a mutant huntingtin fragment. *Proc Natl Acad Sci USA* **2002**, *99*, 727–732.

51 C. Landles, G. P. Bates, Huntingtin and the molecular pathogenesis of Huntington's disease. Fourth in molecular medicine review series. *EMBO Rep* **2004**, *5*, 958–963.

52 P. Harjes, E. E. Wanker, The hunt for huntingtin function: interaction partners tell many different stories. *Trends Biochem Sci* **2003**, *28*, 425–433.

53 S. H. Li, X. J. Li, Huntingtin and its role in neuronal degeneration. *Neuroscientist* **2004**, *10*, 467–475.

54 D. C. Rubinsztein, Lessons from animal models of Huntington's disease. *Trends Genet* **2002**, *18*, 202–209.

55 H. Goehler, et al., A protein interaction network links GIT1, an enhancer of huntingtin aggregation, to Huntington's disease. *Mol Cell* **2004**, *15*, 853–865.

56 J. H. Cha, Transcriptional dysregulation in Huntington's disease. *Trends Neurosci* **2000**, *23*, 387–392.

57 K. L. Sugars, D. C. Rubinsztein, Transcriptional abnormalities in Huntington disease. *Trends Genet* **2003**, *19*, 233–238.

58 H. Sakahira, P. Breuer, M. K. Hayer-Hartl, F. U. Hartl, Molecular chaperones as modulators of polyglutamine protein aggregation and toxicity. *Proc Natl Acad Sci USA* **2002**, *99 (Suppl 4)*, 16412–16418.

59 A. Ciechanover, P. Brundin, The ubiquitin proteasome system in neurodegenerative diseases: sometimes the chicken, sometimes the egg. *Neuron* **2003**, *40*, 427–446.

60 T. Schmelzle, M. N. Hall, TOR, a central controller of cell growth. *Cell* **2000**, *103*, 253–262.

61 A. Petersen, K. E. Larsen, G. G. Behr, N. Romero, S. Przedborski, P. Brundin, D. Sulzer, Expanded CAG repeats in exon 1 of the Huntington's disease gene stimulate dopamine-mediated striatal neuron autophagy and degeneration. *Hum Mol Genet* **2001**, *10*, 1243–1254.

62 S. B. Prusiner, M. P. McKinley, K. A. Bowman, D. C. Bolton, P. E. Bendheim, D. F. Groth, G. G. Glenner, Scrapie prions aggregate to form amyloid-like birefringent rods. *Cell* **1983**, *35*, 349–358.

63 M. Jeffrey, I. A. Goodbrand, C. M. Goodsir, Pathology of the transmissible spongiform encephalopathies with special emphasis on ultrastructure. *Micron* **1995**, *26*, 277–298.

64 S. B. Prusiner, Novel proteinaceous infectious particles cause scrapie. *Science* **1982**, *216*, 136–144.

65 S. B. Prusiner, Molecular biology of prion diseases. *Science* **1991**, *252*, 1515–1522.

66 J. Collinge, Prion diseases of humans and animals: their causes and molecular basis. *Annu Rev Neurosci* **2001**, *24*, 519–550.

67 K. M. Luhr, E. K. Nordstrom, P. Low, K. Kristensson, Cathepsin B and L are involved in degradation of prions in GT1-1 neuronal cells. *Neuroreport* **2004**, *15*, 1663–1667.

68 P. P. Liberski, B. Sikorska, J. Bratosiewicz-Wasik, D. C. Gajdusek, P. Brown, Neuronal cell death in transmissible spongiform encephalopathies (prion diseases) revisited: from apoptosis to autophagy. *Int J Biochem Cell Biol* **2004**, *36*, 2473–2490.

69 B. Sikorska, P. P. Liberski, P. Giraud, N. Kopp, P. Brown, Autophagy is a part of ultrastructural synaptic pathology in Creutzfeldt–Jakob disease: a brain biopsy study. *Int J Biochem Cell Biol* **2004**, *36*, 2563–2573.

70 D. Jesionek-Kupnicka, R. Kordek, J. Buczynski, P. P. Liberski, Apoptosis in relation to neuronal loss in experimental Creutzfeldt–Jakob disease in mice. *Acta Neurobiol Exp (Wars)* **2001**, *61*, 13–19.

71 M. B. Graeber, L. B. Moran, Mechanisms of cell death in neurodegenerative diseases: fashion, fiction, and facts. *Brain Pathol* **2002**, *12*, 385–390.

72 P. G. Pentchev, M. E. Comly, H. S. Kruth, M. T. Vanier, D. A. Wenger, S. Patel, R. O. Brady, A defect in cholesterol esterification in Niemann–Pick disease (type C) patients. *Proc Natl Acad Sci USA* **1985**, *82*, 8247–8251.

73 L. Liscum, S. L. Sturley, Intracellular trafficking of Niemann–Pick C proteins 1 and 2: obligate components of subcellu-

lar lipid transport. *Biochim Biophys Acta* **2004**, *1685*, 22–27.

74 N. Ohgami, D.C. Ko, M. Thomas, M.P. Scott, C.C. Chang, T.Y. Chang, Binding between the Niemann–Pick C1 protein and a photoactivatable cholesterol analog requires a functional sterol-sensing domain. *Proc Natl Acad Sci USA* **2004**, *101*, 12473–12478.

75 Y.A. Ioannou, The structure and function of the Niemann–Pick C1 protein. *Mol Genet Metab* **2000**, *71*, 175–181.

76 M.E. Higgins, J.P. Davies, F.W. Chen, Y.A. Ioannou, Niemann–Pick C1 is a late endosome-resident protein that transiently associates with lysosomes and the *trans*-Golgi network. *Mol Genet Metab* **1999**, *68*, 1–13.

77 E.B. Neufeld, et al., The Niemann–Pick C1 protein resides in a vesicular compartment linked to retrograde transport of multiple lysosomal cargo. *J Biol Chem* **1999**, *274*, 9627–9635.

78 W.S. Garver, R.A. Heidenreich, R.P. Erickson, M.A. Thomas, J.M. Wilson, Localization of the murine Niemann–Pick C1 protein to two distinct intracellular compartments. *J Lipid Res* **2000**, *41*, 673–687.

79 T.S. Blom, M.D. Linder, K. Snow, H. Pihko, M.W. Hess, E. Jokitalo, V. Veckman, A.C. Syvanen, E. Ikonen, Defective endocytic trafficking of NPC1 and NPC2 underlying infantile Niemann–Pick type C disease. *Hum Mol Genet* **2003**, *12*, 257–272.

80 S. Mukherjee, F.R. Maxfield, Lipid and cholesterol trafficking in NPC. *Biochim Biophys Acta* **2004**, *1685*, 28–37.

81 B. Karten, D.E. Vance, R.B. Campenot, J.E. Vance, Cholesterol accumulates in cell bodies, but is decreased in distal axons, of Niemann–Pick C1-deficient neurons. *J Neurochem* **2002**, *83*, 1154–1163.

82 B. Karten, D.E. Vance, R.B. Campenot, J.E. Vance, Trafficking of cholesterol from cell bodies to distal axons in Niemann Pick C1-deficient neurons. *J Biol Chem* **2003**, *278*, 4168–4175.

83 D.C. Ko, L. Milenkovic, S.M. Beier, H. Manuel, J. Buchanan, M.P. Scott, Cell-autonomous death of cerebellar purkinje neurons with autophagy in niemann-pick type C disease. *PLoS Genet* **2005**, *1*, e7.

84 N. Mizushima, Y. Ohsumi, T. Yoshimori, Autophagosome formation in mammalian cells. *Cell Struct Funct* **2002**, *27*, 421–429.

85 E. Eskelinen, D. Schmidt, S. Neu, I.M, G. Fuertes, N. Salvador, Y. Tanaka, R. Lullmann-Rauch, D. Hartmann, J. Heeren, K. von Figura, E. Knecht, P. Saftig, Disturbed cholesterol traffic but normal proteolytic function in LAMP-1/LAMP-2 double deficient fibroblasts. *Mol Biol Cell* **2004**, *15*, 3132–3145.

86 J. Gunn, M.G. Clark, S.E. Knowles, M.F. Hopgood, F.J. Ballard, Reduced rates of proteolysis in transformed cells. *Nature* **1977**, *266*, 58–60.

87 G. Kisen, L. Tessitore, P. Costelli, P.B. Gordon, P.E. Schwarze, F.M. Baccino, P.O. Seglen, Reduced autophagic activity in primary rat hepatocellular carcinoma and ascites hepatoma cells. *Carcinogenesis* **1993**, *14*, 2501–2505.

88 P. Schwarze, P.O. Seglen, Reduced autophagic activity, improved protein balance and enhanced *in vitro* survival of hepatocytes isolated from carcinogen-treated rats. *Exp Cell Res* **1985**, *157*, 15–28.

89 H. Ito, S. Daido, T. Kanzawa, S. Kondo, Y. Kondo, Radiation-induced autophagy is associated with LC3 and its inhibition sensitizes malignant glioma cells. *Int J Oncol* **2005**, *26*, 1401–1410.

90 T. Proikas-Cezanne, S. Waddell, A. Gaugel, T. Frickey, A. Lupas, A. Nordheim, WIPI-1alpha (WIPI49), a member of the novel 7-bladed WIPI protein family, is aberrantly expressed in human cancer and is linked to starvation-induced autophagy. *Oncogene* **2004**, *23*, 9314–9325.

91 E. Ogier-Denis, J.J. Houri, C. Bauvy, P. Codogno, Guanine nucleotide exchange on heterotrimeric GI3 protein controls autophagic sequestration in HT-29 cells. *J Biol Chem* **1996**, *271*, 28593–28600.

92 H. Lee, R.T. Jones, R.A. Myers, L. Marzella, Regulation of protein degradation in normal and transformed human bronchial epithelial cells in culture. *Arch Biochem Biophys* **1992**, *296*, 271–278.

93 A. Alva, S.H. Gultekin, E.H. Baehrecke, Autophagy in human tumors: cell survival or death? *Cell Death Differ* **2004**, *11*, 1046–1048.

94 D. Gozuacik, A. Kimchi, Autophagy as a cell death and tumor suppressor mechanism. *Oncogene* **2004**, *23*, 2891–2906.

95 S. Paglin, T. Hollister, T. Delohery, N. Hackett, M. McMahill, E. Sphicas, D. Domingo, J. Yahalom, A novel response of cancer cells to radiation involves autophagy and formation of acidic vesicles. *Cancer Res* **2001**, *61*, 439–444.

96 K. Yao, T. Komata, Y. Kondo, T. Kanzawa, S. Kondo, I.M. Germano, Molecular response of human glioblastoma multiforme cells to ionizing radiation: cell cycle arrest, modulation of the expression of cyclic-dependent kinase inhibitors, and autophagy. *J Neurosurg* **2003**, *98*, 378–384.

97 W. Bursch, A. Ellinger, H. Kienzl, L. Torok, S. Pandey, M. Sikorska, R. Walker, R.S. Hermann, Active cell death induced by the anti-estrogens tamoxifen and ICI 164 384 in human mammary carcinoma cells (MCF-7) in culture: the role of autophagy. *Carcinogenesis* **1996**, *17*, 1595–1607.

98 H. Takeuchi, Y. Kondo, K. Fujiwara, T. Kanzawa, H. Aoki, G.B. Mills, S. Kondo, Synergistic augmentation of rapamycin-induced autophagy in malignant glioma cells by phosphatidylinositol 3-kinase/protein kinase B inhibitors. *Cancer Res* **2005**, *15*, 3336–3346.

99 T. Kanazawa, I.M. Germano, T. Komata, H. Ito, Y. Kondo, S. Kondo, Role of autophagy in temozolomide-induced cytotoxicity for malignant glioma cells. *Cell Death Differ* **2004**, *11*, 448–457.

100 T. Kanazawa, Y. Kondo, H. Ito, S. Kondo, I. Germano, Induction of autophagic cell death in malignant glioma cells by arsenic trioxide. *Cancer Res* **2003**, *63*.

101 T. Kanzawa, L. Zhang, L. Xiao, I.M. Germano, Y. Kondo, S. Kondo, Arsenic trioxide induces autophagic cell death in malignant glioma cells by upregulation of mitochondrial cell death protein BNIP3. *Oncogene* **2004**, *24*, 980–991.

102 A.J. Opipari, L. Tan, A.E. Boitano, D.R. Sorenson, A. Aurora, J.R. Liu, Resveratrol-induced autophagocytosis in ovarian cancer cells. *Cancer Res* **2004**, *15*, 696–703.

103 A. Ellington, M. Berhow, K.W. Singletary, Induction of macroautophagy in human colon cancer cells by soybean B-group triterpenoid saponins. *Carcinogenesis* **2005**, *26*, 159–167.

104 S. Pattingre, A. Tassa, X. Qu, R. Garuti, X. Huan Liang, N. Mizushima, M. Packer, M. Scheneider, B. Levine, Bcl-2 antiapoptotic proteins inhibit beclin 1-dependent autophagy. *Cell* **2005**, *122*, 927–939.

105 X. Liang, S. Jackson, M. Seaman, K. Brown, B. Kempkes, H. Hibshoosh, B. Levine, Induction of autophagy and inhibition of tumorigenesis by beclin 1. *Nature* **1999**, *402*, 672–676.

106 X. Qu, G. Bhagat, N. Furuya, H. Hibshoosh, A. Troxel, J. Rosen, E.L. Eskelinen, N. Mizushima, Y. Ohsumi, G. Cattoretti, B. Levine, Promotion of tumorigenesis by heterozygous disruption of the *beclin 1* autophagy gene. *J Clin Invest* **2003**, *112*, 1809–1820.

107 Z. Yue, S. Jin, C. Yang, A.J. Levine, N. Heintz, Beclin 1, an autophagy gene essential for early embryonic development, is a haploinsufficient tumor suppressor. *Proc Natl Acad Sci USA* **2003**, *100*, 15077–15082.

108 A. Kihara, Y. Kabeya, Y. Ohsumi, T. Yoshimori, Beclin–phosphatidylinositol 3-kinase complex functions at the *trans*-Golgi network. *EMBO Rep* **2001**, *2*, 330–335.

109 W. Brown, D.B. DeWald, S.D. Emr, H. Plutner, W.E. Balch, Role for phosphatidylinositol 3-kinase in the sorting and transport of newly synthesized lysosomal enzymes in mammalian cells. *J Cell Biol* **1995**, *130*, 781–796.

110 X. Liang, L.K. Kleeman, H.H. Jiang, G. Gordon, J.E. Goldman, G. Berry, B. Herman, B. Levine, Protection against fatal Sindbis virus encephalitis by beclin, a novel Bcl-2-interacting protein. *J Virol* **1998**, *72*, 8586–8596.

111 R. Loewith, E. Jacinto, S. Wullschleger, A. Lorberg, J.L. Crespo, D. Bonenfant, W. Oppliger, P. Jenoe, M.N. Hall, Two TOR complexes, only one of which is rapamycin sensitive, have distinct roles in cell growth control. *Mol Cell* **2002**, *10*, 457–468.

112 E. Jacinto, R. Loewith, A. Schmidt, S. Lin, M.A. Ruegg, A. Hall, M.N. Hall, Mammalian TOR complex 2 controls the actin cytoskeleton and is rapamycin insensitive. *Nat Cell Biol* **2004**, *6*, 1122–1128.

113 S. Chan, Targeting the mammalian target of rapamycin (mTOR): a new approach to treating cancer. *Br J Cancer* **2004**, *91*, 1420–1424.

114 A. Gingras, B. Raught, N. Sonenberg, Regulation of translation initiation by FRAP/mTOR. *Genes Dev* **2001**, *15*, 807–826.

115 D. Sarbassov, D.A. Guertin, S.M. Ali, D.M. Sabatini, Phosphorylation and regulation of Akt/PKB by the rictor–mTOR complex. *Science* **2005**, *307*, 1098–1101.

116 D. Guertin, D.M. Sabatini, An expanding role for mTOR in cancer. *Trends Mol Med* **2005**, *11*, 353–361.

117 S. Arico, A. Petiot, C. Bauvy, P.F. Dubbelhuis, A.J. Meijer, P. Codogno, E. Ogier-Denis, The tumor suppressor PTEN positively regulates macroautophagy by inhibiting the phosphatidylinositol 3-kinase/protein kinase B pathway. *J Biol Chem* **2001**, *276*, 35243–35246.

118 P. Steck, M.A. Pershouse, S.A. Jasser, W.K. Yung, H. Lin, A.H. Ligon, L.A. Langford, M.L. Baumgard, T. Hattier, T. Davis, C. Frye, R. Hu, B. Swedlund, D.H. Teng, S.V. Tavtigian, Identification of a candidate tumour suppressor gene, MMAC1, at chromosome 10q23.3 that is mutated in multiple advanced cancers. *Nat Genet* **1997**, *15*, 356–362.

119 E. Ogier-Denis, P. Codogno, Autophagy: a barrier or an adaptive response to cancer. *Biochim Biophys Acta* **2003**, *1603*, 113–1128.

120 Y. Kondo, T. Kanzawa, R. Sawaya, S. Kondo, The role of autophagy in cancer development and response to therapy. *Nat Rev Cancer* **2005**, *5*, 726–734.

121 B. Inbal, S. Bialik, I. Sabanay, G. Shani, A. Kimchi, DAP kinase and DRP-1 mediate membrane blebbing and the formation of autophagic vesicles during programmed cell death. *J Cell Biol* **2002**, *157*, 455–468.

122 M. Davies, S.J. Kim, N.U. Parikh, Z. Dong, C.D. Bucana, G.E. Gallick, Adenoviral-mediated expression of MMAC/PTEN inhibits proliferation and metastasis of human prostate cancer cells. *Clin Cancer Res* **2002**, *8*, 1904–1914.

123 W. Mondesire, W. Jian, H. Zhang, J. Ensor, M.C. Hung, G.B. Mills, F. Meric-Bernstam, Targeting mammalian target of rapamycin synergistically enhances chemotherapy-induced cytotoxicity in breast cancer cells. *Clin Cancer Res* **2004**, *10*, 7031–7042.

124 N. Mizushima, A. Yamamoto, M. Matsui, T. Yoshimori, Y. Ohsumi, *In vivo* analysis of autophagy in response to nutrient starvation using transgenic mice expressing a fluorescent autophagosome marker. *Mol Biol Cell* **2004**, *15*, 1101–1111.

125 A. Tassa, M.P. Roux, D. Attaix, D.M. Bechet, Class III phosphoinositide 3-kinase–Beclin1 complex mediates the amino acid-dependent regulation of autophagy in C2C12 myotubes. *Biochem J* **2003**, *376*, 577–586.

126 S. Mordier, C. Deval, D. Bechet, A. Tassa, M. Ferrara, Leucine limitation induces autophagy and activation of lysosome-dependent proteolysis in C2C12 myotubes through a mammalian target of rapamycin-independent signaling pathway. *J Biol Chem* **2000**, *275*, 29900–29906.

127 I. Nishino, Autophagic vacuolar myopathies. *Curr Neurol Neurosci Rep* **2003**, *3*, 64–69.

128 M.J. Danon, S.J. Oh, S. DiMauro, J.R. Manaligod, A. Eastwood, S. Naidu, L.H. Schliselfeld, Lysosomal glycogen storage disease with normal acid maltase. *Neurology* **1981**, *31*, 51–57.

129 I. Nishino, J. Fu, K. Tanji, T. Yamada, S. Shimojo, T. Koori, M. Mora, J.E. Riggs, S.J. Oh, Y. Koga, C.M. Sue, A. Yamamoto, N. Murakami, S. Shanske, E. Byrne, E. Bonilla, I. Nonaka, S. DiMauro, M. Hirano, Primary LAMP-2 deficiency causes X-linked vacuolar cardiomyopathy and myopathy (Danon disease). *Nature* **2000**, *406*, 906–910.

130 Y. Tanaka, G. Guhde, A. Suter, E.L. Eskelinen, D. Hartmann, R. Lullmann-

Rauch, P. Saftig, Accumulation of autophagic vacuoles and cardiomyopathy in LAMP-2-deficient mice. *Nature* **2000**, *406*, 902–906.

131 J. Horvath, U. P. Ketelsen, A. Geibel-Zehender, N. Boehm, H. Olbrich, R. Korinthenberg, H. Omran, Identification of a novel LAMP2 mutation responsible for X-chromosomal dominant Danon disease. *Neuropediatrics* **2003**, *34*, 270–273.

132 K. Sugie, S. Noguchi, Y. Kozuka, E. Arikawa-Hirasawa, M. Tanaka, C. Yan, P. Saftig, K. von Figura, M. Hirano, S. Ueno, I. Nonaka, I. Nishino, Autophagic vacuoles with sarcolemmal features delineate Danon disease and related myopathies. *J Neuropathol Exp Neurol* **2005**, *64*, 513–522.

133 L. Villard, V. des Portes, N. Levy, L. J. Pouboutin, D. Recan, M. Coquet, et al., Linkage of X-linked myopathy with excessive autophagy (XMEA) to Xq28. *Eur J Hum Genet* **2000**, *8*, 125–129.

134 I. Nonaka, N. Murakami, Y. Suzuki, M. Kawai, Distal myopathy with rimmed vacuoles. *Neuromuscul Disord* **1998**, *8*, 333–337.

135 P. O. Seglen, Inhibitors of lysosomal function. *Methods Enzymol* **1983**, *96*, 737–764.

136 T. Suzuki, M. Nakagawa, A. Yoshikawa, N. Sasagawa, T. Yoshimori, Y. Ohsumi, et al., The first molecular evidence that autophagy relates rimmed vacuole formation in chloroquine myopathy. *J Biochem* **2002**, *131*, 647–651.

137 A. Yamamoto, Y. Morisawa, A. Verloes, N. Murakami, N. M. I. Hirano, et al., Infantile autophagic vacuolar myopathy is distinct from Danon disease. *Neurology* **2001**, *57*, 903–905.

138 D. Kaneda, K. Sugie, A. Yamamoto, M. H. atsumoto, T. Kato, I. Nonaka, et al., A novel form of autophagic vacuolar myopathy with late-onset and multiorgan involvement. *Neurology* **2003**, *61*, 128–131.

139 A. Buj-Bello, V. Laugel, N. Messaddeq, H. Zahreddine, J. Laporte, J. F. Pellissier, J. L. Mandel, The lipid phosphatase myotubularin is essential for skeletal muscle maintenance but not for myogenesis in mice. *Proc Natl Acad Sci USA* **2002**, *99*, 15060–15065.

140 H. Shimomura, F. Terasaki, H. Tayashi, Y. Kitaura, T. Isomura, H. Suma, Autophagic degeneration as a possible mechanism of myocardial cell death in dilated cardiomyopathy. *Jpn Circ J* **2001**, *65*, 965–968.

141 M. Knaapen, M. J. Davies, M. De Bie, A. J. Haven, W. Martinet, M. M. Kockx, Apoptotic versus autophagic cell death in heart failure. *Cardiovasc Res* **2001**, *51*, 304–312.

142 S. Kostin, L. Pool, A. Elsasser, S. Hein, H. C. Drexler, E. Arnon, Y. Hayakawa, R. Zimmermann, E. Bauer, W. P. Klovekorn, J. Schaper, Myocytes die by multiple mechanisms in failing human hearts. *Circ Res* **2003**, *18*, 715–724.

143 M. Saijo, G. Takemura, M. Koda, H. Okada, et al., Cardiomyopathy with prominent autophagic degeneration, accompanied by an elevated plasma brain natriuretic peptide level despite the lack of overt heart failure. *Intern Med* **2004**, *43*, 700–703.

144 L. Yan, D. E. Vatner, S. J. Kim, H. Ge, M. Masurekar, W. H. Massover, G. Yang, Y. Matsui, J. Sadoshima, S. F. Vatner, Autophagy in chronically ischemic myocardium. *Proc Natl Acad Sci USA* **2005**, *102*, 13807–13812.

145 J. Teckman, J. K. An, K. Blomenkamp, B. Schmidt, D. Perlmutter, Mitochondrial autophagy and injury in the liver in alpha 1-antitrypsin deficiency. *Am J Physiol Gastrointest Liver Physiol* **2004**, *286*, G851–G862.

146 J.-O. Jeppsson, Amino acid substitution Glu to Lys in alpha1antitrypsin PiZ. *FEBS Lett* **1976**, *65*, 195–197.

147 B. Schmidt, D. H. Perlmutter, Grp78, Grp94, and Grp170 interact with alpha1-antitrypsin mutants that are retained in the endoplasmic reticulum. *Am J Physiol Gastrointest Liver Physiol* **2005**, *289*, G444–G455.

148 E. Werner, J. L. Brodsky, A. A. McCracken, Proteasome-dependent endoplasmic reticulum-associated protein degradation: an unconventional route to a familiar fate. *Proc Natl Acad Sci USA* **1996**, *93*, 13797–13801.

149 D. Qu, J.H. Teckman, S. Omura, D.H. Perlmutter, Degradation of mutant secretory protein alpha1antitrypsin Z, in the endoplasmic reticulum requires proteasome activity. *J Biol Chem* **1996**, *271*, 22791–22795.

150 J. Teckman, D. Perlmutter, Retention of mutant alpha1antitrypsin Z in endoplasmic reticulum is associated with an autophagic response. *Am J Physiol Gastrointest Liver Physiol* **2000**, *279*, G961–G974.

151 D. Perlmutter, Liver injury in alpha1antitrypsin deficiency: an aggregated protein induces mitochondrial injury. *J Clin Invest* **2002**, *110*, 1579–1583.

152 M. Komatsu, S. Waguri, T. Ueno, J. Iwata, S. Murata, I. Tanida, J. Ezaki, N. Mizushima, Y. Ohsumi, Y. Uchiyama, E. Kominami, K. Tanaka, T. Chiba, Impairment of starvation-induced and constitutive autophagy in Atg7-deficient mice. *J Cell Biol* **2005**, *169*, 425–434.

153 U. Pfeifer, Inhibition by insulin of the physiological autophagic breakdown of cell organelles. *Acta Biol Med Ger* **1977**, *36*, 1691–1694.

154 U. Pfeifer, Inhibition by insulin of the formation of autophagic vacuoles in rat liver. A morphometric approach to the kinetics of intracellular degradation by autophagy. *J Cell Biol* **1978**, *78*, 152–167.

155 G. Mortimore, C.E. Mondon, Inhibition by insulin of valine turnover in liver. *J Biol Chem* **1970**, *245*, 2375–2383.

156 M. Hopgood, C.M.G. Lark, F.J. Ballard, Protein degradation in hepatocyte monolayers. Effect of glucagon, adenosine 3′:5′-cyclic monophosphate and insulin. *Biochem J* **1980**, *186*, 71–79.

157 U. Pfeifer, M. Warmuth-Metz, Inhibition by insulin of cellular autophagy in proximal tubular cells of rat kidney. *Am J Physiol* **1983**, *244*, E109–E114.

158 S. Lenk, D. Bhat, W. Blakeney, W.A. Dunn, Jr., Effects of strptozotocin-induced diabetes on rough endoplasmic reticulum and lysosomes of rat liver. *Am J Physiol* **1992**, *263*, E856–E862.

159 K. Han, H. Zhou, U. Pfeifer, Inhibition and restimulation by insulin of cellular autophagy in distal tubular cells of the kidney in early diabetic rats. *Kidney Blood Press Res* **1997**, *20*, 258–263.

160 T. Kanazawa, I. Taneike, R. Akaishi, F. Yoshizawa, N. Furuya, S. Fujimura, M. Kadowaki, Amino acids and insulin control autophagic proteolysis through different signaling pathways in relation to mTOR in isolated rat hepatocytes. *J Biol Chem* **2004**, *279*, 8452–8459.

161 R. Towns, Y. Kabeya, T. Yoshimori, G. Guo, Y. Shangguan, S. Hong, M. Kaplan, D. Klionsky, J. Wiley, Sera from patients with type 2 diabetes and neuropathy induce autophagy and colocalization with mitochondria in SY5Y cells. *Autophagy* **2005**, *1*, 141–145.

162 J. Davies, D. Murphy, Autophagy in hypothalamic neurons of rats expressing a familial neurohypophysial diabetes insipidus transgene. *J Neuroendocrinol* **2002**, *14*, 629–637.

163 R. Castino, J. Davies, S. Beaucourt, C. Isidoro, D. Murphy, Autophagy is a prosurvival mechanism in cells expressing an autosomal dominant familial neurohypophyseal diabetes insipidus mutant vasopressin transgene. *FASEB J* **2005**, *19*, 1021–1023.

164 H.A. Franch, S. Sooparb, J. Du, N.S. Brown, A mechanism regulating proteolysis of specific proteins during renal tubular cell growth. *J Biol Chem* **2001**, *276*, 19126–19131.

165 S. Sooparb, S.R. Price, J. Shaoguang, H.A. Franch, Suppression of chaperone-mediated autophagy in the renal cortex during acute diabetes mellitus. *Kidney Int* **2004**, *65*, 2135–2144.

166 M. Martinez-Vicente, G. Sovak, A.M. Cuervo, Protein degradation and aging. *Exp Gerontol* **2005**, *40*, 622–633.

167 A.M. Cuervo, E. Bergamini, U.T. Brunk, W. Droge, M. Ffrench, A. Terman, Autophagy and aging: the importance of maintaining "clean" cells. *Autophagy* **2005**, *1*, 131–140.

168 A. Reznick, D. Gershon, The effect of age on the protein degradation system in the nematode *Turbatrix aceti*. *Mech Ageing Dev* **1979**, *11*, 403–415.

169 L. Lavie, A.Z. Reznick, D. Gershon, Decreased protein and puromycinylpeptide degradation in livers of senescent mice. *Biochem J* **1982**, *202*, 47–51.

170 J.N. Keller, E. Dimayuga, Q. Chen, J. Thorpe, J. Gee, Q. Ding, Autophagy,

proteasomes, lipofuscin, and oxidative stress in the aging brain. *Int J Biochem Cell Biol* **2004**, *36*, 2376–2391.

171 E. Stadtman, Protein oxidation in aging and age-related diseases. *Ann NY Acad Sci* **2001**, *928*, 22–38.

172 A. Ryazanov, B. Nefsky, Protein turnover plays a key role in aging. *Mech Ageing Dev* **2002**, *123*, 207–213.

173 A. Terman, U.T. Brunk, Lipofuscin. *Int J Biochem Cell Biol* **2004**, *36*, 1400–1404.

174 A. Terman, U.T. Brunk, Aging as a catabolic malfunction. *Int J Biochem Cell Biol* **2004**, *36*, 2365–2375.

175 W. Ward, Protein degradation in the aging organism. *Prog Mol Subcell Biol* **2002**, *29*, 35–42.

176 A. Del Roso, S. Vittorini, G. Cavallini, A. Donati, Z. Gori, M. Masini, M. Pollera, E. Bergamini, Ageing-related changes in the *in vivo* function of rat liver macroautophagy and proteolysis. *Exp Gerontol* **2003**, *38*, 519–527.

177 A. Donati, G. Cavallini, C. Paradiso, S. Vittorini, M. Pollera, Z. Gori, E. Bergamini, Age-related changes in the autophagic proteolysis of rat isolated liver cells: effects of antiaging dietary restrictions. *J Gerontol A Biol Sci Med Sci* **2001**, *56*, B375–B383.

178 E. Schmid, J. El Benna, D. Galter, G. Klein, W. Dröge, Redox priming of the insulin receptor-chain associated with altered tyrosine kinase activity and insulin responsiveness in the absence of tyrosine autophosphorylation. *FASEB J.* **1998**, *12*, 863–870.

179 A. Melendez, Z. Talloczy, M. Seaman, E. L. Eskelinen, D. H. Hall, B. Levine, Autophagy genes are essential for dauer development and life-span extension in *C. elegans*. *Science* **2003**, *301*, 1387–1391.

180 J. Dice, Altered degradation of proteins microinjected into senescent human fibroblasts. *J Biol Chem* **1982**, *257*, 14624–14627.

181 A. M. Cuervo, J. F. Dice, Age-related decline in chaperone-mediated autophagy. *J Biol Chem* **2000**, *275*, 31505–31513.

182 A. Terman, Garbage catastrophe theory of aging: imperfect removal of oxidative damage? *Redox Rep* **2001**, *6*, 15–26.

183 R. Kiffin, C. Christian, E. Knecht, A. Cuervo, Activation of chaperone-mediated autophagy during oxidative stress. *Mol Biol Cell* **2004**, *15*, 4829–40.

184 A. Cuervo, A. Palmer, A. Rivett, E. Knecht, Degradation of proteasomes by lysosomes in rat liver. *Eur J Biochem* **1995**, *227*, 792–800.

185 L.-M. Gerland, L. Genestier, S. Peyrol, M. C. Michallet, S. Hayette, I. Urbanowicz, P. Ffrench, J. P. Magaud, M. Ffrench, Autolysosomes accumulate during *in vitro* CD8⁺ T-lymphocyte aging and may participate in induced death sensitization of senescent cells. *Exp Gerontol* **2004**, *39*, 789–800.

186 E. Bergamini, Targets for antiageing drugs. *Expert Opin Ther Targets* **2005**, *9*, 77–82.

187 K. Hauer, W. Hildebrandt, Y. Sehl, L. Edler, P. Oster, W. Dröge, Improvement in muscular performance and decrease in tumor necrosis factor level in old age after antioxidant treatment. *J Mol Med* **2003**, *81*, 118–125.

188 A. C. Massey, S. Kaushik, G. Sovak, R. Kiffin, A. M. Cuervo, Consequence of the selective bloackade of chaperone-mediated autophagy. *Proc Nat Acad Sci* **2006**, *103*, 5805–5810.

5
The Dual Roles for Autophagy in Cell Death and Survival

Jayanta Debnath and Christopher Fung

5.1
Introduction

Macroautophagy (commonly called autophagy) is a fundamental catabolic process where the cytoplasmic contents of a cell are sequestered within double-membrane vacuoles (autophagosomes) and subsequently delivered to the lysosome for degradation. In recent years, autophagy has received increased attention as a potential nonapoptotic or "alternative" cell death mechanism [1–4]. Indeed, autophagic death has historically been classified as type 2 programmed cell death based on morphological grounds [5, 6]. However, the functional contribution of autophagy to programmed cell death has been a matter of intense debate, because autophagy is well recognized as a survival mechanism in all eukaryotic cells during conditions of nutrient limitation.

Basal levels of autophagy occur in eukaryotic cells as a housekeeping mechanism of cytoplasmic and organelle turnover. During starvation, autophagy is utilized to generate both nutrients and energy through the bulk degradation of cytoplasmic material [7–9]. Under these circumstances, autophagy is critical for maintaining cell viability. Moreover, recent evidence indicates that autophagy is utilized to degrade deleterious contents that accumulate in the cytoplasm, such as toxic protein aggregates [10]. Accordingly, the presence of autophagy in dying cells has been proposed to function as an adaptive cytoprotective response that prolongs viability, rather than regulates autophagic (type 2) cell death [11].

This chapter presents an overview of the dual role of autophagy as a contributor to programmed cell death and as a cytoprotective mechanism. Importantly, over the past decade, landmark studies in yeast have identified the genes and proteins constituting the core machinery of the autophagic degradation process; many of these genes, now called *atg* genes for autophagy genes, are conserved in higher organisms [12]. These studies have facilitated the investigation of how autophagy functionally contributes to a myriad of key physiological and pathological processes in higher organisms, including programmed cell death. Indeed, several loss-of-function studies of yeast *atg* orthologs that directly address

Autophagy in Immunity and Infection. Edited by Vojo Deretic
Copyright © 2006 WILEY-VCH Verlag GmbH & Co. KGaA, Weinheim
ISBN: 3-527-31450-4

the role of autophagy during both cell death and survival have been published recently.

5.2
Types of Programmed Cell Death

Cell death is traditionally classified into two groups – necrosis and programmed cell death. Necrosis is a passive, uncontrolled way to die that elicits a strong inflammatory response, whereas programmed cell death represents a highly regulated process with defined cellular pathways. Programmed cell death is further classified into three categories, which are largely defined on the basis of morphological features and the role of lysosomes – apoptotic (type 1), autophagic (type 2) and nonlysosomal (type 3) cell death [5, 6]. Although apoptosis (type 1) is the most commonly observed of these morphologies, the morphological features of autophagic (type 2) cell death are also frequently observed both *in vitro* and *in vivo*; consequently, it has often been classified as a nonapoptotic or "alternative" form of programmed cell death. The morphological features of these types of programmed cell death are summarized below.

5.2.1
Type 1 Programmed Cell Death

The characteristic features of apoptosis are cell shrinkage, condensation of the nuclear chromatin, nuclear fragmentation and membrane blebbing [13]. The clearance of dying cells is mediated by the engulfment by neighboring phagocytic cells, commonly termed heterophagy [14, 15]. During apoptotic cell death, the death signals elicited by various stimuli ultimately converge upon two fundamental biochemical events: (a) mitochondrial depolarization and the release of death-inducing factors (e.g. cytochrome *c*) and (b) the activation of cysteine proteases called caspases [16]. Caspases serve to dismantle the cell during apoptosis. Certain caspases, such as caspase-8 and caspase-9, act as upstream initiators in response to death signals, whereas others, like caspase-3 and caspase-7, act as downstream executioners in an apoptotic cell. Caspase activation involves a rapidly amplified proteolytic cascade, ultimately resulting in the degradation of a wide spectrum of substrates as the cell dies [17]. The absence of caspase activation is usually viewed as a requisite in order to categorize a death process as being "alternative" [2, 18].

5.2.2
Type 2 Programmed Cell Death

In contrast to apoptosis, autophagic cell death is notable for the presence of numerous autophagic vacuoles (AVs) in the dying cell that are utilized for self-degradation [5, 6]. One may speculate that the level of autophagic degradation observed during type 2 cell death is far more extensive than the rate of autophagy observed during the turnover of organelles in normal cell homeostasis. As a result, autophagy ultimately leads to cell demise because the cell literally "eats itself" to death. As discussed in greater detail below, dying cells that exhibit characteristic type 2 morphological features cell frequently demonstrate concomitant caspase activation; hence, elements of the classical apoptotic machinery may still serve critical functions in cell death even when extensive autophagic degradation exists. Similar to apoptosis, but unlike necrosis, autophagic death is not thought to evoke a tissue inflammatory response.

Unlike apoptosis, there are limited tools for the detection or quantification of autophagy, especially during programmed cell death [19]. The gold standard has been the demonstration of the autophagic vesicles by electron microscopy; however, this method requires considerable skill. More recently, monitoring the intracellular location of Atg proteins, using Green Fluorescent Protein (GFP)-tagged molecules, has emerged as an effective and reliable method to monitor autophagy [19–21]. During autophagy, Atg8 is modified with the lipid phosphatidylethanolamine (PE) through a ubiquitin-like conjugation process, upon which it specifically relocates to early autophagosomes in yeast [22, 23]. In mammalian cells, evaluating similar changes in the intracellular location of the mammalian Atg8 ortholog, microtubule-associated protein 1 light chain 3 (LC3), along with measuring changes in LC3 electrophoretic mobility upon PE lipid conjugation, now provide important molecular markers for the detection and quantification of autophagic activity [20, 21]. These new techniques for monitoring autophagy are already facilitating the investigation of autophagy during programmed cell death.

5.2.3
Type 3 Programmed Cell Death

Nonlysosomal (type 3) cell death is notable for organelle swelling and the formation of empty cytoplasmic spaces. Overall, several features of type 3 cell death resemble those associated with necrosis. However, type 3 cell death is rarely observed in physiological situations compared to the type 1 and type 2 morphologies [6].

5.2.4
Other Types of Programmed Cell Death

Recent evidence has indicated that necrosis, commonly considered a passive, unregulated form of cell death, may actually be "programmed"; in other words, it is regulated by defined cellular pathways [1, 24]. Like autophagy, programmed necrosis is viewed as an alternative form of programmed cell death, which has been observed in apoptosis-resistant transformed cells treated with DNA-damaging agents [25], human myeloid leukemia cells treated with the Abl tyrosine kinase inhibitor imatinib (Gleevec) [26] and in virally infected cells treated with the inflammatory cytokine tumor necrosis factor (TNF) [27]. The distinct and interconnected roles of autophagy and necrosis as alternative death processes remain a subject for further study. Finally, several other alternative forms of cell death have also been described based on studies of derived cell lines, but the physiological relevance of these morphologies remains unclear [18].

5.3
The Contribution of Autophagy to Programmed Cell Death

Autophagic cell death has been commonly observed in dying cells from animals of diverse species including insects, amphibians and mammals [6]. In mammals, programmed cell death with an autophagic or type 2 morphology is observed throughout development, including the regression of the corpus luteum, the involution of mammary and prostate glands, and the regression of Mullerian duct structures during male genital development [28–31]. In these situations, extensive autophagy is most commonly associated with large-scale tissue remodeling. For example, the post-lactational mammary gland is notable for rapid, widespread tissue destruction, where physical barriers may be imposed on professional phagocytes due to an intact epithelial basement membrane or myoepithelial cell layer [29]. These results intimate that self-degradation by autophagy serves as an important cell clearance mechanism when phagocytes are absent or when widespread tissue histolysis overwhelms the professional phagocytic machinery.

The presence of AVs in many types of dying cultured cells also suggests that autophagy might play an active role in the regulation of type 2 programmed cell death. Treatment of MCF-7 mammary cancer cells with 4-hydroxytamoxifen, a selective estrogen antagonist, triggers cells to die over a period of several days; these cells possess massive amounts of AVs [32]. Because MCF-7 cells lack critical apoptosis regulators, such as caspase-3, it is possible that autophagy or autophagic degradation can compensate for defects in the apoptotic machinery [33]. Type 2 programmed cell death has also been described in a variety of cancer cells upon treatment with chemotherapeutic agents *in vitro*, which may have important clinical implications for the treatment of cancers with apoptotic defects [34–38]. While plenty of evidence demonstrates the correlation between autophagy and cell death, the lack of causative evidence in most of these situations has led to the proposal that the extensive autophagy observed in dying cells serves as an adaptive response to the stress produced by death-inducing stimuli [11].

5.3.1
Death Processes That Require *atg* Genes

Until recently, a functional link between autophagy and programmed cell death has usually been based on experiments using 3-methyladenine (3-MA) – a pharmacological compound that inhibits the early sequestration events in autophagy

[39]. 3-MA has been shown to prevent the formation of AVs as well as the eventual death of cells in various systems, such as the tamoxifen-induced death of MCF-7 mammary carcinoma cells [32, 40], neuronal cell death upon withdrawal from nerve growth factor (NGF) [41] and the death of T lymphoblastic leukemia cells upon treatment with TNF-α [42]. However, conclusions drawn from these experiments have important caveats. First, 3-MA is a general inhibitor of phosphatidylinositol-3-kinases, enzymes that are involved in a diverse array of cellular processes in addition to autophagy [43]. Second, 3-MA is not a pharmacologically potent compound and is commonly utilized at millimolar concentrations; at these concentrations, additional effects have been reported, including the inhibition of Jun N-terminal kinase (JNK) and p38 kinase, both of which are involved in the regulation of stress-induced cell death [44, 45]. Clearly, the need for more specific loss-of-function approaches to interrogate the role of various *atg* genes in cell death has been long overdue.

Accordingly, several recent studies have indicated that *atg* genes are required to initiate cell death in cells unable to undergo apoptosis. Relatively high numbers of AVs are observed in murine L929 fibrosarcoma cells when they undergo cell death during caspase inhibition [46, 47]. This autophagic death of L929 cells depends on the function of the receptor-interacting protein (RIP), which has been described in previous studies of a necrosis-like cell death process [48]. L929 autophagic cell death also requires components of the JNK pathway including MKK7 and c-Jun [47]. Significantly, the induction of this type 2 autophagic cell death and formation of AVs requires the function of the autophagy genes *atg7* and *atg6/beclin 1*.

The Bcl-2 family proteins function as important regulators of apoptosis. Mouse embryonic fibroblasts which lack the Bcl-2 family members Bax and Bak, two proapoptotic proteins critical for mitochondrial permeabilization, are resistant to apoptosis induced by numerous agents [49]. A recent study indicates that cells doubly deficient for Bax and Bak are capable of dying with a nonapoptotic morphology in response to apoptosis-inducing agents. Notably, these dying cells possess AVs and elevated levels of Atg proteins. Furthermore, this autophagic cell death requires the function of two autophagy genes, *atg5* and *atg6/beclin 1* [50]. Similar to the results observed in L929 cells, these studies of Bax/Bak double-knockout cells indicate that caspase activity and autophagic degradation may serve complementary roles that regulate distinct forms of cell death.

Overall these studies support the idea that autophagy can contribute to death in a caspase-independent manner and operate as a nonapoptotic death mechanism in certain contexts [47, 50]. These alternative death pathways may function as an important failsafe response in tissue homeostasis or in diseases where caspase function has been compromised. Such a function may also have important implications for the chemotherapeutic killing of cancer cells with apoptotic defects [32, 34–38]. Moreover, because viruses commonly possess gene products that inhibit caspase activation, autophagic death may serve as an important back-up mechanism to elicit nonapoptotic cell death in infected cells, and thus, prevent further pathogenic effects. Hence, loss-of-function studies of *atg* genes in response to other death stimuli are warranted in order to discover additional situations where autophagy is required in caspase-independent cell death. Studies are also needed to clarify whether a cell dies by type 2 programmed cell death, programmed necrosis or another process in these situations.

The use of gene-targeted mice lacking various components of the apoptotic machinery may clarify these questions *in vivo*. Initial evidence does indicate that alternative pathways compensate when apoptotic gene function is abolished, e.g. cell death with a vacuolated or necrosis-like morphology is observed in mice null lacking the apoptotic molecules, Apaf-1, caspase-3 or caspase-9 [51, 52]. However, the functional contribution of autophagy to cell death in these models remains unclear until loss-of-function studies of *atg* genes are performed. Further investigation is needed to determine if autophagy can regulate a caspase-independent cell death *in vivo*.

5.4
The Combined Activation of Autophagy and Apoptosis during Programmed Cell Death

While increased autophagy activity is observed in cells undergoing cell death in many situations, few studies have postulated a causative link between programmed cell death and the autophagic pathway. Furthermore, dying cells with characteristic type 2 morphological features cell frequently demonstrate the cardinal features of classical apoptosis (type 1 programmed cell death), such as caspase activation and mitochondrial membrane permeability; hence, delineating the true function of autophagy during programmed cell death in higher organisms has been difficult. Conversely, the fact that autophagy is concurrently activated with apoptosis has led to the idea that autophagy functions as a cellular response to the stress of programmed cell death rather that as an alternative death process. Nonetheless, understanding the function of autophagy in programmed cell death during circumstances where caspases are simultaneously activated remains an important subject for further investigation.

The larval salivary glands of the fruit fly *Drosophila melanogaster* have served as a particularly useful model for studying the relationship between autophagy and cell death under physiological conditions [53]. An increase in the steroid hormone, 20-hydroxyecdysone (ecdysone), triggers a series of morphological and biochemical changes that culminate in removal of salivary glands 6 h later. These morphological and biochemical changes occur 2–4 h after the rise in ecdysone, and include DNA fragmentation, changes in structural protein localization and dynamic changes in vacuole structure, including the formation of AVs [54–56].

Importantly, the expression of the caspase inhibitor p35 prevents destruction of salivary glands [54], and salivary glands of animals with mutations in the caspase *Dronc* also fail to die [57]. Indeed, this requirement for caspase function raises questions about the functional role of autophagy in these dying cells. From a kinetic standpoint, caspase activation destroys a cell more rapidly than degradation via autophagy; hence, it is possible that apoptotic processes represent the predominant death mechanism when both processes take place simultaneously, even if extensive autophagic degradation exists. However, caspase inhibition does not completely block vacuolar changes following the rise in ecdysone, raising the possibility that autophagy might play an active role in cell removal and remodeling of larval salivary glands [55]. This hypothesis is supported by data indicating that caspases are upregulated at the same time as *atg* genes and prior to the formation of AVs in these cells [56]. Although it is feasible that autophagy is being utilized to maintain cell viability after the activation of caspases, it seems counter-productive that the same steroid signal simultaneously promotes both the transcription of cell death regulators as well as *atg* gene transcription for survival; this does not appear to be an efficient use of energy and resources. It appears more likely that autophagy participates in the death or degradation of salivary glands in *Drosophila*.

The concurrent activation of autophagic and apoptotic pathways has also been observed in mammalian cells. The deprivation of NGF from cultured sympathetic neurons elicits cell death that involves both apoptosis and autophagy; based on pharmacological inhibition studies, autophagic degradation has been proposed to be triggered during the early stages of cell death and to function upstream of principal apoptotic events, including cytochrome *c* release and caspase activation [41]. In addition, serum deprivation of PC12 cells induces cell death that exhibits both apoptotic and autophagic features [58]. Autophagy also appears to be required for death induced by the cytokine interferon (IFN)-γ. In HeLa cervical carcinoma cells, the antisense downregulation of Atg5 suppresses both AV formation and cell death mediated by IFN-γ. Conversely, the overexpression of Atg5 induces both cell death and autophagy. Interestingly, Atg5 interacts with FADD both *in vitro* and *in vivo*, and downregulation of FADD inhibits both IFN-γ- and Atg5-mediated cell death without affecting vacuole formation. Hence, the death-promoting activity of Atg5 requires FADD as a downstream mediator. Notably, a pan-caspase inhibitor (zVAD-fmk) blocks IFN-γ-induced cell death, but not vacuole formation [59]. Overall, these results favor the hypothesis in which autophagy ensues during the early stages of cell death and subsequently triggers the activation of caspases.

Finally, the characteristic features of autophagy and apoptosis have also been observed during lumen formation of mammary epithelial acini grown in three-dimensional culture. When cultured on reconstituted basement membrane, MCF-10A cells, a cell line derived from normal mammary epithelium, form acini – spherical structures in which a layer of polarized epithelial cells surround a hollow lumen; these acini resemble glandular epithelium *in vivo* [60]. Interestingly, lumen formation involves the selective death of centrally located cells; both AVs and activated caspase-3 are present in the dying cells occupying the developing acinar lumen. Preventing apoptosis by the ectopic expression of either Bcl-2 or Bcl-X$_L$ delays the clearance of cells, but does not completely inhibit lumen formation in this system; notably, the central cells in Bcl-2-expressing acini also exhibit extensive autophagy [61]. On the other hand, inhibiting downstream signaling by the TNF-related apoptosis inducing ligand (TRAIL) reduces the formation of vacuoles in the central cells, but does not prevent luminal apoptosis. However, the combined expression of Bcl-X$_L$ and inhibition of TRAIL signaling results in the stronger defects in lumen formation, intimating that both caspase activation and autophagy contribute to proper clearance during lumen formation [61, 62]. Although these results may support the notion that autophagy functions as an alternative cell death pathway upon apoptotic inhibition, they do not exclude the possibility that autophagy could function as a stress response mechanism in dying cells occupying the lumen. Loss-of-function studies of *atg* orthologs during acini development are needed to clearly establish the role of autophagy during cell death or clearance in the lumen. Nevertheless, these studies in mammalian neuronal and epithelial cells, similar to data from studies of *Drosophila* larval salivary gland destruction, suggest that programmed cell death could involve two degradation mechanisms – caspase activation and autophagy – that either function in parallel or through interconnected regulatory pathways.

5.5
Emerging Relationships between Apoptosis and Autophagy

As the above studies suggest, the current state of knowledge strongly argues against discretely classifying apoptosis and autophagy as distinct mechanisms of programmed cell death. The terms "type 1" and "type 2" essentially describe morphological features reflective of the death-initiating influence or the environmental context of the dying cell. They do necessarily not define discrete, mutually exclusive functional programs. The wealth of correlative evidence presented in this chapter suggests that some relationship between the two exists. Additionally, gene expression studies of *Drosophila* salivary gland death [63, 64], death of sympathetic neurons during NGF withdrawal [41, 44] and loss-of-function studies of *atg* genes during nutrient depletion all strongly suggest that coordination between these two fundamental cellular processes exists [65, 66].

Beclin 1 was originally identified as a Bcl-2-interacting protein, suggesting an interesting association between molecules regulating autophagy and cell death [67]. In recent studies, the overexpression of Bcl-2 was demonstrated to inhibit beclin 1-dependent autophagy upon nutrient deprivation in a variety of cell types as well as in mouse cardiac muscle [68]. Additionally, the ectopic expression of mutant versions of beclin 1 that were unable to bind Bcl-2 induced higher levels of autophagy in these cells and promoted cell death. Significantly, the silencing of Atg5 using RNA interference (RNAi) blocked the cell death mediated by these beclin 1 mutants. Based on the these results, the interaction between Bcl-2 and beclin 1 has been proposed to act as an important control mechanism in maintaining homeostatic levels of autophagy compatible with cell survival [68]. However, the inhibitory effects of Bcl-2 overexpression on autophagy may be dependent on cell type and stimulus. In immortalized mouse embryonic fibroblasts treated with the cytotoxic agent etoposide, the overexpression of Bcl-2 promotes autophagy and autophagy gene-dependent cell death [50]. Additional studies are required to determine how the interaction between beclin 1 and Bcl-2 regulates autophagy in various contexts.

Several additional molecules may participate in the crosstalk between apoptosis and autophagy. Two related Ca^{2+}-caldmodulin dependent kinases, death-associated protein kinase (DAPK) and DAPK-related protein-1 (DRP-1), regulate membrane blebbing during apoptosis, but can also promote AV formation in dying cells [69]. Ectopic expression of the human ortholog of *Drosophila Spinster* (hSpin1) in cancer cells causes a necrosis-like cell death notable for increased acidic compartments [70]. hSpin1 binds to both Bcl-2 and Bcl-X_L proteins, and the expression of Bcl-X_L can inhibit hSpin1-induced cell death. Interestingly, Spinster localizes to the late endosome and lysosome, suggesting that it may function in the late stages in autophagy [70, 71]. Future studies of these pathways should provide greater insight into the relationships between apoptosis and autophagy during programmed cell death.

5.6
Autophagy and Cell Survival

As discussed above, certain studies have identified the requirement for *atg* orthologs in certain situations, whereas others have revealed the opposite – that autophagy is not required for cell death or, alternatively, actually promotes the survival of cells in response to death stimuli.

Two groups recently generated mice deficient for beclin 1, the mammalian ortholog of the yeast Atg6 protein [72, 73]. Complete loss of *beclin 1* is embryonic lethal and is notable for widespread cell death throughout the embryo, measured by staining with the vital dye, Acridine Orange. Furthermore, embryonic stem (ES) cell lines genetically null for beclin 1 displayed no differences in classical apoptosis in response to ultraviolet radiation or to serum withdrawal, when compared to control wild-type cells. Hence, beclin 1 does not play a critical role in certain forms of programmed cell death [73]. Importantly, the heterozygous loss of *beclin 1* in mice results in the spontaneous formation of both hematopoietic and epithelial tumors; in these tumors, the wild-type allele was not deleted, corroborating previous work that beclin 1 can act as a haploinsufficient tumor suppressor [72, 73]. Although further proof is necessary, one interesting speculation has been that retention of the wild-type allele in *beclin*$^{+/-}$ tumors takes place because minimal levels of autophagy are required for tumor initiation or maintenance; if so, this would argue against autophagy serving as an alternative cell death mechanism that limits tumor progression [1, 74].

In the social amoeba *Dictyostelium discoidium*, programmed cell death is caspase independent and notable for an autophagic vacuolar morphology, reminiscent of a type 2 process [75–77]. This cell death can be induced in *Dictyostelium* through the combined effects of starvation and the production of a morphogen called differentiation-inducing factor (DIF). However, studies in a mutant strain unable to produce DIF demonstrate that starvation alone can continue to induce autophagy without triggering cell death. Furthermore, upon inactivation of the *Dictyostelium atg1* autophagy gene in cells via homologous recombination, both autophagy and vacuolization are suppressed, yet cell death still proceeds with a striking nonvacuolar and centrally condensed morphology. Hence, developmental cell death in *Dictyostelium* does not require autophagic vacuolization [78].

5.6.1
Autophagy is Cytoprotective during Nutrient Depletion in Mammalian Cells

The critical role of autophagy in maintaining cell viability is well-documented in yeast; through the bulk degradation of cytoplasmic material, autophagy is utilized to generate both nutrients and energy when external nutritional sources are lacking [7, 8]. A protective role for autophagy has also been proposed in mammalian cells through the turnover and elimination of excess or damaged organelles like peroxisomes and mitochondria. Depolarized mitochondria are rapidly eliminated by autophagy in primary hepatocytes, corroborating that autophagy may be protective against apoptosis by sequestering death promoting molecules [79, 80].

Moreover, recent work in cultured mammalian cells has indicated that the inhibition of autophagy during nutrient depletion can actually sensitize cells to cell death by apoptosis. HeLa carcinoma cells treated with lysosomotrophic agents, such as hydroxychloroquine, exhibit the typical morphological features associated with type 2 cell death including the accumulation of early AVs; however, follow-up work has delineated that autophagic degradation is actually inhibited in these cells due to the lack of fusion between autophagosomes and lysosomes [65, 81]. Cells exhibiting this morphology can recover from this stress, indicating that this vacuolization is not necessarily lethal; instead, a subset of these cells reach a point of no return and activate classical apoptotic programs, such as caspase activation and mitochondrial membrane permeabilization [65]. Moreover, when autophagy is inhibited, either pharmacologically or by small interfering (si) RNA-mediated silencing of several autophagy genes (including *atg5*, *atg6*, *atg10* and *atg12*), cells exhibit increased sensitivity to cell death upon nutrient depletion; this death is apoptotic in nature because it is reduced by caspase inhibition and by overexpression of Bcl-2. Cell death due to the combined effects of nutrient depletion and autophagy inhibition is also reduced in mouse fibroblasts doubly deficient for Bax and Bak. These results in mammalian cells substantiate a cytoprotective role for autophagy during nutrient depletion that prevents the activation of classical apoptotic death pathways [65]. Similarly, the siRNA-mediated knockdown of lysosome-associated membrane protein 2 (LAMP-2) can sensitize cells to cell death induced by nutrient depletion. The loss of LAMP-2, which is required for the fusion of lysosomes with autophagosomes, leads to an accumulation of AVs *in vivo* [82, 83]. Accordingly, the combined depletion of nutrients and LAMP-2 in cells *in vitro* leads to the acquisition of a morphology associated with type 2 cell death, which subsequently is followed by the hallmarks of apoptosis, including loss of the mitochondrial trans-membrane potential, release of cytochrome *c*, activation of caspase-3 and nuclear chromatin condensation [84].

The cytoprotective role of autophagy is also important during growth factor withdrawal. Trophic factor withdrawal has been associated with high rates of autophagy as well as cell death; an important question is whether autophagy contributes to programmed cell death in these cells or, instead, represents a response to stress. Recent work using hematopoietic cell lines dependent on the growth factor interleukin (IL)-3 for cell survival has provided valuable insight into this question. Upon withdrawal of IL-3, the cell surface expression of multiple nutrient transporters is reduced; as a result, these cells cannot take up the abundant nutrients present in their extracellular milieu and, thus, rapidly undergo apoptosis [85, 86]. In cells unable to undergo apoptosis because they are doubly deficient for Bax and Bak, a progressive atrophy ensues due to the lack of nutrient uptake; these cells are viable for several weeks. Further examination reveals that these growth factor-deprived cells survive because they utilize autophagy to digest intracellular components, providing a source of energy to maintain ATP production. If IL-3 is re-added to the cultures, these cells do recover from their atrophic state and begin to exhibit normal cell growth and proliferation. Importantly, the inhibition of autophagy in these growth factor deprived Bax/Bak double-knockout cells, either by knockdown of Atg5 or Atg7 or by treatment with 3-MA, results in rapid cell death that is associated with compromised bioenergetics and reduced ATP levels. The results once again illustrate a critical role for autophagy in maintaining proper bioenergetics and promoting cell viability [66].

Notably, these studies also suggest that autophagy is self-limiting as a survival mechanism in cells unable to undergo apoptosis; the nutrient and energy supplies obtained through autophagy are ultimately depleted, upon which cells do die, albeit after several weeks [66]. Based on these results, one can construe that cells notable for extensive autophagic vacuolation are probably predisposed to die and that, under these circumstances, the cell death would likely exhibit the morphological features ascribed to type 2 programmed cell death. Still, the major question that requires further investigation is determining if cell death in these situations is due to the autophagic process gone awry, or whether other mechanisms, such as apoptosis or necrosis, are the primary initiators of death.

5.6.2
Autophagy and Neuroprotection

Autophagic vacuoles have been observed in the degenerating neurons of patients with Parkinson, Alzheimer and Huntington diseases, and in neuronal cell lines expressing toxic proteins [87–92]. It is possible that autophagy plays a role in the death of these cells during neurodegeneration; in fact, autophagic death has been described in the cerebeller Purkinje neurons of *lurcher* mice, a naturally occurring mouse strain that exhibits neurodegeneration due to a mutation that results in its constitutive activation of the glutamate receptor, GluRdelta2 [93]. However, recent studies of Huntington disease have poignantly illustrated that autophagy can be used as a cytoprotective mechanism by removing accumulating protein aggregates in cells. Huntington disease is an autosomal dominant neurodegenerative disorder cause by mutations in the *huntingtin* gene that result in the expansion of a polyglutamine (poly-Q) tract in the primary amino acid sequence of the huntingtin (Htt) protein; as a result, the mutant Htt is misfolded and accumulates in the affected neurons as insoluble aggregates. However, the pharmacological induction of autophagy using the mammalian target of rapamycin (mTOR) inhibitor rapamycin facilitates the clearance of mutant Htt aggregates in cultured cells, and protects against neurotoxicity in transgenic *Drosophila* and mouse models of Huntington disease [94, 95]. Remarkably, although the Htt aggregates are highly ubiquitinated, these mutant poly-Q proteins appear to be poor substrates for the proteosomal degradation machinery [96]. Accordingly, autophagy is induced in response to impaired proteasomal activity and may represent a critical pathway to degrade toxic misfolded poly-Q-containing aggregates in the cytoplasm of cells [97, 98]. Additional studies suggest that the cytoprotective effects of autophagy may extend to other neurodegenerative diseases, either by clearing toxic intracellular aggregates or by protecting against the other noxious factors that contribute to neurotoxicity [99, 100].

5.6.3
Cytoprotective Roles of Autophagy in the Response to Infectious Pathogens

The role of autophagy in limiting infection in response to various pathogens has recently been uncovered in numerous recent studies. These reports have indicated that autophagy can promote cell survival by sequestering intracellular pathogens and, hence, limiting the infectious load. IFN-γ, a cytokine secreted in response to intracellular pathogens, has been shown to induce autophagy in nonprofessional phagocytic cells as well as in macrophages [59, 101]. Notably, the induction of autophagy by IFN-γ has been shown to circumvent the phagosome trafficking defect produced by *Mycobacterium tuberculosis* and suppress the survival of *M. tuberculosis*. Similar results were obtained when autophagy was induced using the pharmacological compound, rapamycin [101]. In addition, *Shigella flexneri*, an intracellular bacterium that normally evades autophagy, was found to secrete a protein IcsB that suppressed autophagy by binding to another *Shigella* protein VirG. Interestingly, it was discovered that in the absence of IcsB, VirG interacted with Atg5 and elicited an autophagic response that resulted in degradation of the *Shigella* bacteria, establishing a direct, molecular link between the host-pathogen response and autophagy [102]. An autophagic mechanism also regulates the sequestration and degradation of the extracellular pathogen group A *Streptococcus* (GAS). Normally able to escape the endosome after internalization, GAS is located in LC3-positive, AVs. The contents of these vacuoles are subsequently degraded by the lysosome, whereas in *atg5*$^{-/-}$ cells, these bacteria were not degraded [103].

Moreover, autophagy has also been observed to negatively regulate programmed cell death during the plant innate immune response. In response to viral infection, plants exhibit a limited form of programmed cell death, called the hypersensitive response (HR) that is largely restricted to the site of infection; in contrast, the adjacent healthy tissue does not die. As a result, during HR, this tight regulation of cell death eliminates diseased tissue in an attempt to protect the whole plant. The silencing of multiple plant *atg* gene orthologs, including *beclin 1*, produces unrestricted programmed cell death throughout the plant during viral infection, suggesting a role for autophagy in suppressing programmed cell death in healthy tissue and regulating this fundamental innate immune response [104]. Autophagy also appears to play a protective role in encephalitis induced by Sindbis virus; enforced expression of *beclin 1* reduces viral titers as well as Sindbis virus-induced apoptosis in mouse brains [67]. Overall, these studies suggest that autophagy serves as an important cell survival mechanism in the innate immune response to a number of pathogens.

Interestingly, situations in which invading pathogens circumvent the host cell autophagic machinery in order to thrive and proliferate have also been documented in recent studies. Three bacterial species, *Legionella pneumophila*, *Brucella abortus* and *Porphyromonas gingivalis*, have all been shown to subvert the autophagic machinery, which allows these organisms to escape degradation in the lysosome [105–108]. *Legionella*, a species of intracellular bacteria that can proliferate after phagocytosis into host macrophages, can replicate within modified autophagosomes, and metabolize amino acids and peptides produced there by the normal autophagic machinery. An unspecified bacterial product whose secretion to the host cell is regulated by a translocation channel controls the trafficking of *Legionella* containing endosomes to early autophagosomes [105, 109]. Successful delivery of the bacterial product to the host cell resulted in upregulation of autophagy and *Legionella* localization to autophagosomes. Blockage of the translocation channel resulted in normal endosome trafficking and destruction of the bacteria in the lysosome [110]. *B. abortus* and *P. gingivalis* utilize similar modifications of endosome trafficking to avoid degradation in the lysosome [107, 108].

Although autophagy appears to play a protective role during the cellular response to a variety of infectious pathogens, the direct induction of autophagic cell death by an infectious pathogen has also been observed. Autophagic cell death has been observed in macrophages upon infection with the intracellular bacteria *Salmonella enterica*. In fact, cultured macrophages exhibit the morphological features of type II cell death in response to *Salmonella* and the secreted *Salmonella* virulence protein SipB under caspase inhibition; in addition, *caspase-$1^{-/-}$* macrophages exhibit a similar cell death morphology upon infection. Nonetheless, these studies do not provide direct evidence that autophagy is functionally responsible macrophage for cell death [111]. Overall, additional studies using autophagy deficient cells and organisms are necessary to establish how autophagy functionally regulates cell survival or death in the host response to infection.

5.7
Autophagy and Organism Survival

The prosurvival role of autophagy in mammalian cells delineated in the above studies argues against autophagy as a mediator of cell death. Furthermore, the recent examination of *atg* orthologs in various model systems strongly indicates that this catabolic process can promote survival in higher organisms *in vivo* by regulating a variety of developmental and systemic programs. When faced with unfavorable environmental conditions, the nematode *Caenorhabditis elegans* undergoes a reversible form of developmental arrest known as the dauer diapause. Recent work using nematodes with a mutation in *daf-2* has illustrated the importance of autophagy in this critical process; *daf-2* encodes for an insulin-like receptor and mutant worms exhibit high rates of dauer entry [112, 113]. Dauer formation is associated with increased autophagy in hypodermal seam cells, a cell type required for multiple aspects of dauer morphogenesis; furthermore, abnormal dauers result as a consequence of loss-of-function mutations or RNAi-mediated silencing of multiple *atg* genes, including *atg1* (*unc51*), *atg6* (*beclin 1*), *atg7*, *atg8* and *atg18/aut10* [113]. These results demonstrate that autophagy functionally contributes to an important environmental stress response program in the worm. Moreover, *beclin 1* is also required for the lifespan extension phenotype associated with *daf-2* mutant nematodes, suggesting that autophagy components contribute to the enhanced survival of a multicellular organism [112, 113].

Autophagy possesses a similar survival function during nutrient starvation in *Drosophila*. In the fly, the fat body is thought to function as a sensor of nutritional status by secreting both nutrients and trophic factors [114, 115]. During starvation, the fat body is the predominant organ to exhibit high rates of autophagy, presumably to maintain adequate nutrient levels for the entire organism [116, 117]. Accordingly, the disruption of autophagy machinery genes, either in the whole organism or specifically in the fat body, decreases survival; when combined with the loss of TOR, an essential regulator of cell growth and nutrient uptake, the deleterious effects of decreased autophagy are exacerbated [116].

Recent studies in mice have also revealed an essential role for autophagy for survival during the neonatal period. Mice lacking Atg5 exhibit reduced autophagy and die in their first day of life; the reduced viability of these mice is independent of their ability to nurse. Instead, mice lacking Atg5 display significantly reduced systemic levels of amino acids, as well as signs of energy depletion, such as activation of the energy sensor, AMP-activated protein kinase (AMPK). Interestingly, the highest rates of autophagy in the newborn are found in tissues that exhibit substantial increases in energy requirements (e.g. heart, diaphragm) or marked environmental changes due to the switch from amniotic fluid to air (e.g. lung alveoli, skin) [118].

Certainly, the aforementioned studies of autophagy gene function in model organisms and in mammalian cells have corroborated that autophagy can serve as a survival mechanism in the primary response to various stresses. They have also indicated that inhibiting autophagy in starving cells can actually activate apoptosis. These results would support the idea that a cell undergoes autophagy in order to keep it alive under stressful conditions. Upon removal of the initiating stress, the cell can resume normal rates of growth and proliferation. In this model, autophagy does not function in cell death during the primary response to stress or death stimuli.

However, these studies do not exclude the possibility that autophagic degradation may execute or promote cell death in tissue-specific contexts. As discussed above, high levels of autophagy have been observed in certain mammalian tissues *in vivo*, such as the involution of the prostate and mammary gland [28–30]. The enormous amounts of autophagy observed in these circumstances have bolstered the case for autophagy as an alternative form of programmed cell death. As a result, determining whether autophagy contributes to cell survival or cell death in specific developmental situations remains an important and intriguing question for further study.

5.8
Concluding Remarks

The studies discussed above highlight the expanding importance of autophagic degradation during programmed cell death as well as cell survival in a variety of physiological and pathological processes. Nonetheless, although programmed cell death was one of the first settings to broach the biological significance of autophagy, much remains to be learned about the functional role of autophagy during programmed cell death. The isolation of the *atg* genes from genetic screens in yeast have provided enormous insight into the mechanisms that regulate autophagy, and are beginning to inform studies of both autophagic cell death and apoptosis. These studies have confirmed that the role of autophagy in cell survival is conserved among organisms. Conversely, the study of the role and regulation of autophagy in programmed cell death requires further investigation in both higher organisms and mammalian cell culture systems.

Acknowledgments

We apologize in advance to the researchers who were not referenced due to space limitations. J. D. is funded by a KO8 Award from the National Cancer Institute and by the Sandler Program in Biomedical Sciences at UCSF.

References

1 Edinger, A.L., Thompson, C.B. Death by design: apoptosis, necrosis and autophagy. *Curr Opin Cell Biol 16*, 663–669, **2004**.

2 Lockshin, R.A., Zakeri, Z. Caspase-independent cell death? *Oncogene 23*, 2766–2773, **2004**.

3 Nelson, D.A., White, E. Exploiting different ways to die. *Genes Dev 18*, 1223–1226, **2004**.

4 Tsujimoto, Y., Shimizu, S. Another way to die: autophagic programmed cell death. *Cell Death Differ 12 (Suppl 2)*, 1528–1534, **2005**.

5 Schweichel, J.U., Merker, H.J. The morphology of various types of cell death in prenatal tissues. *Teratology 7*, 253–266, **1973**.

6 Clarke, P.G. Developmental cell death: morphological diversity and multiple mechanisms. *Anat Embryol (Berl) 181*, 195–213, **1990**.

7 Levine, B., Klionsky, D.J. Development by self-digestion: molecular mechanisms and biological functions of autophagy. *Dev Cell 6*, 463–477, **2004**.

8 Ohsumi, Y. Molecular dissection of autophagy: two ubiquitin-like systems. *Nat Rev Mol Cell Biol 2*, 211–216, **2001**.

9 Klionsky, D.J., Emr, S.D. Autophagy as a regulated pathway of cellular degradation. *Science 290*, 1717–1721, **2000**.

10 Thoreen, C.C., Sabatini, D.M. Huntingtin aggregates ask to be eaten. *Nat Genet 36*, 553–554, **2004**.

11 Levine, B., Yuan, J. Autophagy in cell death: an innocent convict? *J Clin Invest 115*, 2679–2688, **2005**.

12 Klionsky, D.J., Cregg, J.M., Dunn, W.A., Jr., Emr, S.D., Sakai, Y., Sandoval, I.V., Sibirny, A., Subramani, S., Thumm, M., Veenhuis, M., Ohsumi, Y. A unified nomenclature for yeast autophagy-related genes. *Dev Cell 5*, 539–545, **2003**.

13 Kerr, J.F., Wyllie, A.H., Currie, A.R. Apoptosis: a basic biological phenomenon with wide-ranging implications in tissue kinetics. *Br J Cancer 26*, 239–257, **1972**.

14 Savill, J., Fadok, V. Corpse clearance defines the meaning of cell death. *Nature 407*, 784–788, **2000**.

15 Hengartner, M.O. Apoptosis: corralling the corpses. *Cell 104*, 325–328, **2001**.

16 Danial, N.N., Korsmeyer, S.J. Cell death: critical control points. *Cell 116*, 205–219, **2004**.

17 Riedl, S.J., Shi, Y. Molecular mechanisms of caspase regulation during apoptosis. *Nat Rev Mol Cell Biol 5*, 897–907, **2004**.

18 Leist, M., Jaattela, M. Four deaths and a funeral: from caspases to alternative mechanisms. *Nat Rev Mol Cell Biol 2*, 589–598, **2001**.

19 Mizushima, N. Methods for monitoring autophagy. *Int J Biochem Cell Biol 36*, 2491–2502, **2004**.

20 Kabeya, Y., Mizushima, N., Ueno, T., Yamamoto, A., Kirisako, T., Noda, T., Kominami, E., Ohsumi, Y., Yoshimori, T. LC3, a mammalian homologue of yeast Apg8p, is localized in autophagosome membranes after processing. *EMBO J 19*, 5720–5728, **2000**.

21 Mizushima, N., Yamamoto, A., Matsui, M., Yoshimori, T., Ohsumi, Y. *In vivo* analysis of autophagy in response to nutrient starvation using transgenic mice expressing a fluorescent autophagosome marker. *Mol Biol Cell 15*, 1101–1111, **2004**.

22 Ichimura, Y., Kirisako, T., Takao, T., Satomi, Y., Shimonishi, Y., Ishihara, N., Mizushima, N., Tanida, I., Kominami, E., Ohsumi, M., Noda, T., Ohsumi, Y. A ubiquitin-like system mediates protein lipidation. *Nature 408*, 488–492, **2000**.

23 Kirisako, T., Baba, M., Ishihara, N., Miyazawa, K., Ohsumi, M., Yoshimori, T., Noda, T., Ohsumi, Y. Formation process of autophagosome is traced with Apg8/Aut7p in yeast. *J Cell Biol 147*, 435–446, **1999**.

24 Yuan, J., Lipinski, M., Degterev, A. Diversity in the mechanisms of neuronal cell death. *Neuron 40*, 401–413, **2003**.

25 Zong, W.X., Ditsworth, D., Bauer, D.E., Wang, Z.Q., Thompson, C.B. Alkylating DNA damage stimulates a regulated form of necrotic cell death. *Genes Dev 18*, 1272–1282, **2004**.

26 Okada, M., Adachi, S., Imai, T., Watanabe, K., Toyokuni, S.Y., Ueno, M., Zer-

vos, A. S., Kroemer, G., Nakahata, T. A novel mechanism for imatinib mesylate-induced cell death of BCR–ABL-positive human leukemic cells: caspase-independent, necrosis-like programmed cell death mediated by serine protease activity. *Blood 103*, 2299–2307, **2004**.

27 Chan, F. K., Shisler, J., Bixby, J. G., Felices, M., Zheng, L., Appel, M., Orenstein, J., Moss, B., Lenardo, M. J. A role for tumor necrosis factor receptor-2 and receptor-interacting protein in programmed necrosis and antiviral responses. *J Biol Chem 278*, 51613–51621, **2003**.

28 Helminen, H. J., Ericsson, J. L. Ultrastructural studies on prostatic involution in the rat. Mechanism of autophagy in epithelial cells, with special reference to the rough-surfaced endoplasmic reticulum. *J Ultrastruct Res 36*, 708–724, **1971**.

29 Helminen, H. J., Ericsson, J. L., Orrenius, S. Studies on mammary gland involution. IV. Histochemical and biochemical observations on alterations in lysosomes and lysosomal enzymes. *J Ultrastruct Res 25*, 240–252, **1968**.

30 Sensibar, J. A., Liu, X. X., Patai, B., Alger, B., Lee, C. Characterization of castration-induced cell death in the rat prostate by immunohistochemical localization of cathepsin D. *Prostate 16*, 263–276, **1990**.

31 Dyche, W. J. A comparative study of the differentiation and involution of the Mullerian duct and Wolffian duct in the male and female fetal mouse. *J Morphol 162*, 175–209, **1979**.

32 Bursch, W., Ellinger, A., Kienzl, H., Torok, L., Pandey, S., Sikorska, M., Walker, R., Hermann, R. S. Active cell death induced by the anti-estrogens tamoxifen and ICI 164 384 in human mammary carcinoma cells (MCF-7) in culture: the role of autophagy. *Carcinogenesis 17*, 1595–1607, **1996**.

33 Janicke, R. U., Sprengart, M. L., Wati, M. R., Porter, A. G. Caspase-3 is required for DNA fragmentation and morphological changes associated with apoptosis. *J Biol Chem 273*, 9357–9360, **1998**.

34 Kanzawa, T., Germano, I. M., Komata, T., Ito, H., Kondo, Y., Kondo, S. Role of autophagy in temozolomide-induced cyto-toxicity for malignant glioma cells. *Cell Death Differ 11*, 448–457, **2004**.

35 Daido, S., Kanzawa, T., Yamamoto, A., Takeuchi, H., Kondo, Y., Kondo, S. Pivotal role of the cell death factor BNIP3 in ceramide-induced autophagic cell death in malignant glioma cells. *Cancer Res 64*, 4286–4293, **2004**.

36 Kanzawa, T., Zhang, L., Xiao, L., Germano, I. M., Kondo, Y., Kondo, S. Arsenic trioxide induces autophagic cell death in malignant glioma cells by upregulation of mitochondrial cell death protein BNIP3. *Oncogene 24*, 980–991, **2005**.

37 Opipari, A. W., Jr., Tan, L., Boitano, A. E., Sorenson, D. R., Aurora, A., Liu, J. R. Resveratrol-induced autophagocytosis in ovarian cancer cells. *Cancer Res 64*, 696–703, **2004**.

38 Shao, Y., Gao, Z., Marks, P. A., Jiang, X. Apoptotic and autophagic cell death induced by histone deacetylase inhibitors. *Proc Natl Acad Sci USA 101*, 18030–18035, **2004**.

39 Seglen, P. O., Gordon, P. B. 3-Methyladenine: specific inhibitor of autophagic/lysosomal protein degradation in isolated rat hepatocytes. *Proc Natl Acad Sci USA 79*, 1889–1892, **1982**.

40 Bursch, W., Hochegger, K., Torok, L., Marian, B., Ellinger, A., Hermann, R. S. Autophagic and apoptotic types of programmed cell death exhibit different fates of cytoskeletal filaments. *J Cell Sci 113*, 1189–1198, **2000**.

41 Xue, L., Fletcher, G. C., Tolkovsky, A. M. Autophagy is activated by apoptotic signalling in sympathetic neurons: an alternative mechanism of death execution. *Mol Cell Neurosci 14*, 180–198, **1999**.

42 Jia, L., Dourmashkin, R. R., Allen, P. D., Gray, A. B., Newland, A. C., Kelsey, S. M. Inhibition of autophagy abrogates tumour necrosis factor alpha induced apoptosis in human T-lymphoblastic leukaemic cells. *Br J Haematol 98*, 673–685, **1997**.

43 Rameh, L. E., Cantley, L. C. The role of phosphoinositide 3-kinase lipid products in cell function. *J Biol Chem 274*, 8347–8350, **1999**.

44 Tolkovsky, A. M., Xue, L., Fletcher, G. C., Borutaite, V. Mitochondrial disappear-

ance from cells: a clue to the role of autophagy in programmed cell death and disease? *Biochimie 84*, 233–240, **2002**.

45 Caro, L. H., Plomp, P. J., Wolvetang, E. J., Kerkhof, C., Meijer, A. J. 3-Methyladenine, an inhibitor of autophagy, has multiple effects on metabolism. *Eur J Biochem 175*, 325–329, **1988**.

46 Vercammen, D., Beyaert, R., Denecker, G., Goossens, V., Van Loo, G., Declercq, W., Grooten, J., Fiers, W., Vandenabeele, P. Inhibition of caspases increases the sensitivity of L929 cells to necrosis mediated by tumor necrosis factor. *J Exp Med 187*, 1477–1485, **1998**.

47 Yu, L., Alva, A., Su, H., Dutt, P., Freundt, E., Welsh, S., Baehrecke, E. H., Lenardo, M. J. Regulation of an ATG7-beclin 1 program of autophagic cell death by caspase-8. *Science 304*, 1500–1502, **2004**.

48 Holler, N., Zaru, R., Micheau, O., Thome, M., Attinger, A., Valitutti, S., Bodmer, J. L., Schneider, P., Seed, B., Tschopp, J. Fas triggers an alternative, caspase-8-independent cell death pathway using the kinase RIP as effector molecule. *Nat Immunol 1*, 489–495, **2000**.

49 Wei, M. C., Zong, W. X., Cheng, E. H., Lindsten, T., Panoutsakopoulou, V., Ross, A. J., Roth, K. A., MacGregor, G. R., Thompson, C. B., Korsmeyer, S. J. Proapoptotic BAX and BAK: a requisite gateway to mitochondrial dysfunction and death. *Science 292*, 727–730, **2001**.

50 Shimizu, S., Kanaseki, T., Mizushima, N., Mizuta, T., Arakawa-Kobayashi, S., Thompson, C. B., Tsujimoto, Y. Role of Bcl-2 family proteins in a non-apoptotic programmed cell death dependent on autophagy genes. *Nat Cell Biol 6*, 1221–1228, **2004**.

51 Chautan, M., Chazal, G., Cecconi, F., Gruss, P., Golstein, P. Interdigital cell death can occur through a necrotic and caspase-independent pathway. *Curr Biol 9*, 967–970, **1999**.

52 Oppenheim, R. W., Flavell, R. A., Vinsant, S., Prevette, D., Kuan, C. Y., Rakic, P. Programmed cell death of developing mammalian neurons after genetic deletion of caspases. *J Neurosci 21*, 4752–4760, **2001**.

53 Baehrecke, E. H. Autophagic programmed cell death in *Drosophila*. *Cell Death Differ 10*, 940–945, **2003**.

54 Jiang, C., Baehrecke, E. H., Thummel, C. S. Steroid regulated programmed cell death during *Drosophila* metamorphosis. *Development 124*, 4673–4683, **1997**.

55 Lee, C. Y., Baehrecke, E. H. Steroid regulation of autophagic programmed cell death during development. *Development 128*, 1443–1455, **2001**.

56 Martin, D. N., Baehrecke, E. H. Caspases function in autophagic programmed cell death in *Drosophila*. *Development 131*, 275–284, **2004**.

57 Daish, T. J., Mills, K., Kumar, S. *Drosophila* caspase DRONC is required for specific developmental cell death pathways and stress-induced apoptosis. *Dev Cell 7*, 909–915, **2004**.

58 Ohsawa, Y., Isahara, K., Kanamori, S., Shibata, M., Kametaka, S., Gotow, T., Watanabe, T., Kominami, E., Uchiyama, Y. An ultrastructural and immunohistochemical study of PC12 cells during apoptosis induced by serum deprivation with special reference to autophagy and lysosomal cathepsins. *Arch Histol Cytol 61*, 395–403, **1998**.

59 Pyo, J. O., Jang, M. H., Kwon, Y. K., Lee, H. J., Jun, J. I., Woo, H. N., Cho, D. H., Choi, B., Lee, H., Kim, J. H., Mizushima, N., Oshumi, Y., Jung, Y. K. Essential roles of Atg5 and FADD in autophagic cell death: dissection of autophagic cell death into vacuole formation and cell death. *J Biol Chem 280*, 20722–20729, **2005**.

60 Debnath, J., Muthuswamy, S. K., Brugge, J. S. Morphogenesis and oncogenesis of MCF-10A mammary epithelial acini grown in three-dimensional basement membrane cultures. *Methods 30*, 256–268, **2003**.

61 Debnath, J., Mills, K. R., Collins, N. L., Reginato, M. J., Muthuswamy, S. K., Brugge, J. S. The role of apoptosis in creating and maintaining luminal space within normal and oncogene-expressing mammary acini. *Cell 111*, 29–40, **2002**.

62 Mills, K. R., Reginato, M., Debnath, J., Queenan, B., Brugge, J. S. Tumor necrosis factor-related apoptosis-inducing

ligand (TRAIL) is required for induction of autophagy during lumen formation *in vitro. Proc Natl Acad Sci USA 101*, 3438–3443, **2004**.

63 Lee, C. Y., Clough, E. A., Yellon, P., Teslovich, T. M., Stephan, D. A., Baehrecke, E. H. Genome-wide analyses of steroid- and radiation-triggered programmed cell death in *Drosophila. Curr Biol 13*, 350–357, **2003**.

64 Gorski, S. M., Chittaranjan, S., Pleasance, E. D., Freeman, J. D., Anderson, C. L., Varhol, R. J., Coughlin, S. M., Zuyderduyn, S. D., Jones, S. J., Marra, M. A. A SAGE approach to discovery of genes involved in autophagic cell death. *Curr Biol 13*, 358–363, **2003**.

65 Boya, P., Gonzalez-Polo, R. A., Casares, N., Perfettini, J. L., Dessen, P., Larochette, N., Metivier, D., Meley, D., Souquere, S., Yoshimori, T., Pierron, G., Codogno, P., Kroemer, G. Inhibition of macroautophagy triggers apoptosis. *Mol Cell Biol 25*, 1025–1040, **2005**.

66 Lum, J. J., Bauer, D. E., Kong, M., Harris, M. H., Li, C., Lindsten, T., Thompson, C. B. Growth factor regulation of autophagy and cell survival in the absence of apoptosis. *Cell 120*, 237–248, **2005**.

67 Liang, X. H., Kleeman, L. K., Jiang, H. H., Gordon, G., Goldman, J. E., Berry, G., Herman, B., Levine, B. Protection against fatal Sindbis virus encephalitis by beclin, a novel Bcl-2-interacting protein. *J Virol 72*, 8586–8596, **1998**.

68 Pattingre, S., Tassa, A., Qu, X., Garuti, R., Liang, X. H., Mizushima, N., Packer, M., Schneider, M. D., Levine, B. Bcl-2 antiapoptotic proteins inhibit Beclin 1-dependent autophagy. *Cell 122*, 927–939, **2005**.

69 Inbal, B., Bialik, S., Sabanay, I., Shani, G., Kimchi, A. DAP kinase and DRP-1 mediate membrane blebbing and the formation of autophagic vesicles during programmed cell death. *J Cell Biol 157*, 455–468, **2002**.

70 Yanagisawa, H., Miyashita, T., Nakano, Y., Yamamoto, D. HSpin1, a transmembrane protein interacting with Bcl-2/Bcl-x$_L$, induces a caspase-independent autophagic cell death. *Cell Death Differ 10*, 798–807, **2003**.

71 Sweeney, S. T., Davis, G. W. Unrestricted synaptic growth in spinster-a late endosomal protein implicated in TGF-beta-mediated synaptic growth regulation. *Neuron 36*, 403–416, **2002**.

72 Qu, X., Yu, J., Bhagat, G., Furuya, N., Hibshoosh, H., Troxel, A., Rosen, J., Eskelinen, E. L., Mizushima, N., Ohsumi, Y., Cattoretti, G., Levine, B. Promotion of tumorigenesis by heterozygous disruption of the beclin 1 autophagy gene. *J Clin Invest 112*, 1809–1820, **2003**.

73 Yue, Z., Jin, S., Yang, C., Levine, A. J., Heintz, N. Beclin 1, an autophagy gene essential for early embryonic development, is a haploinsufficient tumor suppressor. *Proc Natl Acad Sci USA 100*, 15077–15082, **2003**.

74 Ogier-Denis, E., Codogno, P. Autophagy: a barrier or an adaptive response to cancer. *Biochim Biophys Acta 1603*, 113–128, **2003**.

75 Cornillon, S., Foa, C., Davoust, J., Buonavista, N., Gross, J. D., Golstein, P. Programmed cell death in *Dictyostelium. J Cell Sci 107*, 2691–2704, **1994**.

76 Roisin-Bouffay, C., Luciani, M. F., Klein, G., Levraud, J. P., Adam, M., Golstein, P. Developmental cell death in *Dictyostelium* does not require paracaspase. *J Biol Chem 279*, 11489–11494, **2004**.

77 Olie, R. A., Durrieu, F., Cornillon, S., Loughran, G., Gross, J., Earnshaw, W. C., Golstein, P. Apparent caspase independence of programmed cell death in *Dictyostelium. Curr Biol 8*, 955–958, **1998**.

78 Kosta, A., Roisin-Bouffay, C., Luciani, M. F., Otto, G. P., Kessin, R. H., Golstein, P. Autophagy gene disruption reveals a non-vacuolar cell death pathway in *Dictyostelium. J Biol Chem 279*, 48404–48409, **2004**.

79 Lemasters, J. J., Nieminen, A. L., Qian, T., Trost, L. C., Elmore, S. P., Nishimura, Y., Crowe, R. A., Cascio, W. E., Bradham, C. A., Brenner, D. A., Herman, B. The mitochondrial permeability transition in cell death: a common mechanism in necrosis, apoptosis and autophagy. *Biochim Biophys Acta, 1366:* 177–196, **1998**.

80 Rodriguez-Enriquez, S., He, L., Lemasters, J. J. Role of mitochondrial permeability transition pores in mitochondrial

autophagy. *Int J Biochem Cell Biol 36*, 2463–2472, **2004**.

81 Boya, P., Andreau, K., Poncet, D., Zamzami, N., Perfettini, J. L., Metivier, D., Ojcius, D. M., Jaattela, M., Kroemer, G. Lysosomal membrane permeabilization induces cell death in a mitochondrion-dependent fashion. *J Exp Med 197*, 1323–1334, **2003**.

82 Eskelinen, E. L., Illert, A. L., Tanaka, Y., Schwarzmann, G., Blanz, J., Von Figura, K., Saftig, P. Role of LAMP-2 in lysosome biogenesis and autophagy. *Mol Biol Cell 13*, 3355–3368, **2002**.

83 Tanaka, Y., Guhde, G., Suter, A., Eskelinen, E. L., Hartmann, D., Lullmann-Rauch, R., Janssen, P. M., Blanz, J., von Figura, K., Saftig, P. Accumulation of autophagic vacuoles and cardiomyopathy in LAMP-2-deficient mice. *Nature 406*, 902–906, **2000**.

84 Gonzalez-Polo, R. A., Boya, P., Pauleau, A., Jalil, A., Larochette, N., Souquere, S., Eskelinen, E. L., Pierron, G., Saftig, P., Kroemer, G. The apoptosis/autophagy paradox. Autophagic vacuolization before apoptotic death. *J Cell Sci 118*, 3091–3102, **2005**.

85 Plas, D. R., Talapatra, S., Edinger, A. L., Rathmell, J. C., Thompson, C. B. Akt and Bcl-x$_L$ promote growth factor-independent survival through distinct effects on mitochondrial physiology. *J Biol Chem 276*, 12041–12048, **2001**.

86 Edinger, A. L., Thompson, C. B. Akt maintains cell size and survival by increasing mTOR-dependent nutrient uptake. *Mol Biol Cell 13*, 2276–2288, **2002**.

87 Nixon, R. A., Wegiel, J., Kumar, A., Yu, W. H., Peterhoff, C., Cataldo, A., Cuervo, A. M. Extensive involvement of autophagy in Alzheimer disease: an immuno-electron microscopy study. *J Neuropathol Exp Neurol 64*, 113–122, **2005**.

88 Yu, W. H., Kumar, A., Peterhoff, C., Shapiro Kulnane, L., Uchiyama, Y., Lamb, B. T., Cuervo, A. M., Nixon, R. A. Autophagic vacuoles are enriched in amyloid precursor protein-secretase activities: implications for beta-amyloid peptide overproduction and localization in Alzheimer's disease. *Int J Biochem Cell Biol 36*, 2531–2540, **2004**.

89 Anglade, P., Vyas, S., Javoy-Agid, F., Herrero, M. T., Michel, P. P., Marquez, J., Mouatt-Prigent, A., Ruberg, M., Hirsch, E. C., Agid, Y. Apoptosis and autophagy in nigral neurons of patients with Parkinson's disease. *Histol Histopathol 12*, 25–31, **1997**.

90 Qin, Z. H., Wang, Y., Kegel, K. B., Kazantsev, A., Apostol, B. L., Thompson, L. M., Yoder, J., Aronin, N., DiFiglia, M. Autophagy regulates the processing of amino terminal huntingtin fragments. *Hum Mol Genet 12*, 3231–3244, **2003**.

91 Kegel, K. B., Kim, M., Sapp, E., McIntyre, C., Castano, J. G., Aronin, N., DiFiglia, M. Huntingtin expression stimulates endosomal-lysosomal activity, endosome tubulation, and autophagy. *J Neurosci 20*, 7268–7278, **2000**.

92 Stefanis, L., Larsen, K. E., Rideout, H. J., Sulzer, D., Greene, L. A. Expression of A53T mutant but not wild-type alpha-synuclein in PC12 cells induces alterations of the ubiquitin-dependent degradation system, loss of dopamine release, and autophagic cell death. *J Neurosci 21*, 9549–9560, **2001**.

93 Yue, Z., Horton, A., Bravin, M., DeJager, P. L., Selimi, F., Heintz, N. A novel protein complex linking the delta 2 glutamate receptor and autophagy: implications for neurodegeneration in lurcher mice. *Neuron 35*, 921–933, **2002**.

94 Ravikumar, B., Vacher, C., Berger, Z., Davies, J. E., Luo, S., Oroz, L. G., Scaravilli, F., Easton, D. F., Duden, R., O'Kane, C. J., Rubinsztein, D. C. Inhibition of mTOR induces autophagy and reduces toxicity of polyglutamine expansions in fly and mouse models of Huntington disease. *Nat Genet 36*, 585–595, **2004**.

95 Ravikumar, B., Duden, R., Rubinsztein, D. C. Aggregate-prone proteins with polyglutamine and polyalanine expansions are degraded by autophagy. *Hum Mol Genet 11*, 1107–1117, **2002**.

96 Venkatraman, P., Wetzel, R., Tanaka, M., Nukina, N., Goldberg, A. L. Eukaryotic proteasomes cannot digest polyglutamine sequences and release them during degradation of polyglutamine-containing proteins. *Mol Cell 14*, 95–104, **2004**.

97 Iwata, A., Riley, B. E., Johnston, J. A., Kopito, R. R. HDAC6 and microtubules are required for autophagic degradation of aggregated huntingtin. *J Biol Chem* 280, 40282–40292, **2005**.

98 Iwata, A., Christianson, J. C., Bucci, M., Ellerby, L. M., Nukina, N., Forno, L. S., Kopito, R. R. Increased susceptibility of cytoplasmic over nuclear polyglutamine aggregates to autophagic degradation. *Proc Natl Acad Sci USA* 102, 13135–13140, **2005**.

99 Jellinger, K. A., Stadelmann, C. Problems of cell death in neurodegeneration and Alzheimer's disease. *J Alzheimers Dis* 3, 31–40, **2001**.

100 Berger, Z., Ravikumar, B., Menzies, F. M., Garcia Oroz, L., Underwood, B. R., Pangalos, M. N., Schmitt, I., Wullner, U., Evert, B. O., O'Kane C. J., Rubinsztein, D. C. Rapamycin alleviates toxicity of different aggregate-prone proteins. *Hum Mol Genet* 15, 433–442, **2006**.

101 Gutierrez, M. G., Master, S. S., Singh, S. B., Taylor, G. A., Colombo, M. I., Deretic, V. Autophagy is a defense mechanism inhibiting BCG and *Mycobacterium tuberculosis* survival in infected macrophages. *Cell* 119, 753–766, **2004**.

102 Ogawa, M., Yoshimori, T., Suzuki, T., Sagara, H., Mizushima, N., Sasakawa, C. Escape of intracellular *Shigella* from autophagy. *Science* 307, 727–731, **2005**.

103 Nakagawa, I., Amano, A., Mizushima, N., Yamamoto, A., Yamaguchi, H., Kamimoto, T., Nara, A., Funao, J., Nakata, M., Tsuda, K., Hamada, S., Yoshimori, T. Autophagy defends cells against invading group A *Streptococcus*. *Science* 306, 1037–1040, **2004**.

104 Liu, Y., Schiff, M., Czymmek, K., Talloczy, Z., Levine, B., Dinesh-Kumar, S. P. Autophagy regulates programmed cell death during the plant innate immune response. *Cell* 121, 567–577, **2005**.

105 Swanson, M. S., Isberg, R. R. Association of *Legionella pneumophila* with the macrophage endoplasmic reticulum. *Infect Immun* 63, 3609–3620, **1995**.

106 Pizarro-Cerda, J., Meresse, S., Parton, R. G., van der Goot, G., Sola-Landa, A., Lopez-Goni, I., Moreno, E., Gorvel, J. P. *Brucella abortus* transits through the autophagic pathway and replicates in the endoplasmic reticulum of non-professional phagocytes. *Infect Immun* 66, 5711–5724, **1998**.

107 Dorn, B. R., Dunn, W. A., Jr., Progulske-Fox, A. *Porphyromonas gingivalis* traffics to autophagosomes in human coronary artery endothelial cells. *Infect Immun* 69, 5698–5708, **2001**.

108 Dorn, B. R., Dunn, W. A., Jr., Progulske-Fox, A. Bacterial interactions with the autophagic pathway. *Cell* Microbiol 4, 1–10, **2002**.

109 Swanson, M. S., Isberg, R. R. Analysis of the intracellular fate of *Legionella pneumophila* mutants. *Ann NY Acad Sci* 797, 8–18, **1996**.

110 Coers, J., Kagan, J. C., Matthews, M., Nagai, H., Zuckman, D. M., Roy, C. R. Identification of Icm protein complexes that play distinct roles in the biogenesis of an organelle permissive for *Legionella pneumophila* intracellular growth. *Mol Microbiol* 38, 719–736, **2000**.

111 Hernandez, L. D., Pypaert, M., Flavell, R. A., Galan, J. E. A *Salmonella* protein causes macrophage cell death by inducing autophagy. *J Cell Biol* 163, 1123–1131, **2003**.

112 Dorman, J. B., Albinder, B., Shroyer, T., Kenyon, C. The *age-1* and *daf-2* genes function in a common pathway to control the lifespan of *Caenorhabditis elegans*. *Genetics* 141, 1399–1406, **1995**.

113 Melendez, A., Talloczy, Z., Seaman, M., Eskelinen, E. L., Hall, D. H., Levine, B. Autophagy genes are essential for dauer development and life-span extension in *C. elegans*. *Science* 301, 1387–1391, **2003**.

114 Colombani, J., Raisin, S., Pantalacci, S., Radimerski, T., Montagne, J., Leopold, P. A nutrient sensor mechanism controls *Drosophila* growth. *Cell* 114, 739–749, **2003**.

115 Britton, J. S., Edgar, B. A. Environmental control of the cell cycle in *Drosophila*: nutrition activates mitotic and endoreplicative cells by distinct mechanisms. *Development* 125, 2149–2158, **1998**.

116 Scott, R. C., Schuldiner, O., Neufeld, T. P. Role and regulation of starvation-induced autophagy in the *Drosophila* fat body. *Dev Cell 7*, 167–178, **2004**.

117 Rusten, T. E., Lindmo, K., Juhasz, G., Sass, M., Seglen, P. O., Brech, A., Stenmark, H. Programmed autophagy in the *Drosophila* fat body is induced by ecdy-sone through regulation of the PI3K pathway. *Dev Cell 7*, 179–192, **2004**.

118 Kuma, A., Hatano, M., Matsui, M., Yamamoto, A., Nakaya, H., Yoshimori, T., Ohsumi, Y., Tokuhisa, T., Mizushima, N. The role of autophagy during the early neonatal starvation period. *Nature 432*, 1032–1036, **2004**.

Part II
Autophagy and Bacteria

6
Autophagy and *Mycobacterium tuberculosis*

James Harris, Sergio De Haro and Vojo Deretic

6.1
Introduction

Responsible for nearly 2 million deaths a year, *Mycobacterium tuberculosis*, the causative agent of tuberculosis (TB), is a pathogen of global significance. Despite the existence of effective treatments for the disease, TB is a growing problem, particularly in countries badly affected by human immunodeficiency virus (HIV). In sub-Saharan Africa, TB notifications have increased 3-fold since the mid-1980s and the death rates with treatment have reached 20%, compared to 5% in the absence of HIV [1]. Spread through airborne droplets, *M. tuberculosis* travels to the distal regions of the lung, where it is engulfed by alveolar macrophages and dendritic cells (DCs). In most individuals, infection leads to a series of innate and adaptive immune responses, resulting in the formation of a granuloma [2]. Consisting of epitheloid macrophages, DCs, T cells, B cells and fibroblasts, the granuloma directs a predominantly T helper (T_h) 1-type immune response that is usually sufficient to contain the infection and prevent active disease, but does not completely eradicate the bacteria. This leaves individuals susceptible to re-activation of tuberculosis when their immune system becomes compromised [2, 3].

As a facultative intracellular pathogen, *M. tuberculosis* is able to survive in macrophages by blocking fusion of the phagosome with lysosomes (phagolysosome biogenesis), avoiding direct bactericidal mechanisms and proteolytic degradation, and preventing efficient antigen processing and presentation [4]. In this chapter, we summarize the mechanisms through which *M. tuberculosis* is able to block phagolysosome biogenesis and how autophagy can overcome this block, representing a previously unrecognized mechanism for elimination of this important human pathogen. We also discuss the recently established link between regulators of the immune response, including interferon (IFN)-γ and autophagy, and the role of a unique GTP-binding protein, LRG-47, in modulating these responses.

Autophagy in Immunity and Infection. Edited by Vojo Deretic
Copyright © 2006 WILEY-VCH Verlag GmbH & Co. KGaA, Weinheim
ISBN: 3-527-31450-4

6.2
M. tuberculosis Blocks Phagolysosome Biogenesis in Macrophages

A key factor in the success of M. tuberculosis as an intracellular pathogen is its ability to survive in macrophages by disrupting phagolysosome biogenesis. The major route of entry for M. tuberculosis into macrophages is still a matter for discussion, but most likely involves a number of surface receptors, including complement receptors and C-type lectins, such as the mannose receptor [5–7]. Upon entry into the macrophage, the nature of the M. tuberculosis phagosome is well understood, and can be characterized by incomplete lumenal acidification and the absence of mature lysosomal hydrolases [8]. The block in phagolysosome biogenesis occurs between the maturation stages controlled by the GTP-binding proteins Rab5 (early endosomal) and Rab7 (late endosomal) [9]. The mycobacterial phagosome maintains access to early endosomal contents, including transferrin-bound iron, through Rab5-mediated processes, but does not acquire late endosomal or lysosomal components [9–11]. While many Rab5 effectors are recruited to the mycobacterial phagosome, early endosomal autoantigen 1 (EEA1) is displaced [12]. EEA1, which interacts directly with Rab5 [13], binds via its FYVE domain to phosphatidylinositol-3-phosphate (PI3P), which is generated on organellar membranes by hVPS34, the only type III phosphatidylinosi-

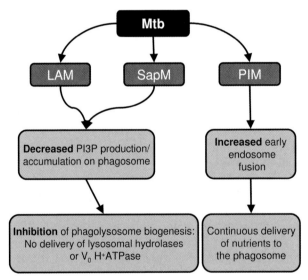

Fig. 6.1 Mycobacterial factors induce phagosome maturation block. PIM enhances homotypic fusion of early endosomes, and fusion between the phagosome and early endosomes, allowing delivery of nutrients, including iron. SapM removes PI3P from the phagosomal membrane, while LAM reduces PI3P production and prevents phagolysosome biogenesis by inhibiting the Ca^{2+}-dependent recruitment of the PI3K, hVPS34 and EEA1, and other effectors to the phagosome. This, in turn, prevents the delivery of hydrolases and H^+ATPase from the trans-Golgi network.

tol-3-kinase (PI3K) [13, 14]. Although retained on phagosomes containing dead mycobacteria, PI3P is continuously eliminated from those containing live bacilli [15] (Fig. 6.1).

Several mycobacterial factors have been shown to regulate different aspects of the phagolysosome biogenesis block and affect pathogenesis of *M. tuberculosis* (Fig. 6.1). The glycophospholipids phosphatidylinositol mannoside (PIM) and lipoarabinomannan (LAM) are able to insert into host endomembranes and traffic within infected cells [16]. Whereas LAM blocks calcium- and PI3K-dependent phagosome maturation [12, 17, 18], PIM stimulates fusion of early endosomes with the mycobacterial phagosome, permitting acquisition of nutrients, such as iron [19]. *M. tuberculosis* also secretes an acid phosphatase, SapM, which removes any PI3P that may be generated past the LAM block [15]. It is likely that additional mycobacterial factors target host cell effectors to influence, either directly or indirectly, the intracellular and extracellular environment [20, 21]. For example, a 24-kDa lipoprotein, LprG, has been proposed to inhibit major histocompatibility complex (MHC) class II antigen processing in human macrophages through a Toll-like receptor 2-dependent mechanism [22].

6.3
Autophagy and the Host Response to *M. tuberculosis*

The fact that macroautophagy (herein referred to as autophagy) is dependent on the generation of PI3P by hVPS34 during initiation and maturation/flux stages led our group to investigate whether the induction of autophagy in *M. tuberculosis*-infected macrophages could influence the phagolysosome biogenesis block. In a model system using *M. tuberculosis* variant *bovis* BCG and the murine macrophage cell line RAW 264.7, induction of autophagy by starvation or with rapamycin increases acidification of mycobacterial phagosomes and promotes maturation of the phagolysosome [23]. These effects can be inhibited by the addition of 3-methyladenine (3-MA) and wortmannin, both inhibitors of PI3Ks commonly used to inhibit autophagosome formation. Following induction of autophagy, BCG can be seen by electron microscopy in vacuoles containing partially degraded internal membranes, a characteristic of maturing autophagosomes [24–26] (Fig. 6.2). In addition, BCG shows increased colocalization with microtubule-associated protein 1 light chain 3 (LC3) and beclin, both associated with autophagy [23]. Moreover, the survival of both BCG and virulent *M. tuberculosis* in macrophages (RAW 246.7 cells and primary murine and human macrophages) is inhibited after induction of autophagy with rapamycin or by starvation [23] (Fig. 6.3). Thus, autophagy promotes the killing of *M. tuberculosis* by diverting the mycobacterial phagosome to an autophagic vacuole, which can, in turn, fuse with lysosomes, overcoming the *M. tuberculosis*-induced phagolysosome biogenesis block.

Control

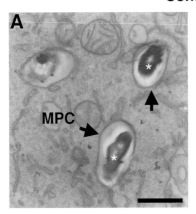

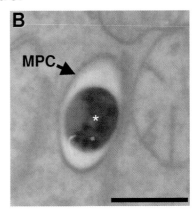

Starvation

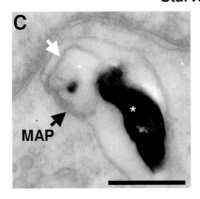

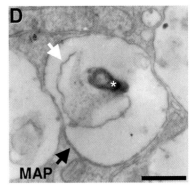

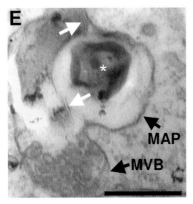

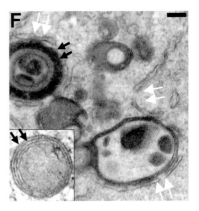

Human Peripheral Blood Monocyte-Derived Macrophages - BCG   **RAW 264.7 - *M. tuberculosis* HR37v**

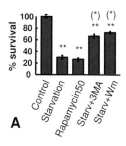
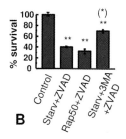
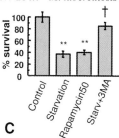

Fig. 6.3 Autophagy inhibits mycobacterial survival in macrophages. (A and B) Human peripheral blood monocyte-derived macrophages were infected with *M. tuberculosis* variant *bovis* BCG. (C) RAW 264.7 cells were infected with virulent *M. tuberculosis* HR37v. Cells were incubated for 2 h in starvation medium, with or without 10 mM 3-MA or 100 nM wortmannin (WM). Full nutrient medium was used for rapamycin (50 µg/ml) treatment. The pan-caspase inhibitor z-VAD-fmk (100 µM) was added where indicated (ZVAD). Cells were lysed and mycobacterial viability (c.f.u.) determined. Data represent means ± SEM from three independent experiments. **$P \leq 0.01$; *$P < 0.05$; †$P \geq 0.05$; all relative to control. (*)$P < 0.05$; relative to single treatment samples. Modified with permission from Gutierrez *et al.* [23].

6.4
Regulation of Autophagy by the Immune System

In activated macrophages the phagolysosome maturation block induced by *M. tuberculosis* can be overcome by IFN-γ – a critical anti-tuberculosis immune mediator [27]. Moreover, IFN-γ can induce or augment autophagy in macrophages and other cells [23, 24, 28]. These findings raise important questions regarding the role of cytokines in regulating this process, especially in the context of T_h1 versus T_h2 responses to pathogens. If IFN-γ can induce autophagy, could other T_h1 cytokines [e.g. interleukin (IL)-12, tumor necrosis factor (TNF)-a] have

◀───────────────────────────────

Fig. 6.2 Ultrastructural analysis of mycobacterial phagosomes upon induction of autophagy. Macrophages (RAW 264.7 cells) were infected with *M. tuberculosis* var. *bovis* BCG (for 1 h), incubated in full nutrient or starvation medium (for 2 h) and processed for electron microscopy. (A and B) Full nutrient medium control. The bacilli were found inside typical phagosomal compartments (MPC). (C–F) Infected macrophages induced for autophagy by starvation. The bacilli were found in altered phagosomes, referred to as mycobacterial autophagosomes (MAP) with internal membranes (C and D). Fusion events between MAP and multivesicular bodies (MVB) were observed (E). Under starvation conditions, numerous onion-like multi lamella structures (inset) or myelin-like figures containing bacterial and other debris were observed (F). Black arrows = phagosomal membrane. Asterisks = bacilli. White arrows = internal membranes within autophagosomes. Double black arrows = onion (myelin)-like multilamellar structures. Double white arrows = ER-like membranes juxtaposed to autophagosomes and multilamellar structures. Reproduced with permission from Gutierrez *et al.* [23].

similar effects and could T$_h$2 cytokines (e.g. IL-4, IL-13 and IL-10) be inhibitory? In HT-29 human colon carcinoma epithelial cells, IL-13 inhibits autophagy through activation of the protein kinase B (PKB)/mammalian inhibitor of rapamycin (mTOR) pathway [29–31], so in macrophages this cytokine could prove to be a potent regulator of the autophagic process. In this context, it is particularly interesting that the hyper-virulent strains of *M. tuberculosis*, HN878 and W/Beijing, promote secretion of IL-4 and IL-13 by infected human monocytes, while the less pathogenic isolate, CDC1551, induces more IL-12 [32]. Like IL-13, IL-4 also signals through the PKB (Akt) pathway, via the shared receptor, IL-4Rα [33] and has similar immunomodulatory effects, so could also be a target for future research. However, it should be noted that these cytokines could influence the response to *M. tuberculosis* (and other pathogens) through other mechanisms, such as inhibition of inducible nitrogen oxide (iNOS) expression [34] or indirectly through inhibition of T$_h$1 responses [20, 35, 36].

A recent study has shown that the removal of a single growth factor can be sufficient to induce autophagy. Using a hematopoietic cell dependent on IL-3 that also lacks the apoptotic regulators Bax and Bak, Lum *et al.* [37] were able to show that upon removal of IL-3 from the media, these cells formed autophagosomes and survived on intracellular substrates. Although these cells did eventually die, when autophagy was blocked by transfecting the cells with short hairpin RNA constructs to Atg5, ATP production declined rapidly and the cells died more quickly, demonstrating that autophagy is a survival mechanism in this case [37]. Could growth factors influence autophagy in macrophages and other cells of the immune system? Numerous cytokines act as growth or survival factors, such as IL-2 in lymphoblasts [38] and macrophage colony-stimulating factor (M-CSF) in macrophages [39]. Murine bone marrow macrophages require growth factors in order to survive and proliferate *in vitro*. The most potent of these is M-CSF, although both granulocyte/macrophage colony-stimulating factor (GM-CSF) and IL-3 can support both survival and proliferation in these cells [40]. All three factors promote survival through activation of PI3K/Akt and their absence leads to apoptosis [40]. How these, and other, growth factors modulate autophagy is of significance in the context of macrophage biology and the immune response.

6.5
p47 GTPases and Autophagy

The IFN-γ-inducible 47-kDa (p47) GTPase family members are among the most potent known protective factors against intracellular parasites [41, 42]. One member, LRG-47, is important in the host response to mycobacteria [42]. Studies *in vivo* have demonstrated that LRG-47 knockout mice are unable to control *M. tuberculosis* infections and that this effect is independent of reactive oxygen and nitrogen intermediates [42, 43]. Although the mechanism of action for these GTPases is largely unknown, LRG-47 may be involved in IFN-γ-induced

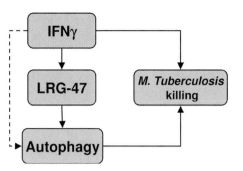

Fig. 6.4 IFN-γ-dependent killing of *M. tuberculosis*. Induction of autophagy, via LRG-47, is one way in which IFN-γ can enhance intracellular killing of *M. tuberculosis* by macrophages. It is not yet clear whether IFN-γ affects autophagy in an LRG-47-independent manner (dotted line).

autophagy [23]. In macrophages, IFN-γ induces the formation of large LC3-positive vacuoles, increases proteolysis of long-lived proteins and increases co-localization between BCG and LC3 [23]. Similarly, formation of large LC3-positive puncta is induced, in the absence of IFN-γ by transfection of macrophages with LRG-47 [23]. Further studies are required to elucidate the mechanisms by which IFN-γ and LRG-47 induce autophagy (Fig. 6.4). It is not yet clear whether autophagy represents a strategy by which macrophages can kill mycobacteria *in vivo* or whether *M. tuberculosis* has evolved mechanisms to modulate the process in the host. In this context, it would be of great value to look at autophagy markers in infected tissue *in situ*, particularly in TB granulomas, where infected macrophages and DCs are concentrated and the immune response is largely controlled by IFN-γ.

6.6
Future Directions

Relatively few studies have been conducted on the process of autophagy in macrophages and related cells. A few early studies concentrated on the role of autophagy in the constitutive degradation of long-lived proteins by macrophages [44, 45], while autophagy in microglia has been linked to disc shedding in the retina [46], and may play in role in neurodegenerative diseases, such as Parkinson's and Alzheimer's [47]. Several studies have also demonstrated a role for macroautophagy in MHC class II antigen processing and presentation by both macrophages and DCs (see Chapter 12).

Whilst the role of autophagy in innate and adaptive immunity is now receiving more attention, very little is known about the dynamics of the autophagic process in macrophages compared with epithelial cells and fibroblasts. As professional phagocytes, and exceptionally active endocytic cells, macrophages (and DCs) may display higher basal levels of autophagy or use this process for specialized functions. Numerous questions arise from the few studies so far presented. Do macrophages require different stimuli to undergo autophagy? Do different stimuli (including different pathogens) influence the specific mechanisms of autophagy and how does this affect the outcome? Does autophagy, under certain circumstances, lead to cell death in these cells and can this be controlled? Further studies on the basic biology of these processes, along with detailed investigation of its role in immune responses to pathogens will allow us to answer these questions and fully understand the importance of autophagy in macrophages and the immune system.

References

1 Nunn, P., Williams, B., Floyd, K., Dye, C., Elzinga, G., Raviglione, M. 2005. Tuberculosis control in the era of HIV. *Nat Rev Immunol 5*, 819–826.

2 Salgame, P. 2005. Host innate and T_h1 responses and the bacterial factors that control *Mycobacterium tuberculosis* infection. *Curr Opin Immunol 17*, 374–380.

3 Co, D.O., Hogan, L.H., Kim, S.I., Sandor, M. 2004. Mycobacterial granulomas: keys to a long-lasting host-pathogen relationship. *Clin Immunol 113*, 130–136.

4 Vergne, I., Chua, J., Singh, S.B., Deretic, V. 2004. Cell biology of *mycobacterium tuberculosis* phagosome. *Annu Rev Cell Dev Biol 20*, 367–394.

5 Ernst, J.D. 1998. Macrophage receptors for *Mycobacterium tuberculosis*. *Infect Immun 66*, 1277–1281.

6 Ferguson, J.S., Weis, J.J., Martin, J.L., Schlesinger, L.S. 2004. Complement protein C3 binding to *Mycobacterium tuberculosis* is initiated by the classical pathway in human bronchoalveolar lavage fluid. *Infect Immun 72*, 2564–2573.

7 Kang, P.B., Azad, A.K., Torrelles, J.B., Kaufman, T.M., Beharka, A., Tibesar, E., DesJardin, L.E., Schlesinger, L.S. 2005. The human macrophage mannose receptor directs *Mycobacterium tuberculosis* lipoarabinomannan-mediated phagosome biogenesis. *J Exp Med 202*, 987–999.

8 Russell, D.G., Mwandumba, H.C., Rhoades, E.E. 2002. Mycobacterium and the coat of many lipids. *J Cell Biol 158*, 421–426.

9 Via, L.E., Deretic, D., Ulmer, R.J., Hibler, N.S., Huber, L.A., Deretic, V. 1997. Arrest of mycobacterial phagosome maturation is caused by a block in vesicle fusion between stages controlled by rab5 and rab7. *J Biol Chem 272*, 13326–13331.

10 Clemens, D.L., Horwitz, M.A. 1996. The *Mycobacterium tuberculosis* phagosome interacts with early endosomes and is accessible to exogenously administered transferrin. *J Exp Med 184*, 1349–1355.

11 Kelley, V.A., Schorey, J.S. 2003. *Mycobacterium's* arrest of phagosome maturation in macrophages requires Rab5 activity and accessibility to iron. *Mol Biol Cell 14*, 3366–3377.

12 Fratti, R.A., Backer, J.M., Gruenberg, J., Corvera, S., Deretic, V. 2001. Role of phosphatidylinositol 3-kinase and Rab5 effectors in phagosomal biogenesis and mycobacterial phagosome maturation arrest. *J Cell Biol 154*, 631–644.

13 Simonsen, A., Lippe, R., Christoforidis, S., Gaullier, J.M., Brech, A., Callaghan, J., Toh, B.H., Murphy, C., Zerial, M., Stenmark, H. 1998. EEA1 links PI(3)K function to Rab5 regulation of endosome fusion. *Nature 394*, 494–498.

14 Christoforidis, S., Miaczynska, M., Ashman, K., Wilm, M., Zhao, L., Yip, S.C., Waterfield, M.D., Backer, J.M., Zerial, M. 1999. Phosphatidylinositol-3-OH kinases are Rab5 effectors. *Nat Cell Biol 1*, 249–252.

15 Vergne, I., Chua, J., Lee, H.H., Lucas, M., Belisle, J., Deretic, V. 2005. Mechanism of phagolysosome biogenesis block by viable *Mycobacterium tuberculosis*. *Proc Natl Acad Sci USA 102*, 4033–4038.

16 Beatty, W.L., Rhoades, E.R., Ullrich, H.J., Chatterjee, D., Heuser, J.E., Russell, D.G. 2000. Trafficking and release of mycobacterial lipids from infected macrophages. *Traffic 1*, 235–247.

17 Fratti, R.A., Chua, J., Vergne, I., Deretic, V. 2003. *Mycobacterium tuberculosis* glycosylated phosphatidylinositol causes pha-

gosome maturation arrest. *Proc Natl Acad Sci USA 100*, 5437–5442.

18 Vergne, I., Chua, J., Deretic, V. **2003**. Tuberculosis toxin blocking phagosome maturation inhibits a novel Ca²⁺/calmodulin–PI3K hVPS34 cascade. *J Exp Med 198*, 653–659.

19 Vergne, I., Fratti, R. A., Hill, P. J., Chua, J., Belisle, J., Deretic, V. **2004**. *Mycobacterium tuberculosis* phagosome maturation arrest: mycobacterial phosphatidylinositol analog phosphatidylinositol mannoside stimulates early endosomal fusion. *Mol Biol Cell 15*, 751–760.

20 Mijatovic, T., Kruys, V., Caput, D., Defrance, P., Huez, G. **1997**. Interleukin-4 and -13 inhibit tumor necrosis factor-alpha mRNA translational activation in lipopolysaccharide-induced mouse macrophages. *J Biol Chem 272*, 14394–14398.

21 Stewart, G. R., Patel, J., Robertson, B. D., Rae, A., Young, D. B. **2005**. Mycobacterial mutants with defective control of phagosomal acidification. *PLoS Pathog 1*, e33.

22 Gehring, A. J., Dobos, K. M., Belisle, J. T., Harding, C. V., Boom, W. H. **2004**. *Mycobacterium tuberculosis* LprG (Rv1411c): a novel TLR-2 ligand that inhibits human macrophage class II MHC antigen processing. *J Immunol 173*, 2660–2668.

23 Gutierrez, M. G., Master, S. S., Singh, S. B., Taylor, G. A., Colombo, M. I., Deretic, V. **2004**. Autophagy is a defense mechanism inhibiting BCG and *Mycobacterium tuberculosis* survival in infected macrophages. *Cell 119*, 753–766.

24 Inbal, B., Bialik, S., Sabanay, I., Shani, G., Kimchi, A. **2002**. DAP kinase and DRP-1 mediate membrane blebbing and the formation of autophagic vesicles during programmed cell death. *J Cell Biol 157*, 455–468.

25 Mizushima, N., Yamamoto, A., Hatano, M., Kobayashi, Y., Kabeya, Y., Suzuki, K., Tokuhisa, T., Ohsumi, Y., Yoshimori, T. **2001**. Dissection of autophagosome formation using Apg5-deficient mouse embryonic stem cells. *J Cell Biol 152*, 657–668.

26 Talloczy, Z., Jiang, W., Virgin, H. W., Leib, D. A., Scheuner, D., Kaufman, R. J., Eskelinen, E. L., Levine, B. **2002**. Regulation of starvation- and virus-induced autophagy by the eIF2alpha kinase signaling pathway. *Proc Natl Acad Sci USA 99*, 190–195.

27 Via, L. E., Fratti, R. A., McFalone, M., Pagan-Ramos, E., Deretic, D., Deretic, V. **1998**. Effects of cytokines on mycobacterial phagosome maturation. *J Cell Sci 111*, 897–905.

28 Pyo, J. O., Jang, M. H., Kwon, Y. K., Lee, H. J., Jun, J. I., Woo, H. N., Cho, D. H., Choi, B., Lee, H., Kim, J. H., Mizushima, N., Oshumi, Y., Jung, Y. K. **2005**. Essential roles of Atg5 and FADD in autophagic cell death: dissection of autophagic cell death into vacuole formation and cell death. *J Biol Chem 280*, 20722–20729.

29 Arico, S., Petiot, A., Bauvy, C., Dubbelhuis, P. F., Meijer, A. J., Codogno, P., Ogier-Denis, E. **2001**. The tumor suppressor PTEN positively regulates macroautophagy by inhibiting the phosphatidylinositol 3-kinase/protein kinase B pathway. *J Biol Chem 276*, 35243–35246.

30 Petiot, A., Ogier-Denis, E., Blommaart, E. F., Meijer, A. J., Codogno, P. **2000**. Distinct classes of phosphatidylinositol 3′-kinases are involved in signaling pathways that control macroautophagy in HT-29 cells. *J Biol Chem 275*, 992–998.

31 Wright, K., Ward, S. G., Kolios, G., Westwick, J. **1997**. Activation of phosphatidylinositol 3-kinase by interleukin-13. An inhibitory signal for inducible nitric-oxide synthase expression in epithelial cell line HT-29. *J Biol Chem 272*, 12626–12633.

32 Manca, C., Reed, M. B., Freeman, S., Mathema, B., Kreiswirth, B., Barry, C. E., 3rd, and Kaplan, G. **2004**. Differential monocyte activation underlies strain-specific *Mycobacterium tuberculosis* pathogenesis. *Infect Immun 72*, 5511–5514.

33 Nelms, K., Keegan, A. D., Zamorano, J., Ryan, J. J., Paul, W. E. **1999**. The IL-4 receptor: signaling mechanisms and biologic functions. *Annu Rev Immunol 17*, 701–738.

34 Bogdan, C., Vodovotz, Y., Paik, J., Xie, Q. W., Nathan, C. **1994**. Mechanism of suppression of nitric oxide synthase expression by interleukin-4 in primary mouse macrophages. *J Leukoc Biol 55*, 227–233.

35 Gordon, S. **2003**. Alternative activation of macrophages. *Nat Rev Immunol 3*, 23–35.

36 O'Garra, A. **1998**. Cytokines induce the development of functionally heterogeneous T helper cell subsets. *Immunity 8*, 275–283.

37 Lum, J.J., Bauer, D.E., Kong, M., Harris, M.H., Li, C., Lindsten, T., Thompson, C.B. **2005**. Growth factor regulation of autophagy and cell survival in the absence of apoptosis. *Cell 120*, 237–248.

38 Duke, R.C., Cohen, J.J. **1986**. IL-2 addiction: withdrawal of growth factor activates a suicide program in dependent T cells. *Lymphokine Res 5*, 289–299.

39 Stanley, E.R., Berg, K.L., Einstein, D.B., Lee, P.S., Pixley, F.J., Wang, Y., Yeung, Y.G. **1997**. Biology and action of colony-stimulating factor-1. *Mol Reprod Dev 46*, 4–10.

40 Comalada, M., Xaus, J., Sanchez, E., Valledor, A.F., Celada, A. **2004**. Macrophage colony-stimulating factor-, granulocyte-macrophage colony-stimulating factor-, or IL-3-dependent survival of macrophages, but not proliferation, requires the expression of p21^{Waf1} through the phosphatidylinositol 3-kinase/Akt pathway. *Eur J Immunol 34*, 2257–2267.

41 MacMicking, J.D. **2005**. Immune control of phagosomal bacteria by p47 GTPases. *Curr Opin Microbiol 8*, 74–82.

42 MacMicking, J.D., Taylor, G.A., McKinney, J.D. **2003**. Immune control of tuberculosis by IFN-γgamma-inducible LRG-47. *Science 302*, 654–659.

43 Feng, C.G., Collazo-Custodio, C.M., Eckhaus, M., Hieny, S., Belkaid, Y., Elkins, K., Jankovic, D., Taylor, G.A., Sher, A. **2004**. Mice deficient in LRG-47 display increased susceptibility to mycobacterial infection associated with the induction of lymphopenia. *J Immunol 172*, 1163–1168.

44 Dean, R.T. **1977**. Lysosomes and protein degradation. *Acta Biol Med Ger 36*, 1815–1820.

45 Thyberg, J., Hedin, U., Stenseth, K. **1982**. Mechanisms of autophagy in resident and thioglycollate-elicited mouse peritoneal macrophages *in vivo* and *in vitro*. *Eur J Cell Biol 29*, 24–33.

46 Long, K.O., Fisher, S.K., Fariss, R.N., Anderson, D.H. **1986**. Disc shedding and autophagy in the cone-dominant ground squirrel retina. *Exp Eye Res 43*, 193–205.

47 Jellinger, K.A. **2001**. Cell death mechanisms in neurodegeneration. *J Cell Mol Med 5*, 1–17.

7
Autophagy Eliminates Group A *Streptococcus* Invading Host Cells

Atsuo Amano and Tamotsu Yoshimori

7.1
Group A *Streptococcus* (GAS; *Streptococcus pyogenes*)

GAS is a Gram-positive extracellular etiological agent, which is one of the most ubiquitous and versatile of human bacterial pathogens [1]. GAS colonizes the throat and skin, and is responsible for a number of suppurative infections and nonsuppurative sequelae. GAS is the most common causes of bacterial pharyngitis as well as scarlet fever and impetigo. Further, GAS is responsible for streptococcal toxic shock syndrome (STSS), and it is recently known as the "flesh-eating" bacterium which invades skin and soft tissues, destroying tissues and limbs. During the 5-year period from 1995 to 1999, there were 9600–9700 cases of invasive GAS infections reported annually, including 1100–1300 deaths [2]. In those, pneumonia, necrotizing fasciitis and central nervous system infections occurred in more than 20%, and STSS occurred in 44.5% of cases. In addition, acute post-streptococcal glomerulonephritis is a leading cause of cardiovascular morbidity and mortality in many developing countries throughout the world [3].

7.2
Adherence to Host Cells by GAS

The initiation of GAS diseases needs the adhesion of GAS to mucosal or cutaneous surfaces. In the process of adhesion, extracellular matrix proteins such as fibronectin and laminin serve as mediators between the bacteria and host cells [4–7]. At least 17 candidates of adhesin (GAS components which promote bacterial adhesion) have been described. Among them, lipoteichoic acid (LTA), surface M protein and fibronectin binding proteins have been extensively studied. LTA adheres to fibronectin on human buccal epithelial cells [8, 9] and passive immunization for LTA markedly protected mice against streptococcal challenge [10]. Thus, LTA is thought to serve as a "first-step" with its hydrophobic interactions. In addition, GAS surface proteins that bind fibronectin are considered as "second-step" factors mediating GAS adherence. Those involve protein F1, also known as SfbI (streptococcal fibronectin-binding protein I) [11, 12], and its related proteins including F2 [13], SfbII [14], FBP54 [15], PFBP [16] and Fba [17]. Such a coordinated strategy likely promotes efficient adhesion to and subsequent invasion of host cells by GAS.

7.3
Invasion of Host Cells by GAS

Various evidence over the past decade suggested that GAS not only adheres to epithelial cells, but is also internalized by them [18]. In fact, GAS has been show to penetrate a variety of cultured human epithelial cells [19–22] (Fig. 7.1). Streptococci were serologically separated based on M protein serotypes and more than 80 M protein serotypes are currently known with a molecular approach to identify the *emm* (M protein) genes [23]. Among those serotypes, M1 protein as well as SfbI are considered components which promote bacterial invasion to human cells, called invasins [24]. The binding of Sfbs to fibronectin [25] also triggers invasion via the attachment of the complex to cellular $\alpha_5\beta_1$ integrin. In addition, laminin has been shown to bind GAS [26] and subsequently induce internalization of M1-GAS, independent of serum and fibronectin. These findings suggest that GAS possesses multiple pathways to penetrate host cells. The intracellular organisms subsequently activate the focal adhesion complex [27] and induce cytoskeletal rearrangements [28], which disable cellular functions such as adhesion, migration and proliferation. The intracellular GAS also induces proinflammatory cytokine production [29] and apoptosis by invaded cells [26, 30]. Although the biological significance of intracellular localization remains to be further elucidated, intracellular localization would allow the pathogen to penetrate to deep tissues and may provide it a nutritionally rich "shelter" protected from the host immune system, including phagocytes, humoral antibodies and antibiotics. Such a hypothesis is partially supported by the findings that GAS were recovered from biopsies of infected tonsils as well as from surgically removed tonsils of streptococcal carriers with no infectious symptoms [31, 32].

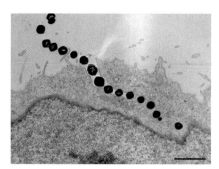

Fig. 7.1 The GAS chain penetrating HeLa cells. The electron micrograph was taken by Hitomi Yamaguchi, Research Institute of Microbial Diseases, Osaka University, Japan. Bar=2 μm.

7.4
Survival of Intracellular Bacteria

A number of pathogenic bacterial species have developed a variety of mechanisms to invade eukaryotic cells and survive inside them, maximizing their potential virulence [33]. Although the molecular mechanisms differ among the bacterial species, most of these bacteria are internalized into phagosomes, thereby maturing into vacuoles with the characteristics of endosomes [34]. Subsequent fusion of these compartments with lysosomes results in their degradation and this harsh environment provides a host defense against intracellular pathogens [34, 35]. However, some bacteria have invented one of two different strategies to avoid this. The first strategy is to modify the endocytic compartments for prevention of fusion with a lysosome or to interfere with its lytic action. For instance, phagosomes containing *Mycobacterium tuberculosis* lack the V-ATPase, thereby preventing acidification [36]. In addition, *Salmonella enterica* serovar *typhimurium* traffics to late endosome-like compartments that do not mature into lysosomes [37]. The second strategy is to escape from the endocytic compartment by lysing it into the cytoplasm of the host cell. *Listeria monocytogenes* normally enters host cells by phagocytosis, after which the bacteria escapes from the phagosomes. These bacteria can replicate within the host cells. Therefore, they are called "intracellular parasites". Thus, escape from the phagocytic/endocytic degradation system has been thought to be a crucial step for the survival of intracellular bacteria [38, 39].

7.5
Streptolysin O (SLO) Enables GAS to Escape
form Phagocytic/Endocytic Degradation

We confirmed that GAS successfully invaded human epithelial cells (HeLa cells) (Fig. 7.1) and further examined the fate of intracellular GAS. For infection experiments, cells were incubated with GAS for 1 h. After washing, the infected cells were cultured for various times in the presence of antibiotics to kill extracellular GAS. At the first setout after invasion of HeLa cells, GAS were found to be colocalized with the FYVE domain of early endosomal autoantigen 1 (EEA1), which specifically binds to early endosomes, demonstrating that GAS first enters endosomes [40]. Then, after 1 h of infection, endosomes containing GAS gradually disappeared, which suggests escape of GAS from endosomes. GAS is known to secrete SLO, a member of a conserved family of cholesterol-dependent pore-forming cytolysins [41]. Although the role of SLO is not clear, one of the family members, listeriolysin O (LLO), which is produced by *L. monocytogenes*, is essential for the escape of the bacterium from endocytic compartments [42, 43]. We found that JRS4Δ*SLO*, an isogenic SLO-deficient mutant of strain JRS4, was not extricated and remained within FYVE-positive endosomes even at 2 h post-infection. The endosomes enclosing JRS4Δ*SLO* grew to larger sizes during incubation than did those enclosing wild-type GAS. GAS seems to escape from endosomes to cytoplasm via a SLO-dependent mechanism.

7.6
Autophagy

Macroautophagy, usually referred to simply as autophagy, is a physiologically important cellular process for the bulk degradation of organelles and cytosolic proteins [44, 45] (chapter 1). Like the endocytic pathway, it is a ubiquitous pathway in eukaryotic cells. Regions of cytoplasm, as well as organelles, are first engulfed by double- or multiple-membrane structures called autophagosomes with diameters of 0.5–1.0 μm, which are formed through elongation and closure of cap-shaped cisternae known as isolation membranes. These autophagosomes eventually fuse with lysosomes to form autolysosomes. The cytoplasm-derived contents of the autophagosomes are degraded by lysosomal hydrolases. This lysosomal degradation system is thought to be required for the nonselective degradation and recycling of cellular proteins, as well as for the removal of nonfunctional or superfluous organelles. Under starvation conditions, autophagy is dramatically enhanced to maintain the intracellular amino acid pool for gluconeogenesis and for the synthesis of proteins essential for survival. In addition, it has been postulated that some intracellular bacteria, e.g. *Rickettsia conorii*, are targeted by this degradation system [46]. However, it has been difficult to substantiate the hypothesis that autophagy degrades intracellular pathogens as well as cellular constituents, since individual components of the autophagic machinery have not been identified.

Recently, we identified the microtubule-associated protein 1 light chain 3 (LC3) as an autophagosome-specific membrane marker in mammalian cells [47–49] (chapter 3). LC3 is the only reliable marker for the compartments at the present time. We also found that a protein complex including Atg5 is involved in autophagosome formation [50, 51]; *atg5* gene-knockout cells are unable to form autophagosome [52] (chapter 3). Using the autophagosome marker LC3 as well as cells lacking autophagic activity, we examined whether autophagosomes trap intracellular GAS following its escape to the cytoplasm [40].

7.7
Intracellular GAS is Trapped by Autophagosome-like Compartments

To investigate the possibility that the autophagic machinery sequesters GAS harboring in the cytoplasm, we infected cells expressing LC3 coupled with Green Fluorescent Protein (GFP–LC3) as a marker for autophagosomal membranes. Confocal microscopy revealed that in uninfected cells cultured under nutrient-rich conditions, the majority of cellular LC3 is found diffused throughout the cytoplasm; under starvation conditions, however, LC3-positive autophagosomes (with diameters of 0.5–1.0 μm) were found in the cytoplasm as punctuate structures. In HeLa

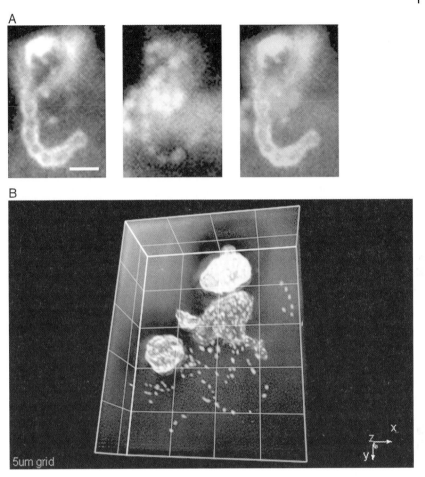

Fig. 7.2 GcAVs. (A) LC3-positive compartments (a left panel and green in a right panel) sequestering intracellular GAS (a middle panel and magenta in a right panel) in HeLa cells expressing enhanced GFP (EGFP)–LC3 at 1 h post-infection [40]. The bacteria were visualized by staining bacterial DNA with propidium iodide. Bar=2 µm. (B) Three-dimensional image of a large GcAV at 3 h post-infection. The image was made by Shunsuke Kimura, Research Institute for Microbial Diseases, Osaka University, Japan. Grid=5 µm. (This figure also appears with the color plates).

cells infected by GAS strain JRS4 cells, we found the large LC3-positive compartments acquiring the bacteria cluster (Fig. 7.2). This was true even under nutrient-rich conditions and was not observed in uninfected cells, showing that GAS induces autophagy and is trapped by autophagosomes. The size and morphology of the GAS-induced LC3-positive compartments were distinct from canonical starvation-induced autophagosomes, whose size is sometimes over 10 µm, so we designated these novel structures GAS-containing LC3-positive autophagosome-like vacuoles (GcAVs). GcAVs appeared within 30 min of infection (in $6.8 \pm 3.4\%$ of infected cells), after which the number of cells bearing GcAVs increased in a time-dependent manner, reaching a maximum level (79.6±7.2% of infected cells) at 2 h post-infection. The area of GcAVs also increased over time to around 10% of the total cytoplasmic area. Finally, about 80% of intracellular GAS was trapped by the LC3-positive compartments, indicating that most intracellular GAS had been trapped by GcAVs. Interestingly, the number of GcAVs in HeLa cells infected with JRS4Δ*SLO* was only 4% of that seen in wild-type GAS-infected cells through infection and post-infection. This suggests that emergence of the bacteria in the cytoplasm triggers autophagic induction to attack the bacteria. SLO seems to be a critical factor for bacterial escape from endosomes, while autophagosomes are likely to be resistant to SLO. There is no explanation for this at this moment but SLO targets cholesterol and the autophagic membranes may contain less or no cholesterol compared to endosomes.

At an early stage of post-infection (around 2 h), LC3 frequently surrounded GAS as a chain-like structure fitting closely around the GAS chain; the LC3-positive membrane may thus be formed contiguously with GAS surfaces (Fig. 7.2 A). By 3 h post-infection, GcAVs had fused together and were larger (5–10 μm). At least 20–30 bacterial cells were associated with a single GcAV (Fig. 7.2 B). Localization of GAS within LC3-positive vacuoles was observed not only for the GAS strain JRS4, but also for other M-type strains. We previously demonstrated that LC3 exists in two molecular forms. LC3-I (18 kDa) is cytosolic, whereas LC3-II (16 kDa) binds to autophagosomes [47, 48]. The amount of LC3-II directly correlates with the number of autophagosomes. We found that the total amount of LC3-II increased in a time-dependent manner during post-infection under nutrient-rich conditions. This induction correlated with the development of GcAVs.

7.8
Atg5 is Required for Capture and Killing of GAS

The results of the LC3 analysis suggest that GAS invasion induces autophagy that specifically traps intracellular GAS. To further substantiate this idea, we next examined GcAV formation in Atg5-deficient cells. Our previous gene-targeting study [52] using mouse embryonic stem (ES) cells demonstrated that Atg5 is required for starvation-induced autophagosome formation, membrane targeting of LC3 and formation of LC3-II in wild-type ES cells. GcAVs similar to those found in HeLa cells were clearly observed in the wild-type ES cells during post-infection. In contrast, no GcAVs were observed in Atg5$^{-/-}$ ES cells. Atg5$^{-/-}$ mouse embryonic fibroblasts (MEFs) [53] also showed no GcAV formation at 3 h post-infection, indicating that Atg5 is required for the formation of GcAV in several cell types. The numbers of wild-type MEFs bearing GcAVs increased with time, reaching a maximum (67.8% of infected cells) at 4 h post-infection. On the other hand, no GcAV formation was observed in Atg5$^{-/-}$ MEF cells even at 4 h post-infection. We concluded that GcAV formation requires an Atg5-mediated mechanism.

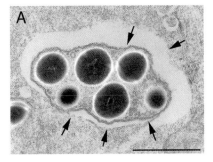
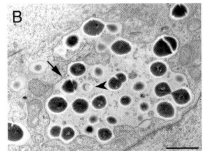
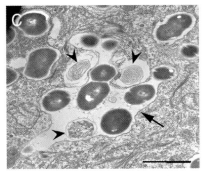

Fig. 7.3 Ultrastructural observation of GAS-infected HeLa cells [40]. (A) An autophagosome-like multiple-membrane-bound compartment (arrows) containing intact cytosol and GAS at 1 h post-infection. (B and C) Autolysosome-like single-membrane-bound compartments (arrows) containing degraded cytosol and GAS (arrowheads) at 4 h post-infection. Bars = 1 µm.

Next, we analyzed the ultrastructure of GcAVs by electron microscopy. In wild-type MEFs infected with GAS we observed characteristic double-membrane cisternae surrounding GAS in the cytoplasm, which resemble, but are larger than, isolation membranes. We also observed double- or multiple-membrane-bound compartments containing intact cytosol and GAS at 1 h post-infection (Fig. 7.3 A), and single-membrane-bound compartments with degraded cytosol and GAS at 4 h post-infection (Fig. 7.3 B). Again, they resemble, but are quite larger than, autophagosomes and autolysosomes, respectively. No GAS were found surrounded by isolation membrane-like structures in Atg5$^{-/-}$ cells. We also examined LC3-II formation in three ES cell lines. In wild-type and WT13 cells, an *atg5* cDNA transformant of Atg5$^{-/-}$ cells, the induction of LC3-II was observed in a time-dependent manner during post-infection. During both starvation conditions and infection with GAS, Atg5$^{-/-}$ cells showed no induction of LC3-II. Similar results were obtained for Atg5$^{-/-}$ MEFs. The results suggest that GcAV formation upon GAS invasion shares common molecular mechanisms with starvation-induced canonical autophagosome formation, in that both GAS-induced increases in LC3-II conversion and GcAV formation require Atg5.

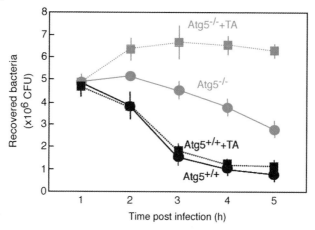

Fig. 7.4 Lack of autophagic ability allows GAS survival in host cells. Viability of intracellular GAS in wild-type MEFs and Atg5$^{-/-}$ MEFs was measured in the presence (squares) and absence (circles) of tannic acid (TA; final concentration 0.5%) [40]. Data are representative of at least three independent experiments.

All data described above are consistent with the idea that GAS is sequestered by autophagic machinery. Our next question was whether the bacteria are killed or survive after entering the compartments. Fluorescence microscopy revealed that the number of intracellular GAS in the two Atg5$^{-/-}$ cell lines was greater than in wild-type cells. To quantify the difference, we directly scored bacterial viability by counting colony-forming units (c.f.u. viability assay) in wild-type and Atg5$^{-/-}$ MEFs (Fig. 7.4). In wild-type MEF cells at 4 h post-infection, intracellular GAS, most of which was within GcAV, were mostly dead (black circle). The numbers of GAS inside the Atg5$^{-/-}$ MEFs, however, slightly increased at 1 h of infection and then decreased, but were still maintained at 60% of the initial level at 4 h of post-infection (grey circle). Thus, many of the intracellular GAS are killed by the Atg5-dependent autophagic system. Then, we wondered why the decrease of GAS was still observed in Atg5$^{-/-}$cells. We considered the possibility that GAS can escape the host cells into the surrounding medium. To explore this, we analyzed the number of intracellular living GAS in the presence of tannic acid (Fig. 7.4, squares). Tannic acid is a cell-impermeable fixative that prevents fusion between secretory vesicles and the plasma membrane, but does not affect intracellular membrane trafficking [54]. Thus, we expected that it also inhibit escape of GAS from inside to outside of cells by penetrating the plasma membrane. As shown in Fig. 7.4, the numbers of intracellular GAS in tannic acid-treated Atg5$^{-/-}$ MEFs increased at 2 h post-infection and maintained the same level through 4 h post-infection (grey squares). In contrast, the numbers of intracellular GAS decreased rapidly in tannic acid-treated Atg5$^{+/+}$ cells as well as that of the untreated Atg5$^{+/+}$ cells (black squares). Thus, GAS were not killed at all in the Atg5$^{-/-}$ cells and some of them were released from the cells. We therefore concluded that the autophagic machinery kills most of intracellular GAS and helps prevent the expansion of GAS infection.

We next assessed whether intracellular GAS actually multiplied in Atg5$^{-/-}$ cells. Host cells were metabolically labeled with [^{35}S]methionine and [^{35}S]cysteine ([^{35}S]Met+Cys) for 48 h, and then infected with GAS. GAS harbored within Atg5$^{-/-}$ cells took up [^{35}S]Met+Cys until 2 h post-infection, while the uptake of [^{35}S]Met+Cys by GAS in the wild-type cells was quite low (about 25% of that in Atg5$^{-/-}$ cells). This supports the idea that the autophagic machinery strongly inhibits the growth of GAS within the host cells.

7.9
GcAVs Fuse with Lysosomes for Degradation

Our results prompted us to inquire if the death of GAS is due to fusion of the GcAVs with lysosomes. We first analyzed the colocalization of lysosome-associated membrane protein 1 (LAMP-1), with GcAVs as an index of fusion between GcAVs and lysosomes. Confocal microscopy revealed that GcAVs formed in GAS-infected HeLa cells did not include LAMP-1 within 2 h post-infection, but clear colocalization of LC3 and LAMP-1 was observed at 2–3 h post-infection. GAS surrounded by LAMP-1, but not LC3, was also found, but these structures represented only 30% of the LAMP-1-positive GcAVs. The fact that GcAVs acquired LAMP-1 indicates fusion with lysosomes subsequent to formation of GcAVs, similar to the canonical autophagic pathway.

Certain bacterial pathogens, such as *Salmonella*, can survive inside lysosomes [55]. We therefore examined whether the viability of GAS is impaired by lysosomal enzymes. We performed a bacterial viability assay in the presence of the lysosomal protease inhibitors leupeptin and E64d. As expected, the decrease of intracellular GAS in wild-type cells was significantly suppressed by treatment with protease inhibitors. In Atg5$^{-/-}$ cells, however, the protease inhibitors did not affect the number of viable intracellular GAS, implying that the decrease in living GAS in wild-type cells was due to proteases provided not by the endocytic pathway, but by the autophagic pathway. Although degradation appeared to be partial in the ultrastructural observation, this may be sufficient to kill GAS because most were not viable at 4 h post-infection in wild-type cells.

7.10
Conclusion and Perspective

As shown schematically in Fig. 7.5, autophagic machinery can act as an important degradation system to eliminate intracellularly invaded GAS. Intracellular GAS was selectively sequestered by LC3-positive compartments. Atg5 was essential for formation of these compartments and elimination of intracellular GAS.

Wild-type cells   Atg5⁻/⁻ cells

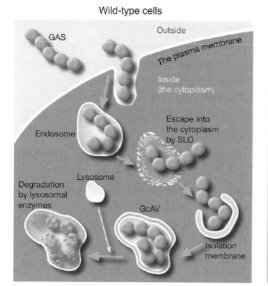
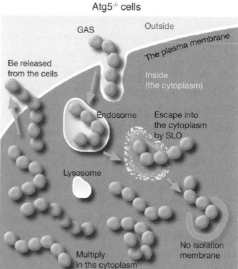

Fig. 7.5 Model for the fate of GAS in host cells with ("Wild-type cells") or without ("Atg5$^{-/-}$ cells") autophagic activity. (This figure also appears wth the color plates).

The compartments eventually fused with lysosomes, where enzymes degraded the GAS. The autophagic machinery apparently has a capacity to trap and degrade intracellularly invading pathogenic bacteria. Does such effective autophagic elimination occur only for GAS? Another common Gram-positive bacterium, *Staphylococcus aureus*, is also sequestered by LC3 compartments and multiplies in Atg5$^{-/-}$ cells (Dr. Ichiro Nakagawa, University of Tokyo, Japan, personal communication). These two species are the most ubiquitous and versatile pathogens, and are etiologically important. Therefore, autophagic elimination might be a general aspect of mammalian human innate defense systems. Autophagic elimination of both pathogens may be a clue as to an effective prevention or new therapeutic methods of overcoming these infectious diseases.

The most important point in our study of GAS is that nonphagocytic cells have the potential for effective elimination of intracellular bacteria that escape from the endocytic pathway to the cytoplasm. This also means that the autophagic machinery may function as a novel hand in innate immunity, which plays an important role in host defense against infection. However, some bacterial species, such as *Shigella* or *Salmonella*, seem to be able to avoid or subvert autophagy [56, 57] (Chapter 8). Secretory proteins produced via the type III secretion system by *Shigella* and *Salmonella* might disable the autophagic as well as endocytic machineries. Autophagy may therefore be the battlefront of the running war between cells and invasive pathogenic bacteria, and its significance in innate immunity is clear.

The autophagic machinery utilized in the defense against GAS is distinct in several points from canonical autophagy. It is specifically induced by the emergence of GAS in the cytoplasm even under nutrient-rich conditions. Canonical autophagy is thought to be nonselective, which represents random and bulk degradation of cytoplasmic contents, whereas the GAS-specific autophagy appears to selectively sequester the bacteria. The autophagosomes engulfing a GAS cluster are extremely larger and live for longer than the canonical autophagosomes. These striking features are reminiscent of the existence of the autophagic machinery specializing in defense against pathogenic bacteria. In fact, in preliminary analysis we identified several molecules essential to the formation of the GAS-specific autophagosomes, but not to that of the canonical autophagosomes. Many critical questions remain cryptic. How does the isolation membrane selectively surround the bacteria? What of the bacteria induces autophagy? How are such large autophagosomes able to be formed beyond the usual limitations? What is the source of the membranes for the large autophagosomes? Extensive work on the cell biological aspects will be required to answer these questions.

References

1 A. Smith, T.L. Lamagni, I. Oliver, A. Efstratiou, R.C. George, J.M. Stuart, *Lancet Infect Dis* **2005**, *5*, 494–500.

2 K.L. O'Brien, B. Beall, N.L. Barrett, P.R. Cieslak, A. Reingold, M.M. Farley, R. Danila, E.R. Zell, R. Facklam, B. Schwartz, A. Schuchat, *Clin Infect Dis* **2002**, *35*, 268–276.

3 J.M. Musser, V. Kapur, J. Szeto, X. Pan, D.S. Swanson, D.R. Martin, *Infect Immun.* **1995**, *63*, 994–1003.

4 A.L. Bisno, *Infect Immun* **1979**, *26*, 1172–1176.

5 D. Cue, P.E. Dombek, H. Lam, P.P. Cleary, *Infect Immun* **1998**, *66*, 4593–4601.

6 D. Cue, S.O. Southern, P.J. Southern, J. Prabhakar, W. Lorelli, J.M. Smallheer, S.A. Mousa, P.P. Cleary, *Proc Natl Acad Sci USA* **2000**, *97*, 2858–2863.

7 H.S. Courtney, D.L. Hasty, J.B. Dale, *Ann Med* **2002**, *34*, 77–87.

8 E.H. Beachey, W.A. Simpson, *Infection* **1982**, *10*, 107–111.

9 I. Ofek, E.H. Beachey, W. Jefferson, G.L. Campbell, *J Exp Med* **1975**, *187*, 1161–1167.

10 J.B. Dale, R.W. Baird, H.S. Courtney, D.L. Hasty, M.S. Bronze, *J Infect Dis* **1994**, *169*, 319–323.

11 S. Sela, A. Aviv, A. Tovi, I. Burstein, M.G. Caparon, E. Hanski, *Mol Microbiol* **1993**, *10*, 1049–55.

12 S.R. Talay, P. Valentin-Weigand, K.N. Timmis, G.S. Chhatwal, *Mol Microbiol* **1994**, *13*, 531–539.

13 J. Jaffe, S. Natanson-Yaron, M.G. Caparon, E. Hanski, *Mol Microbiol* **1996**, *21*, 373–384.

14 B. Kreikemeyer, S. Talay, G.S. Chhatwal, *Mol Microbiol* **1995**, *17*, 137–145.

15 H.S. Courtney, Y. Li, J.B. Dale, D.L. Hasty, *Infect Immun* **1994**, *62*, 3937–3946

16 C.L. Rocha, V.A. Fischetti, *Infect Immun* **1999**, *67*, 2720–2728.

17 Y. Terao, S. Kawabata, E. Kunitomo, J. Murakami, I. Nakagawa, S. Hamada, *Mol Microbiol* **2001**, *42*, 75–86.

18 D. LaPenta, C. Rubens, E. Chi, P.P. Cleary, *Proc Natl Acad Sci USA* **1994**, *91*, 12115–12119.

19 P.P. Cleary, L. McLandsborough, L. Ikeda, D. Cue, J. Krawczak, H. Lam, *Mol Microbiol* **1998**, *28*, 157–167.

20 S. Kawabata, H. Kuwata, I. Nakagawa, S. Morimatsu, K. Sano, S. Hamada, *Microb Pathog* **1999**, *27*, 71–80.

21 G. Molinari, M. Rohde, C.A. Guzman, G.S. Chhatwal, *Cell Microbiol* **2000**, *2*, 145–154.

22 I. Nakagawa, M. Nakata, S. Kawabata, S. Hamada, *Cell Microbiol* **2001**, *3*, 395–405.

23 A.L. Bisno, M.O. Brito, C.M. Collins, *Lancet Infect Dis* **2003**, *3*, 191–200.

24 P. Dombek, D. Cue, J. Sedgewick, H. Lam, B. Ruschkowski, B. Finlay, P. Cleary, *Mol Microbiol* **1999**, *31*, 859–870.

25 V. Ozeri, I. Rosenshine, D. F. Mosher, R. Fassler, E. Hanski, *Mol Microbiol* **1998**, *30*, 625–637.

26 Y. Terao, S. Kawabata, E. Kunitomo, I. Nakagawa, S. Hamada, *Infect Immun.* **2002**, *70*, 993–997.

27 B. R. Tomasini-Johansson, N. R. Kaufman, M. G. Ensenberger, V. Ozeri, E. Hanski, D. F. Mosher, *J Biol Chem* **2001**, *276*, 23430–23439.

28 C. Cywes, M. R. Wessels, *Nature* **2001**, *414*, 648–652.

29 I. Hagakawa, M. Nakata, S. Kawabata, S. Hamada, *Cell Microbiol* **2004**, *6*, 939–952.

30 P. J. Tsai, Y. S. Kin, C. F. Kuo, H. Y. Lei, J. J. Wu, *Infect Immun* **1999**, *67*, 4334–4339.

31 A. Osterlund, L. Engstrand, *Acta Otolaryngol* **1997**, *117*, 883–888.

32 J. Jadoun, O. Eyal, S. Sela, *Infect Immun* **2002**, *70*, 462–469.

33 S. Falkow, *Cell* **1991**, *65*, 1099–1102.

34 J. Tjelle, T. Lovdal, T. Berg, *BioEssays* **2000**, *22*, 255–263.

35 M. Desjardins, N N. Nzala, R. Chorsini, C. Rondeau, *J Cell Sci* **1997**, *110*, 2303–2314.

36 S. Stugill-Koszycki, P. H. Schlesinger, P. Chakraborty, P. L. Haddix, H. L. Collins, A. K. Fok, *Science* **1994**, *263*, 678–681.

37 S. Meresse, O. Steele-Mortimer, B. B. Finlay, J. P. Gorvel, *EMBO J* **1999**, *18*, 4394–4403.

38 C. M. Rosenberger, B. B. Finlay, *Nat Rev Mol Cell Biol* **2003**, *4*, 385–396.

39 A. O. Amer, M. S. Swanson, *Curr Opin Microbiol* **2002**, *5*, 56–61.

40 I. Nakagawa, A. Amano, N. Mizushima, A.Yamamoto, H. Yamaguchi, T. Kamimoto, A. Nara, J. Funao, M. Nakata, K. Tsuda, S. Hamada, T. Yoshimori, *Science* **2004**, *306*, 1037–1040.

41 R. K. Tweten, *Virulence Mechanisms of Bacterial Pathogens*. ASM Press, Washington, DC, **1995**.

42 K. E. Beauregard, K. D. Lee, R. J. Collier, J. A. Swanson, *J Exp Med* **1997**, *186*, 1159.

43 J. Bielecki, P. Youngman, P. Connelly, D. A. Portnoy, *Nature* **1990**, *345*, 175.

44 N. Mizushima, Y. Ohsumi, T. Yoshimori, *Cell Struct Funct* **2002**, *27*, 421–429.

45 T. Yoshimori, *Biochem Biophys Res Commun* **2004**, *313*, 453–458.

46 K. Kirkegaard, M. P. Taylor, W. T. Jackson, *Nat Rev Microbiol* **2004**, *2*, 301.

47 Y. Kabeya, N. Mizushima, T. Ueno, A. Yamamoto, T. Kirisako, T. Noda, E. Kominami, Y. Ohsumi, T. Yoshimori, *EMBO J* **2000**, *19*, 5720–5728.

48 N. Mizushima, A. Yamamoto, M. Matsui, T. Yoshimori, Y. Ohsumi, *Mol Biol Cell.* **2004**, *15*, 1101–1111.

49 Y. Kabeya, N. Mizushima, S. Oshitani-Okamoto, Y. Ohsumi, T. Yoshimori, *J Cell Sci* **2004**, *117*, 2805–2812.

50 N. Mizushima, T. Noda, T. Yoshimori, Y. Tanaka, T. Ishii, M. D. George, D. J. Klionsky, M. Ohsumi, Y. Ohsumi, *Nature* **1998**, *395*, 395–398.

51 N. Mizushima, H. Sugita, T. Yoshimori, Y. Ohsumi, *J Biol Chem* **1998**, *273*, 33889–33892.

52 N. Mizushima, A. Yamamoto, M. Hatano, Y. Kobayashi, Y. Kabeya, K. Suzuki, T. Tokuhisa, Y. Ohsumi, T. Yoshimori, *J Cell Biol* **2001**, *152*, 657–668.

53 A. Kuma, M. Hatano, M. Matsui, A. Yamamoto, H. Nakaya, T. Yoshimori, Y. Ohsumi, T. Tokuhisa, N. Mizushima, *Nature* **2004**, *432*, 1032–1036.

54 R. Polishchuk, A. Di Pentima, J. Lippencott-Schwartz, *Nat Cell Biol* **2004**, *6*, 297.

55 Y. K. Oh, C. Alpuche-Aranda, E. Berthiaume, T. Jinks, S. I. Miller, J. A. Swanson, *Infect Immun* **1996**, *64*, 3877–3883.

56 P. J. Sansonetti, *FEMS Microbiol Rev* **2001**, *25*, 3–14.

57 L. D.Hernandez, M. Pypaert, R. A. Flavell, J. E. Galan, *J Cell Biol* **2003**, *163*, 1123–1131.

8
Shigella Invasion of Host Cells and Escape from Autophagy

Michinaga Ogawa and Chihiro Sasakawa

8.1
Shigella Invasion of Epithelia

Shigella are highly adapted human pathogens and the cause of bacillary dysentery (shigellosis) – a disease manifested by severe bloody and mucous diarrhea. *Shigella* are one of the leading infectious killers of children in developing countries, where more than a hundred million people develop shigellosis annually and a million children die of the disease. Although *Shigella* are members of the pathogenic *Escherichia coli* family, unlike other pathogenic *E. coli* they have no adhesion molecules on their surface and are therefore unable to attach to the intestinal epithelium. Thus, when ingested via the fecal–oral route, *Shigella* eventually reach the colon and rectum, where they translocate through the epithelial barrier via the M cells that overlie solitary lymphoid nodules (Fig. 8.1). Once they reach the M cells, they invade the resident macrophages which lie underneath M cells. However, the infecting bacteria escape from the phagosomes into the cytoplasm, where they multiply and induce rapid cell death [1–4]. *Shigella* released by the dead macrophages immediately enter the surrounding enterocytes via their basolateral surface by inducing macropinocytic events; as soon as a bacterium is surrounded by the vacuole membrane of infected epithelial cells, it disrupts the vacuole membrane and escapes into the cytoplasm. *Shigella* multiply in the cytoplasm, where they move by forming actin comet tails at one pole of the bacterium (Fig. 8.1).

Shigella invade epithelial cells by a special method of entry, called "the trigger mechanism of entry", which is characterized by macropinocytic events that allow cells to trap several bacterial cells simultaneously [5]. When *Shigella* come into contact with epithelial cells, the type III secretion system (TTSS) is activated, and delivers effectors into the host cells and space around the bacteria. The secreted effectors are capable of modulating various host functions engaged in remodeling the surface architecture of host cells and escaping from the innate defense system of the host. *Shigella* effectors secreted via the TTSS, such as IpaA, IpaB, IpaC, IpgB1, IpgD and VirA, are involved in stimulating the reor-

Autophagy in Immunity and Infection. Edited by Vojo Deretic
Copyright © 2006 WILEY-VCH Verlag GmbH & Co. KGaA, Weinheim
ISBN: 3-527-31450-4

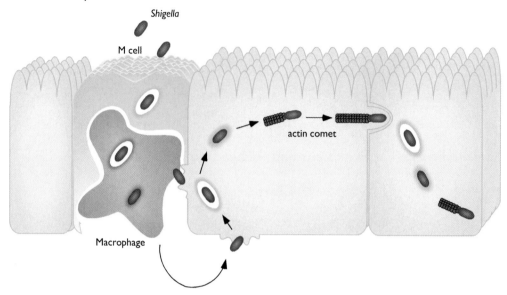

Fig. 8.1 Strategy used by *Shigella* to invade intestinal epithelial cells. Simplified illustration of infection of colonic epithelial cells by *Shigella*. *Shigella* are able to multiply in the cytosol of host cells and move into neighboring cells. (This figure also appears with the color plates).

ganization of F-actin and microtubule cytoskeletons, which trigger bacterial uptake by host cells [6]. Recent studies have indicated that synergism between activities arising from the interplay between *Shigella* effectors and target host proteins orchestrated by Rho-GTPases and tyrosine kinases plays the central role in inducing the phagocytic event [7–10].

8.2
Shigella Disseminate among Epithelial Cells

Some bacteria that invade the cytosol, such as *Shigella*, *Listeria monocytogenes*, *Rickettsia*, *Mycobacterium marinum* and *Burkholderia pseudomallei*, are capable of directing local actin polymerization at one end of the bacterium and moving within the host cells as well as into neighboring cells [5, 11, 12]. The actin-based motility of *Shigella* is dependent on VirG (IcsA), an outer membrane protein exposed on the surface of bacteria [13–15]. When *Shigella* divide, VirG is gradually distributed asymmetrically along the bacterial cell and eventually accumulates at one pole of the bacterium [16]. VirG is composed of 1102 amino acids and contains three distinctive domains: an N-terminal signal sequence (residues 1–52), a 706-amino-acid α-domain (residues 53–758) and a 344-amino-acid C-terminal β-core (residues 759–1102) [17, 18]. The α-domain is exposed on the surface, whereas the β-core is embedded in the outer membrane, where it forms the

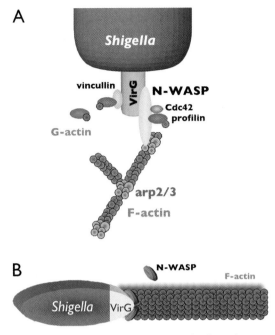

Fig. 8.2 The VirG–N-WASP–Arp2/3 complex formed at one pole of the bacterium mediates local actin nucleation and elongation. (This figure also appears with the color plates).

membrane pore [17, 18]. The unipolar distribution of VirG is a prerequisite for the polar movement of *Shigella* within epithelial cells [16]. VirG is capable of interacting with vinculin and N-WASP [17, 19] – a member of the WASP family that is required to mediate actin polymerization by interacting with Arp2/3 complex [20]. The VirG expressed on the bacterial surface in host cells is capable of directly recruiting and activating N-WASP with the aid of Cdc42, and the activated N-WASP in turn recruits and activates the Arp2/3 complex [19, 21, 22]. This enables the VirG–N-WASP–Arp2/3 complex formed at one pole of the bacterium to mediate local actin nucleation and elongation (Fig. 8.2). In addition, profilin, which interacts with N-WASP and Arp2/3, and some other host proteins, is required to sustain rapid bacterial movement [23]. In this way *Shigella* acquire a propulsive force in the host cytoplasm. Some motile bacteria impinge on the host cytoplasmic membrane, causing the membrane to protrude and penetrate neighboring cells. The processes allow *Shigella* to move into adjacent epithelial cells by disrupting their double membranes, and they multiply within the cytoplasm of the new cells [24–26] (Fig. 8.1).

8.3
Shigella Infection Elicits an Inflammatory Response

Shigella multiplication in the cytoplasm of the macrophages induces cell death 1–2 h after infection. The macrophage cell death is achieved by two distinctive pathways: one is through activation of caspase-1 by IpaB protein secreted via the TTSS, which in turn leads to the maturation and release of interleukin (IL)-1β [1, 2, 27], and the other pathway is through translocation of lipid A from cytosolic *Shigella* and occurs independently of caspase-1 and Toll-like receptor 4 activity [28]. Thus *Shigella* infection of macrophages leads to severe inflammation of intestinal tissue and *Shigella* infection of epithelial cells induces a strong inflammatory response. During their multiplication in epithelial cells, the bacteria release peptidoglycan (PGN) and a muropeptide composed of the diaminopimelate (DAP)-containing *N*-acetylglucosamine-*N*-acetylmuramic acid dipeptide, which is recognized by Nod1 protein, which results in the formation of the IKK–RICK complex required for activation of NF-κB [29–32]. In this way, invasion of epithelial cells by *Shigella* results in the production of pro-inflammatory chemokines and cytokines, such as IL-1β, IL-6, IL-8 and tumor necrosis factor (TNF)-α, which, in turn, induce neutrophil infiltration of intestinal tissue. The migrating neutrophils exacerbate the infection, because they open up epithelial cell–cell junctions as they move toward the lumen, thereby permitting *Shigella* further access to the basolateral surface of the enterocytes. However, eventually the activated neutrophils efficiently kill *Shigella* and that is assumed to be an important step in resolution of the infection [33–35]. It is therefore argued that *Shigella* possess some activity that modulates or optimizes host inflammatory responses and circumvents bacterial killing in the early stage of infection [36–39].

8.4
Shigella that do not Produce IcsB Undergo Autophagic Degradation

IcsB is one of the effectors secreted by intracellular *Shigella* via the TTSS and has recently been shown to play a pivotal role in their escape from autophagy [40–42]. IcsB is a 52-kDa protein composed of 494 amino acids encoded by the *icsB* gene, which is located upstream from the *ipaBCD* genes on the large 220-kb plasmid of *Shigella* [43, 44]. Our previous study showed that an *icsB*-deleted mutant of *S. flexneri* is fully invasive and capable of escaping from the vacuole, but ultimately becomes deficient in its ability to multiply within host cells [42, 43]. For example, the *icsB* mutant multiplies and moves normally for about 3 h in BHK cells, but intracellular multiplication eventually plateaus 4 h after invasion [42]. At that stage the *icsB* mutant is colocalized with markers for acidic lysosomes (LysoTracker) and autophagosomes [monodansylcadaverine and microtubule-associated protein 1 light chain 3 (LC3)] [45], and the morphology of the mutant is less distinct than that of the wild-type. In the infection stage, around 40% of intracellular *icsB* mutants and 8% of intracellular wild-type *Shigella* are associated with LC3 signals, implying that the *icsB* mutant phenotype with deficient ability to multiply intracellularly is associated with autophagy [42]. Autophagy can be triggered by nutrient-starved conditions and is a response that is highly conserved from yeast cells to mammalian cells. When MDCK cells expressing Green Fluorescent Protein (GFP)–LC3 (MDCK/pGFP-LC3) are infected with the *icsB* mutant or the wild-type under amino-acid-starved conditions, the population of LC3-positive bacteria in the MDCK cells significantly increases in response to amino acid depletion. Conversely, when MDCK cells are treated with known inhibitors of autophagy or lysosomes, such as wortmannin [an inhibitor of phosphatidylinositol-3-kinase (PI3K)], 3-methyladenine (an inhibitor of class III PI3K) or bafilomycin-A1 (Baf-A1; a v-ATPase inhibitor), the LC3-positive *icsB* population decreases markedly. The inability of the *icsB* mutant to circumvent autophagy can also be demonstrated by infecting *atg5* knockout mouse embryonic fibroblasts (*atg5*$^{-/-}$ MEFs), which are defective in autophagy (3.3.1 and [46]). Although the LC3-positive *icsB* mutant is detectable in normal MEF cells (*atg5*$^{+/+}$ MEFs) expressing GFP-LC3, hardly any signals are detected in *atg5* knockout MEFs expressing GFP-LC3 infected with the *icsB* mutant. As expected, intracellular multiplication of the *icsB* mutant is restored to the level of wild type in *atg5*$^{-/-}$ MEFs.

When examined with a thin-section electron microscope, at the initial stage of infection, 1–2 h after infection, the *icsB* mutant, the same as the wild-type, can be seen to be free of the vacuoles and they sometimes possess an actin tail, indicating the presence of motile bacteria. However, 3–4 h after infection the *icsB* mutant is frequently enclosed by lamellar membranous structures in striking contrast to the phagocytic membrane surrounding the invading bacterium. In the infection stage, bacteria enclosed by lamellar membranes are occasionally associated with onion-skin-like structures (Fig. 8.3). At that stage, wild-type *Shigella* generally lack lamellar membranes and most have long actin tails instead. Immunogold electron microscopy with anti-GFP antibody has shown that the lamellar membrane around the *icsB* mutant in MDCK/pGFP-LC3 cells is specifically labeled with immunogold and the onion-skin-like membranous structures

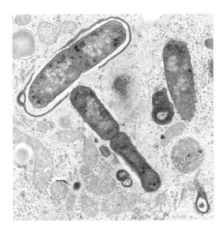

Fig. 8.3 Autophagy induced by the *Shigella icsB* mutant. Electron micrograph of autophagosomes engulfing the *Shigella icsB* mutant. MDCK cells infected with *Shigella icsB* mutant for 4 h were examined by transmission electron microscopy. Bacteria enclosed by lamellar membranes are occasionally associated with onion-skin-like structures.

are also strongly labeled with immunogold. Thus, the intracellular behavior of the *icsB* mutant strongly suggests that *Shigella* that does not produce IcsB readily succumb to autophagy [42].

8.5
Shigella VirG is a Target for Autophagy, but is Camouflaged by its IcsB

Intriguingly, the LC3 and Atg5 signals in the vicinity of intracellular *Shigella*, such as the *icsB* mutant, are occasionally distributed asymmetrically and the signals are accumulated at one pole of the bacterium [16]. Similarly, electron microscopy has revealed that the lamellar membranes associated with the *icsB* mutant are frequently located at one end of the bacterium, suggesting that some *Shigella* surface component(s) is targeted by autophagy. The asymmetric distribution of the autophagic signals on the *icsB* mutant is reminiscent of the asymmetric distribution of VirG on *Shigella* (see above), raising the possibility of VirG being a target for autophagy. When cell lysates prepared from COS-7 cells expressing GFP–Atg5, GFP–beclin, GFP–LC3 or Myc–Vps34 are pulled down with GST–VirGα (the surface-exposed VirG portion) (see above), Atg5 is reproducibly precipitated with GST–VirGα, but none of the autophagic proteins at all is precipitated with GST–IcsB. Indeed, when BHK cells expressing GFP–Atg5 (BHK/pGFP-Atg5) are infected with the *icsB* mutant for 4 h, GFP–Atg5 signals are occasionally confined to one pole of the bacterium, but no association with the Atg5 signals is observed in the BHK/pGFP-Atg5 cells infected with the *virG* mutant. Furthermore, when BHK cells expressing Myc–LC3 (BHK/pMyc-LC3) are infected with an *E. coli* K-12 strain expressing VirG plus GFP together with *virG/icsB* mutant (invasion-positive, autophagy-negative *Shigella* used as the carrier, which allow *E. coli* to move into the host cytoplasm), the Myc–LC3 signals are merged with the VirG signals [42]. These findings strongly indicate that VirG, which is essential to mediating the actin-based motility of *Shigella* (see above), is a target for autophagy and suggest that Atg5 plays a pivotal role as the "seed" in the formation of isolation membranes, which leads to the maturation of autophagosomes.

Both IcsB and Atg5 proteins have been shown to have some ability to bind VirG *in vitro*, and IcsB and Atg5 share the same binding region on VirG. It is noteworthy that in pull-down experiments IcsB shows stronger binding affinity for VirG than Atg5 does, suggesting that IcsB acts as an anti-Atg5 binding protein. Examination of IcsB binding to VirG in a pull-down assay with GST–VirGα in the presence of Atg5–Myc actually showed that Atg5–Myc was precipitated only in the absence of IcsB. The extent of Atg5 binding to VirGα decreased as IcsB was added, supporting the notion that Atg5 binding to VirG is competitively inhibited by IcsB. Attempts to specify the location of the VirG site required for IcsB and Atg5 binding suggested that VirG internal amino acid residues 319–507 are responsible for interacting with both proteins, and that the binding site is located in the C-terminal end of the sequence involved in the interaction with N-WASP [19, 42]. This implies that the IcsB binding to VirG still permits the pathogen to move by mediating actin poly-

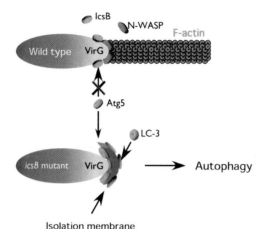

Fig. 8.4 Proposed model for the camouflage against autophagic recognition during *Shigella* infection. In the presence of IcsB (wild type), IcsB binds to VirG and competitively inhibits the binding of Atg5 to VirG. In the absence of IcsB (*icsB* mutant), Atg5 can bind to the VirG accumulated at one pole of the bacterium, and intracellular Shigella are recognized by autophagy. (This figure also appears with the color plates).

merization within the cell's cytoplasm. Thus, the interaction between IcsB and VirG in the vicinity of the bacterial surface seems to be the mechanism of the camouflage that protects against autophagic recognition (Fig. 8.4).

8.6
Perspective

Clearly, *Shigella* possess highly sophisticated systems for adapting to the human intestinal cell environment that allow them to multiply and disseminate. In this chapter we have discussed how *Shigella* use secreted proteins to gain a foothold within the intestinal epithelium that permits them to survive within the host and avoid recognition by the innate immune system, mediated, for example, by autophagy. However, several questions need to be addressed. How and when did *Shigella* acquire the *icsB* gene? Is VirG sufficient to induce autophagy in epithelial cell cytoplasm? Is Atg5 binding to VirG in the cytoplasm alone sufficient to induce autophagy? Further study is needed to answer the questions and should also provide deeper insight into the escape mechanisms of cytosol-invading pathogens from autophagy and the mechanisms underlying autophagy.

References

1 Chen Y, Smith MR, Thirumalai K, Zychlinsky A. A bacterial invasin induces macrophage apoptosis by binding directly to ICE. *EMBO J* 1996, *15*, 3853–3860.

2 Weinrauch Y, Zychlinsky A. The induction of apoptosis by bacterial pathogens. *Annu Rev Microbiol* 1999, *53*, 155–187.

3 Zychlinsky A, Prevost MC, Sansonetti PJ. *Shigella flexneri* induces apoptosis in infected macrophages. *Nature* 1992, *358*, 167–169.

4 Sansonetti PJ. War and peace at mucosal surfaces. *Nat Rev Immunol* 2004, *4*, 953–964.

5 Cossart P, Sansonetti PJ. Bacterial invasion: the paradigms of enteroinvasive pathogens. *Science* 2004, *304*, 242–248.

6 Tran Van Nhieu G, Enninga J, Sansonetti PJ, Grompone G. Tyrosine kinase signaling and type III effectors orchestrating *Shigella* invasion. *Curr Opin Microbiol* 2005, *8*, 16–20.

7 Burton EA, Plattner R, Pendergast AM. Abl tyrosine kinases are required for infection by *Shigella flexneri*. *EMBO J* 2003, *22*, 5471–5479.

8 Lafont F, Tran Van Nhieu G, Hanada K, Sansonetti P, van der Goot FG. Initial steps of *Shigella* infection depend on the cholesterol/sphingolipid raft-mediated CD44–IpaB interaction. *EMBO J* 2002, *21*, 4449–4457.

9 Ohya K, Handa Y, Ogawa M, Suzuki M, Sasakawa C. IpgB1 is a novel *Shigella* effector protein involved in bacterial invasion of host cells: its activity to promote membrane ruffling via RAC1 and CDC42 activation. *J Biol Chem* 2005, *280*, 24022–24034.

10 Yoshida S, Katayama E, Kuwae A, Mimuro H, Suzuki T, Sasakawa C. *Shigella* deliver an effector protein to trigger host microtubule destabilization, which promotes Rac1 activity and efficient bacterial internalization. *EMBO J* 2002, *21*, 2923–2935.

11 Gouin E, Welch MD, Cossart P. Actin-based motility of intracellular pathogens. *Curr Opin Microbiol* 2005, *8*, 35–45.

12 Sinai AP, Joiner KA. Safe haven: the cell biology of nonfusogenic pathogen vacuoles. *Annu Rev Microbiol* 1997, *51*, 415–462.

13 Bernardini ML, Mounier J, d'Hauteville H, Coquis-Rondon M, Sansonetti PJ. Identification of icsA, a plasmid locus of *Shigella flexneri* that governs bacterial intra- and intercellular spread through interaction with F-actin. *Proc Natl Acad Sci USA* 1989, *86*, 3867–3871.

14 Lett MC, Sasakawa C, Okada N, Sakai T, Makino S, Yamada M, Komatsu K, Yoshikawa M. *virG*, a plasmid-coded virulence gene of *Shigella flexneri*: identification of the *virG* protein and determination of the complete coding sequence. *J Bacteriol* 1989, *171*, 353–359.

15 Makino S, Sasakawa C, Kamata K, Kurata T, Yoshikawa M. A genetic determinant required for continuous reinfection of adjacent cells on large plasmid in *S. flexneri* 2a. *Cell* 1986, *46*, 551–555.

16 Goldberg MB, Barzu O, Parsot C, Sansonetti PJ. Unipolar localization and ATPase activity of IcsA, a *Shigella flexneri* protein involved in intracellular movement. *J Bacteriol* 1993, *175*, 2189–2196.

17 Suzuki T, Saga S, Sasakawa C. Functional analysis of *Shigella* VirG domains essential for interaction with vinculin and actin-based motility. *J Biol Chem* 1996, *271*, 21878–21885.

18 Suzuki T, Sasakawa C. Molecular basis of the intracellular spreading of *Shigella*. *Infect Immun* 2001, *69*, 5959–5966.

19 Suzuki T, Miki H, Takenawa T, Sasakawa C. Neural Wiskott–Aldrich syndrome protein is implicated in the actin-based motility of *Shigella flexneri*. *EMBO J* 1998, *17*, 2767–2776.

20 Rohatgi R, Ma L, Miki H, Lopez M, Kirchhausen T, Takenawa T, Kirschner MW. The interaction between N-WASP and the Arp2/3 complex links Cdc42-dependent signals to actin assembly. *Cell* 1999, *97*, 221–231.

21 Suzuki T, Mimuro H, Miki H, Takenawa T, Sasaki T, Nakanishi H, Takai Y, Sasakawa C. Rho family GTPase Cdc42 is essential for the actin-based motility of *Shigella* in mammalian cells. *J Exp Med* 2000, *191*, 1905–1920.

22 Egile C, Loisel TP, Laurent V, Li R, Pantaloni D, Sansonetti PJ, Carlier MF. Activation of the CDC42 effector N-WASP by the *Shigella flexneri* IcsA protein promotes actin nucleation by Arp2/3 complex and bacterial actin-based motility. *J Cell Biol* **1999**, *146*, 1319–1332.

23 Mimuro H, Suzuki T, Suetsugu S, Miki H, Takenawa T, Sasakawa C. Profilin is required for sustaining efficient intra- and intercellular spreading of *Shigella flexneri*. *J Biol Chem* **2000**, *275*, 28893–28901.

24 Rathman M, Jouirhi N, Allaoui A, Sansonetti P, Parsot C, Tran Van Nhieu G. The development of a FACS-based strategy for the isolation of *Shigella flexneri* mutants that are deficient in intercellular spread. *Mol Microbiol* **2000**, *35*, 974–990.

25 Page AL, Ohayon H, Sansonetti PJ, Parsot C. The secreted IpaB and IpaC invasins and their cytoplasmic chaperone IpgC are required for intercellular dissemination of *Shigella flexneri*. *Cell Microbiol* **1999**, *1*, 183–193.

26 Schuch R, Sandlin RC, Maurelli AT. A system for identifying post-invasion functions of invasion genes: requirements for the Mxi–Spa type III secretion pathway of *Shigella flexneri* in intercellular dissemination. *Mol Microbiol* **1999**, *34*, 675–689.

27 Navarre WW, Zychlinsky A. Pathogen-induced apoptosis of macrophages: a common end for different pathogenic strategies. *Cell Microbiol* **2000**, *2*, 265–273.

28 Suzuki T, Nakanishi K, Tsutsui H, Iwai H, Akira S, Inohara N, Chamaillard M, Nunez G, Sasakawa C. A novel caspase-1/toll-like receptor 4-independent pathway of cell death induced by cytosolic *Shigella* in infected macrophages. *J Biol Chem* **2005**, *280*, 14042–14050.

29 Inohara N, Koseki T, del Peso L, Hu Y, Yee C, Chen S, Carrio R, Merino J, Liu D, Ni J, Nunez G. Nod1, an Apaf-1-like activator of caspase-9 and nuclear factor-kappaB. *J Biol Chem* **1999**, *274*, 14560–14567.

30 Inohara N, Nunez G. NODs: intracellular proteins involved in inflammation and apoptosis. *Nat Rev Immunol* **2003**, *3*, 371–382.

31 Girardin SE, Boneca IG, Carneiro LA, Antignac A, Jehanno M, Viala J, Tedin K, Taha MK, Labigne A, Zahringer U, Coyle AJ, DiStefano PS, Bertin J, Sansonetti PJ, Philpott D. J. Nod1 detects a unique muropeptide from gram-negative bacterial peptidoglycan. *Science* **2003**, *300*, 1584–1587.

32 Girardin SE, Tournebize R, Mavris M, Page AL, Li X, Stark GR, Bertin J, DiStefano PS, Yaniv M, Sansonetti PJ, Philpott DJ. CARD4/Nod1 mediates NF-kappaB and JNK activation by invasive *Shigella flexneri*. *EMBO Rep* **2001**, *2*, 736–742.

33 Le-Barillec K, Magalhaes JG, Corcuff E, Thuizat A, Sansonetti PJ, Phalipon A, Di Santo JP. Roles for T and NK cells in the innate immune response to *Shigella flexneri*. *J Immunol* **2005**, *175*, 1735–1740.

34 Brinkmann V, Reichard U, Goosmann C, Fauler B, Uhlemann Y, Weiss DS, Weinrauch Y, Zychlinsky A. Neutrophil extracellular traps kill bacteria. *Science* **2004**, *303*, 1532–1535.

35 Mandic-Mulec I, Weiss J, Zychlinsky A. *Shigella flexneri* is trapped in polymorphonuclear leukocyte vacuoles and efficiently killed. *Infect Immun* **1997**, *65*, 110–115.

36 Kim DW, Lenzen G, Page AL, Legrain P, Sansonetti PJ, Parsot C. The *Shigella flexneri* effector OspG interferes with innate immune responses by targeting ubiquitin-conjugating enzymes. *Proc Natl Acad Sci USA* **2005**, *102*, 14046–14051.

37 Toyotome T, Suzuki T, Kuwae A, Nonaka T, Fukuda H, Imajoh-Ohmi S, Toyofuku T, Hori M, Sasakawa C. *Shigella* protein IpaH(9.8) is secreted from bacteria within mammalian cells and transported to the nucleus. *J Biol Chem* **2001**, *276*, 32071–32079.

38 Okuda J, Toyotome T, Kataoka N, Ohno M, Abe H, Shimura Y, Seyedarabi A, Pickersgill R, Sasakawa C. *Shigella* effector IpaH9.8 binds to a splicing factor U2AF^{35} to modulate host immune responses. *Biochem Biophys Res Commun* **2005**, *333*, 531–539.

39 Haraga A, Miller SI. A *Salmonella enterica* serovar *typhimurium* translocated leucine-rich repeat effector protein inhibits

NF-kappa B-dependent gene expression. *Infect Immun* **2003**, *71*, 4052–4058.

40 Ogawa M, Sasakawa C. Bacterial evasion of the autophagic defense system. *Curr Opin Microbiol*, **2006**, *9*, 62–68.

41 Ogawa M, Sasakawa C. Intracellular survival of *Shigella*. *Cell Microbiol*, **2006**, *8*, 177–184.

42 Ogawa M, Yoshimori T, Suzuki T, Sagara H, Mizushima N, Sasakawa C. Escape of intracellular *Shigella* from autophagy. *Science* **2005**, *307*, 727–731.

43 Ogawa M, Suzuki T, Tatsuno I, Abe H, Sasakawa C. IcsB, secreted via the type III secretion system, is chaperoned by IpgA and required at the post-invasion stage of *Shigella* pathogenicity. *Mol Microbiol* **2003**, *48*, 913–931.

44 Allaoui A, Mounier J, Prevost MC, Sansonetti PJ, Parsot C. *icsB*: a *Shigella flexneri* virulence gene necessary for the lysis of protrusions during intercellular spread. *Mol Microbiol* **1992**, *6*, 1605–1616.

45 Kabeya Y, Mizushima N, Ueno T, Yamamoto A, Kirisako T, Noda T, Kominami E, Ohsumi Y, Yoshimori T. LC3, a mammalian homologue of yeast Apg8p, is localized in autophagosome membranes after processing. *EMBO J* **2000**, *19*, 5720–5728.

46 Kuma A, Hatano M, Matsui M, Yamamoto A, Nakaya H, Yoshimori T, Ohsumi Y, Tokuhisa T, Mizushima N. The role of autophagy during the early neonatal starvation period. *Nature* **2004**, *432*, 1032–1036.

9
Listeria monocytogenes:
A Model System for Studying Autophagy

Kathryn A. Rich and Paul Webster

9.1
Listeriosis

Listeriosis is an economically important bacterial zoonosis that is caused by the intracellular pathogen *Listeria monocytogenes*. It is a food-borne disease that causes severe infections in farm animals and susceptible human hosts. In cows and sheep the disease can cause spontaneous abortion, encephalitis, meningitis and septicemia. Dairy cattle can develop mastitis as a result of a listerial infection. Consumption of foods contaminated with *L. monocytogenes* (e.g. dairy products and especially cheese) can lead to the development of listeriosis in susceptible humans. Such susceptible humans primarily include newborn children, pregnant women, the immunocompromised and the elderly. Consequences of a listeriosis infection in humans are very similar to those exhibited by farm animals and in widespread outbreaks of human listeriosis the mortality rates in susceptible populations can be as high as 30%.

The organism responsible for this disease, *L. monocytogenes*, is a Gram-positive facultative intracellular bacterium. The entry process and subsequent intracellular life in mammalian cells is well-documented (for a recent review, see Dussurget et al., 2004). *In vitro*, *L. monocytogenes* can enter and replicate in a variety of cell lines, including macrophages and epithelial cells. The extent of our knowledge of how this organism interacts with host cells is in part due to the establishment of the organism as a model system to study intracellular parasitism (Cossart and Mengaud, 1989).

9.2
Invasion of Mammalian Cells by *L. monocytogenes*

Many bacteria have evolved mechanisms for intracellular survival after invasion of mammalian cells. Following phagocytosis, bacteria become sequestered in membrane-bound organelles, called phagosomes, which undergo a process of

Autophagy in Immunity and Infection. Edited by Vojo Deretic maturation that involves acidification and several fusion events. The phagocytic pathway is recognized to be an important component of host defense against microorganisms, and many microbes remain in this compartment where they are either are killed and degraded or delivered to lysosomes for subsequent destruction. However, some intracellular pathogens block or alter the maturation of the phagosome and remain in vacuoles that neither acidify nor fuse with lysosomes, while others escape into the cytoplasm (Sinai and Joiner, 1997). Avirulent mutants of these organisms do not generally escape from the phagosome, and are degraded by the phagolysosomal pathway.

Along with *Shigella* and *Rickettsiae*, *L. monocytogenes* belongs to a group of bacteria that are known to replicate in the cytoplasm (Meresse et al., 1999; Andrews and Webster, 1991). *L. monocytogenes* uses the product of the *hly* gene, a thiol-activated hemolysin [listeriolysin O (LLO)], to disrupt the phagosome membrane and enter the cytoplasm. In the cytoplasm, the bacteria use a protein product of the *actA* gene to form comet-shaped tails using host actin (Beauregard et al., 1997; Tilney and Portnoy, 1989). These tails give the cytoplasmic bacteria the ability to move around within the host cell and eventually into adjacent cells. Mutant forms of *L. monocytogenes*, lacking the *actA* gene, have the ability to escape from the phagosome and multiply within host cells. However, they are defective in intra- and inter-cell spread and have reduced virulence *in vivo* (Brundage et al., 1993).

Once the *L. monocytogenes* escape from the phagosome they begin to multiply in the host cell cytoplasm. Tilney et al. (1990) found that treating Listeria-infected J774 cells (a mouse macrophage cell line) with chloramphenicol (a specific inhibitor of bacterial protein synthesis) could inhibit bacterial growth and the process of actin filament polymerization. While examining cytoplasmic *L. monocytogenes* treated with the antibiotic chloramphenicol, we discovered that over time the cytoplasmic bacteria eventually became re-enclosed within vacuoles. Using the model system illustrated in Fig. 9.1, we have demonstrated that this internalization of metabolically arrested cytoplasmic *L. monocytogenes* into vacuoles represents an autophagic process. Our claim was based on the observations that the vacuoles containing the bacteria have double membranes and the process can be inhibited by classic autophagy inhibitors (Rich et al., 2003). In addition, these vacuoles fuse with the endocytic pathway (Fig. 9.2), re-internalized bacteria are eventually degraded (Fig. 9.3) and the time course of internalization is accelerated under starvation conditions (Rich et al., 2003). Our study was the first clear demonstration that cytoplasmic bacteria could be targets for the autophagic pathway. Below, we propose the interactions of *L. monocytogenes* with mammalian cells as an important model system for studying autophagy.

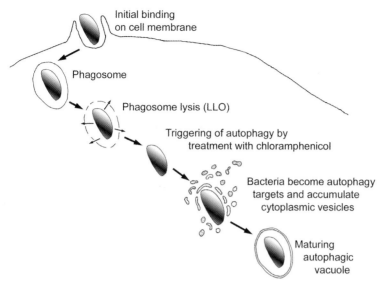

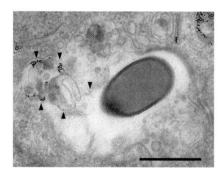

Fig. 9.1 Schematic diagram to illustrate the model system for studying AV formation using *L. monocytogenes* as a cytoplasmic target. Phagocytosed *L. monocytogenes* escape into the cell cytoplasm by secreting LLO.

Treatment of the cells with chloramphenicol arrests the metabolic activity of the bacteria, making them targets for autophagy. Bacteria in AVs are then delivered to lysosomes for degradation.

Fig. 9.2 A transmission electron microscopy (TEM) image of an *L. monocytogenes*-containing autophagosome that appears to be fusing with an endocytic organelle. J774 cells containing cytoplasmic *L. monocytogenes* were incubated for 24 h in the presence of chloramphenicol. The cells were exposed to a suspension of colloidal gold particles

coupled to protein A during the last 30 min of this incubation and then processed for TEM examination. The Protein A–gold particles (arrowheads), taken up by endocytic organelles can be observed in an organelle that is fusing with the *L. monocytogenes*-containing autophagosome. Scale bar = 500 nm.

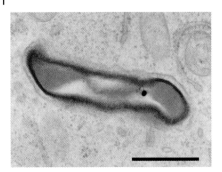

Fig. 9.3 An intracellular *L. monocytogenes* bacterium after 3 days inside a J774 cell. To examine the eventual fate of the *L. monocytogenes* that had entered the autophagic pathway, we incubated cells containing bacteria for 3 days with chloramphenicol. In addition to there being lower numbers of bacteria in the cells than at the start of the time-course, there were many bacterial profiles, such as this one, that were enclosed by a single membrane and appeared to be in the process of being degraded. Scale bar = 500 nm.

9.3
Autophagy

Autophagy is a bulk protein degradation process where cytoplasmic components, including organelles, become enclosed by double-membrane structures termed autophagic vacuoles (AVs) (Yorimitsu and Klionsky, 2005; Yoshimori, 2004). These fuse with the endocytic pathway (Seglen and Bohley, 1992; Dunn, 1990) and their contents are eventually degraded in lysosomes. Morphological and biochemical studies have shown that autophagy is a multistep process. The first step is the sequestration of various organelles (including mitochondria and peroxisomes) and cytoplasmic components into a double-membrane AV. This sequestration can be inhibited by 3-methyladenine (3-MA) (Blommaart et al., 1997b; Seglen and Gordon, 1982), which has been shown to inhibit the class III phosphatidylinositol-3-kinase (PI3K), and by wortmannin, a general inhibitor of PI3Ks (Petiot et al., 2000). AVs then fuse with lysosomes, the inner membrane is degraded and, finally, the sequestered cellular components are also degraded (Punnonen et al., 1993).

The autophagic pathway is understood to be ATP dependent, and is tightly regulated by amino acids and hormones, being dramatically stimulated by starvation (Blommaart et al., 1997a). Autophagy is involved in a number of physiological cellular processes such as lifespan extension and cellular development/differentiation. It may also have a protective role against the progression of some human diseases, including cancer, muscular disorders and neurodegeneration. However, there is evidence that autophagy is involved in some forms of programmed cell death and might contribute to the pathology of some diseases (Cuervo, 2004; Levine and Klionsky, 2004; Shintani and Klionsky, 2004; Bursch, 2001). In addition, the role of autophagy as a cellular defense mechanism against infection by certain pathogenic bacteria and viruses is gaining acceptance (for recent reviews, see Colombo, 2005; Levine, 2005; Mizushima, 2005; Kirkegaard et al., 2004; Shintani and Klionsky, 2004 s).

Prominent gaps remain in our understanding of autophagy in mammalian cells. For example, the origin of the membranes used to assemble AVs is still unresolved. Various sources have been proposed, including the rough endoplasmic reticulum (ER) (Dunn, 1990 a), a post-Golgi compartment (Yamamoto et al., 1990) or a novel compartment (Stromhaug et al., 1998). In addition, there is no definitive information on how cells recognize specific cargo as targets for autophagy, partially a result of the difficulty in identifying newly forming AVs in the cell cytoplasm. The process by which cells are able to identify targets for autophagy was once thought to be a nonselective process. However, it is clear that in some instances cells can selectively target superfluous or damaged organelles and aberrant protein aggregates for degradation (Lemasters, 2005; Reggiori and Klionsky, 2005; Yu et al., 2005; Kissova et al., 2004). Indeed, autophagy is the only mechanism for the turnover of organelles including mitochondria and peroxisomes.

Since autophagy is a degradative pathway, it is not unreasonable to propose that cells might also use this pathway as a defensive mechanism against invading pathogens, in addition to the well-characterized phagolysosomal route. While it is becoming increasingly clear that autophagy has been underestimated as an innate immunity pathway, the term xenophagy has been proposed in order to distinguish use of the autophagy machinery for the degradation of intracellular pathogens from classical autophagy (Levine, 2005). However, throughout this chapter we have continued to use the term autophagy to describe the process by which cells engulf cytoplasmic bacteria into double-membrane organelles.

9.4
The Ideal Target for Studying the Early Stages of Autophagy

Early autophagic events in mammalian cells are difficult to follow, both biochemically and morphologically. Biochemical experiments that look for protein degradation are prone to complications resulting from the fact that there are many degradative processes in cells that target endogenous proteins. Morphologically, the main problem is that early AVs are very similar to the surrounding cytoplasm. Even so, there are many examples in the literature where inspired and creative methods have been used to examine these early structures (Rabouille et al., 1993; Seglen et al., 1986). Examination of late autophagosomes is less problematic as they have the characteristic morphology of a double-enclosing membrane and have unique marker proteins associated with them.

Currently, the earliest stage that can be readily examined in mammalian cells during the AV assembly process is when the Atg8 ortholog microtubule-associated protein 1 light chain 3 (LC3) associates with membranes after it is conjugated to a lipid molecule and post-translationally processed (Mizushima, 2004). LC3 is a general marker for autophagic membranes, and labels both cup and ring-shaped structures by confocal microscopy, probably representing fully formed AVs and bent isolation membranes. Other markers, such as Atg5 and Atg7 could potentially be used to label isolation membranes, but the signal is very transient (Amer and Swanson, 2005; Mizushima et al., 2001).

The identification of the later autophagy stages is possible in mammalian cells because of the unique characteristics associated with this process. Autophagy is generally considered to be occurring if the following criteria are met: (a) if cytoplasmic components are taken into double-membrane vacuoles (Stromhaug et al., 1998), (b) if the process is slowed or stopped using inhibitors of autophagy (Seglen and Gordon, 1982), (c) if the process is stimulated by starvation (Susan and Dunn, 2001; Mortimore et al., 1989) and (d) if vacuoles containing engulfed cytoplasmic material fuse with the endocytic pathway (Liou et al., 1997; Lucocq and Walker, 1997; Punnonen et al., 1993; Tooze et al., 1990).

To make the study of the early events possible requires the availability of an easily identifiable cargo in the cell cytoplasm that can be triggered to become a target for autophagy. A trigger mechanism is essential if synchronous time-courses are to be studied. Such a cargo would have easy access to the cell cytoplasm and remain in the cytoplasm until triggered to become an autophagy target. The cargo should be large enough to apply simple subcellular purification protocols, should have unique morphological features that make it distinct from the surrounding cell cytoplasm and be easily recognizable at the ultrastructural level.

We believe that L. monocytogenes in conjunction with inhibitors of bacterial protein synthesis could be a good model system for studying the early stages of the autophagic process. While it may seem unusual to use bacteria as targets for autophagy, the application of nontraditional targets to study constitutive cellular processes has provided extremely valuable information in other systems. Latex beads have been used as targets for phagocytosis (Defacque et al., 2002; Gagnon et al., 2002; Desjardins et al., 1994), viruses have been important tools for elucidating the mechanisms of the synthetic and secretory pathways (Sodeik and Krijnse-Locker, 2002; Stegmann, 2000; Doms et al., 1993), and protozoans have revealed much about lysosome function (Andrews, 2002).

We performed our initial experiments demonstrating that the internalization of metabolically arrested L. monocytogenes is a form of autophagy using wild-type bacteria. Since it has been established that cell-to-cell spread of wild-type Listeria also results in the formation of double-membrane structures around the bacteria (Tilney and Portnoy, 1989), we needed to carefully evaluate our observations. Although the structures surrounding bacteria that enter cells from adjacent, infected cells are morphologically similar to autophagosomes, they are actually vacuoles consisting of plasma membrane from two cells (Gedde et al., 2000). While internalization of metabolically arrested L. monocytogenes into double-membrane vacuoles was observed with wild-type bacteria (Fig. 9.4), to ensure that we were not examining the unrelated process of cell-to-cell spread, we chose to focus our studies on the actA mutant (Fig. 9.4), the form that multiplies normally within host cells but is defective in actin recruitment and intra- and inter-cellular spread (Brundage et al., 1993).

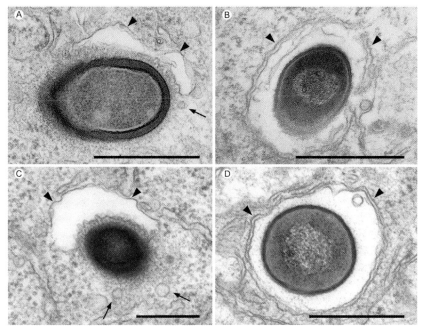

Fig. 9.4 Autophagic vacuoles form around intracellular *wt* and *actA⁻ L. monocytogenes*. Double-membrane structures can form around *wt* (A and B) and *actA⁻ L. monocytogenes* (C and D) 12 h after chloramphenicol treatment. (A) The *wt* bacteria are first observed associated with a large double-membrane structure (arrowheads) that is in close proximity to the bacterial outer membrane. Small vesicles (arrow) are also often observed around the bacteria. (B) The *wt* bacteria are also found completely enclosed within double membrane vacuoles (arrowheads). (C) Cytoplasmic *actA⁻* bacteria also associate with a membrane structure (arrowheads) that is found in close proximity to the bacterial outer membrane. Small vesicles around the bacteria can be seen (arrows). (D) The *actA⁻* bacteria are also enclosed within double-membrane vacuoles (arrowheads). Scale bars = 500 nm.

9.5
Why other Organisms may not be as Useful to Study the Autophagic Process

Recently, a number of studies have provided substantial evidence that stimulation of autophagy can be part of the host response to pathogen invasion, not only to those that access the cytoplasm, but also to those that remain in the phagosomal compartment. The most obvious alternative organism that could be considered as a suitable target for autophagy is *Shigella flexneri* (Ogawa et al., 2005). It is more economically important, being a human pathogen that is endemic worldwide with a broader host pool than *L. monocytogenes*. *S. flexneri* is also able to enter and survive within host cell cytoplasm and, like *L. monocytogenes*, is able to recruit host cell actin to accomplish cell-to-cell spread. VirG which is required for actin-based intracellular motility of *Shigella* bacteria, interacts with the host autophagy protein Atg5, and the VirG–Atg5 interaction is competitively inhibited by IcsB, suggesting a potential mechanism whereby wild-type *Shigella* can avoid autophagy (Ogawa et al., 2005). Prevention of cell-to-cell spread, an essential step if we are to examine bacteria within double-membrane organelles, could be accomplished by deleting virG. However, deletion of this gene may result in the mutant forms of *S. flexneri* becoming unrecognizable by the autophagic pathway when in the host cell cytoplasm. Thus, timed, sequential studies of AV assembly may not be possible using these mutants.

Other organisms that have recently been found to interact with the autophagic pathway include group A *Streptococcus* (GAS), *Mycobacterium tuberculosis, Legionella pneumophila, Brucella abortus, Porphyromonas gingivalis* and *Coxiella burnetti*. Although these findings are important for understanding the pathogenesis of each organism, none of them are suitable on which to base a model system for studying the early stages of AV formation – the reason being that none of these organisms have an extended cytoplasmic stage that can be disturbed by an external trigger such as inhibition of protein synthesis to turn them into autophagy targets.

GAS, upon escaping into the cytosol, are immediately sequestered by autophagosome-like structures which eventually fuse with lysosomes (Nakagawa et al., 2004). *M. tuberculosis* normally resides in a phagosome compartment within macrophages, but if autophagy is stimulated by starvation or rapamycin the mycobacteria are instead transported to lysosomes and degraded (Gutierrez et al., 2004). Similar findings have been reported for *L. pneumophila* (Amer et al., 2005; Amer and Swanson, 2005). Another group of vacuolar pathogens including *L. pneumophila* (Swanson and Isberg, 1995), *B. abortus* (Pizarro-Cerda et al., 1998b; Pizarro-Cerda et al., 1998a), *P. gingivalis* (Dorn et al., 2001) and *C. burnetti* (Gutierrez et al., 2005) appears to infiltrate the autophagic pathway in order to avoid host cell defenses and, in some instances, to gain access to the ER. Once within the ER, all of these bacteria may modify the pathway, perhaps by preventing maturation of autophagosomes, in order to establish an environment necessary for replication and survival. The intracellular route of these organisms is difficult to control so that they can be easily studied. Moreover, recent evidence suggests that *L. pneumophila* manipulate different pathways to gain eventual access to its replicative niche in the rough ER (Kagan and Roy, 2002; Tilney et al., 2001).

In contrast, *L. monocytogenes* is an organism that is not normally associated with the autophagic pathway. Once the bacteria have gained access to the host cell cytoplasm, they are able to continue growing with no apparent damage being inflicted by the host cells. This extended cytoplasmic phase makes it easy to manipulate the bacteria, especially if we use the *actA*-deficient mutant forms that remain trapped inside the cell they invaded. The cells can be incubated with bacteria until all, or most, of the bacteria become cytoplasmic and the cytoplasmic bacteria can be left to multiply within the cell cytoplasm. Most importantly, the cytoplasmic bacteria can be synchronously triggered to become targets for autophagy by treating the infected cells with inhibitors of bacterial protein synthesis. In our initial experiments we used chloramphenicol to inhibit protein synthesis of the cytoplasmic bacteria. However, we have also found that tetracycline treatment will also cause cytoplasmic *L. monocytogenes* to become targets for autophagy.

9.6
Assembly of AVs may Result from Fusion of Cytoplasmic Membrane Structures

Treatment of infected cells with either 3-MA or wortmannin, two established inhibitors of autophagy (Petiot et al., 2000), led to a marked reduction in the percentage of internalized cytoplasmic bacteria (Rich et al., 2003). Interestingly, numerous small vesicular structures were observed around the cytoplasmic bacteria when cells were treated with a low concentration of wortmannin. Although not as obvious, similar observations were made in cultures without inhibitor after about 3 h of chloramphenicol treatment (Tilney and Portnoy, 1989). It is known that the effects of wortmannin are reversible (Blommaart et al., 1997 b). Thus, it is possible that the lower concentration of wortmannin results in a partial inhibition of AV formation and that the small vesicular structures represent an early stage of AV formation. Combined, these observations suggest that AVs may assemble by fusion of small cytoplasmic vesicles around the target destined for autophagic sequestration, rather from any preformed organelle.

Previous studies have described a lipid-rich cup-shaped isolation membrane in the cytoplasm of cells undergoing autophagy, and suggested that double-membrane AVs may form by elongation and closure of this isolation membrane, thus sequestering cytoplasmic components (Mizushima et al., 2001; Seglen et al., 1996). On the other hand, Dunn (1994) proposed that AVs might form by invagination of cisternae from the ER, while Stromhaug et al. (1998) proposed that AVs were derived from preformed structures called phagophores. Our observations, suggesting assembly of AVs by fusion of small cytoplasmic vesicles, could be envisioned to fit with these previously proposed mechanisms of AV formation, simply representing a much earlier stage in the process than has been examined before. Furthermore, our finding that membranes around cytoplasmic bacteria could be immunolabeled with anti-protein disulfide isomerase (PDI) antibodies (Rich et al., 2003) supports the concept that AV membranes may originate from elements of the rough ER. In addition, by demonstrating that the association of Atg7 with *Legionella* vacuoles was reduced when vesicular traffic was inhibited by brefeldin A, Amer and Swanson (2005) provide another example where the ER appears to be the source of AV membranes. However, these data do not provide conclusive proof that the ER membranes are involved in autophagosome formation and more studies are required to elucidate the origin of autophagosome membranes.

The changing composition of the AVs is suggested by our finding that PDI (an accepted ER marker) is absent from fully formed AVs (Rich et al., 2003). This result is consistent with previous suggestions that ER markers may rapidly disappear from AV membranes (Dunn, 1994; Furuno et al., 1990; Reunanen and Hirsimaki, 1983). Rapid loss of ER markers is a process that has been extensively documented in viral systems (e.g. vaccinia virus; Sodeik et al., 1993). The changing composition of AV membranes may also explain the controversy over the origin of AV membranes. It is possible that, with no ability to identify the very beginning of the AV assembly process, investigators have studied AVs in different stages of assembly. It is also possible that AVs are assembled from membranes derived from several sources or that autophagy differs between tissues and cell types (Mizushima et al., 2002; Dunn, 1994).

Interestingly, a similar phenomenon of aggregates of small vesicles forming part of a pre-autophagosomal structure (PAS) has been implicated in yeast autophagy (Kim et al., 2002; Suzuki et al., 2001; Kirisako et al., 1999). Signals that might trigger transport of such cytoplasmic vesicles to sites of AV assembly are unknown at present. However, as a general theme, intracellular membrane trafficking commonly involves the transportation of small vesicles around the cell (e.g. during endocytosis, secretion and receptor recycling as well as during the reassembly of the Golgi and ER following mitosis). A requirement for the actin cytoskeleton in AV formation (Aplin et al., 1992) further suggests that vesicular transport mechanisms may have an important role in autophagy. Interestingly, it has recently been shown in yeast that actin plays a role in selective autophagy, but not in the nonselective, bulk process (Reggiori et al., 2005).

In a recent study, Amer and Swanson (Amer and Swanson, 2005) suggest that biogenesis of the *Legionella* double-membrane replication vacuole forms by homotypic fusion of numerous small vesicles derived from the secretory pathway analogous to the nucleation-assembly elongation process proposed for autophagosome formation in yeast (Noda et al., 2002). However, their model, which requires secretory vesicles to attach to the cytoplasmic face of *L. pneumophila* phagosomes to form a second membrane around the pathogen phagosome should be evaluated in the context of the soluble *N*-ethylmalemide-sensitive fusion protein (NSF) attachment protein/soluble NSF attachment receptor (SNAP/SNARE) hypothesis (for recent review, see Ungermann and Langosch, 2005). The fusion of all vesicles with target membranes is initiated by an ATP-dependent process that involves a v-SNARE on the vesicle membrane binding to a ubiquitous fusion protein, called SNAP25, and an integral membrane protein on the target membrane (t-SNARE). In this way, membranes are fused with each other to form larger membrane structures as part of a complete vesicle. It is unlikely that membranes will fuse to form unstable exposed ends that are not closed to form complete vesicles. The prediction that autophagosome formation during selective autophagy involves expansion of the sequestering membrane would appear to be a more likely scenario since the cargo probably plays an active role in directing membrane elongation (Reggiori and Klionsky, 2005).

9.7
Pathogenic Cytoplasmic Bacteria can Avoid the Autophagic Pathway

Intracellular pathogens are known to modify their environment in multiple ways to avoid destruction by innate host cell defense mechanisms, while exploiting the natural cellular physiology in their favor. The most successful pathogens are those that strongly interfere with host cell functions. Less virulent variants fail to thrive and are presumably overcome by host cell defense mechanisms. It is not surprising that successful pathogens have evolved strategies to avoid autophagy or to actively subvert its components. Bacteria that are normally never found in the cell cytoplasm fail to replicate when microinjected directly into the cytoplasm (O'Riordan and Portnoy, 2002; O'Riordan et al., 2002; Goetz et al., 2001). Other bacteria, able to exploit the cell cytoplasm, thrive by subjugating host cell processes for their own use (O'Riordan and Portnoy, 2002; O'Riordan et al., 2002; Goetz et al., 2001). Ogawa et al. (2005) reported that a *Shigella* mutant lacking the secretory protein IcsB is sequestered by autophagosome-like structures in contrast to the wild-type which can avoid autophagy. They further found that VirG, which is required for actin-based intracellular motility of *Shigella* bacteria, interacts with the host autophagy protein Atg5 and that the VirG–Atg5 interaction is competitively inhibited by IcsB, suggesting a potential mechanism whereby wild-type *Shigella* can avoid autophagy.

L. *monocytogenes* are not normally a target for autophagy – a fact that implies they have an avoidance mechanism. However, the mechanism proposed for *Shigella* cannot be applied to L. *monocytogenes*, because although the actA protein is functionally equivalent to VirG in being required for actin-based motility, we have shown that both the *actA* mutant and the wild-type bacteria are equally capable of avoiding autophagy. The question remains whether all cytoplasmic bacteria have evolved avoidance mechanisms in order to survive in the cytoplasm. We have shown that if cytoplasmic bacteria are prevented from adapting to the intracellular niche (e.g. by antibiotic treatment), this may enable the host cell to access the autophagic pathway as a route for their removal from the cytoplasm and subsequent delivery to the endocytic pathway for degradation. Analysis of such avoidance mechanisms offers a unique opportunity to study autophagy from the perspective of the bacteria.

In order to examine whether the phenomenon observed with *Listeria* represents a general cellular mechanism, we have experimented with a *Bacillus subtilis* strain that has been engineered to express the *hly* gene from *Listeria* (Portnoy et al., 1992; Bielecki et al., 1990). B. *subtilis* is a free-living, Gram-positive bacterium that is nonpathogenic both in humans and animals. When the *hly* gene is present, the bacterium, which can be phagocytosed by macrophages, is able to enter the cell cytoplasm by disrupting the phagosome membrane (Fig. 9.4). Although initially promising in that we were able to place non-pathogenic bacteria into the cell cytoplasm, the experiment revealed something we had overlooked. When the modified B. *subtilis* bacteria entered the cell cytoplasm they immediately became targets for autophagy, a fact revealed by our inability to ac-

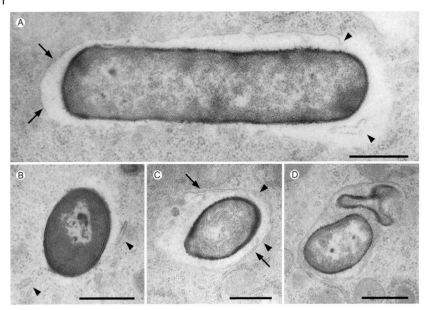

Fig. 9.5 Intracellular B. subtilis mutants that express the hly gene were incubated for 3 h with J774 cells. (A) The B. subtilis bacteria are internalized by J774 cells and are found in many intracellular sites. The B. subtilis, which is approximately 3 times longer than the L. monocytogenes bacteria, can be found partially enclosed within a membrane (arrows). Parts of the membrane appear absent or disrupted (arrowheads). (B) In the same cell population are bacteria that are completely free within the cytoplasm with only small membrane profiles in close proximity (arrowhead). (C) Sometimes the intracellular bacteria are found within vacuoles that have double-membrane profiles in some regions (arrows) with incomplete inner membranes. Some regions of the inner membrane (arrowheads) appear to be in the process of degrading. (D) Other intracellular bacteria are enclosed within double-membrane organelles and have empty or collapsed profiles, suggesting that they are being destroyed. Scale bars = 500 nm.

curately quantify cytoplasmic bacteria. Instead of all the phagocytosed bacteria entering and remaining in the cell cytoplasm, as is the case with L. monocytogenes, the modified B. subtilis could be found in many sites within the cell, even after short incubation times with cells (Fig. 9.5). In addition to finding B. subtilis escaping from phagosomes and being cytoplasmic, we also observed bacteria enclosed within single and double membrane-bounded structures showing various stages of morphological damage (Fig. 9.5), suggesting that they were undergoing digestion by the cell.

These experiments with B. subtilis, a nonpathogenic bacteria, also highlight the differing abilities of different bacteria to avoid autophagy. While L. monocytogenes is capable of actively avoiding autophagy, B. subtilis appears to have no such ability and the host response to these bacteria in the cytoplasm is very rapid. On the other hand, even after L. monocytogenes is metabolically inhibited with chloramphenicol, the time course for the bacteria to become enveloped in an AV is much slower than normal cellular physiological processes, in which the half-life of autophagosomes is only around 10 min (Yoshimori, 2004). However, for *L. monocytogenes* the process takes 4–6 h. The difference between pathogenic and nonpathogenic bacteria and their ability to avoid autophagy is likely to depend on virulence factors. Indeed, a number of pathogenic Gram-negative and Gram-positive bacteria are known to use type III and IV secretion systems to influence host responses (Abe et al., 2005; Christie et al., 2005; Mota and Cornelis, 2005; Segal et al., 2005). Amer and Swanson (2005) demonstrated that a soluble factor from *Legionella* bacteria stimulates macrophages to activate the autophagy pathway for their own benefit.

Although the *hly*-modified forms of *B. subtilis* may not be suitable for experiments that avoid protein synthesis inhibitors such as chloramphenicol, it is possible that this organism may be useful if used with inhibitors of autophagy. For example, if the *B. subtilis* are added to cells in the presence of 3-MA, then the bacteria should accumulate within the host cell cytoplasm. Using a reversible inhibitor such as wortmannin would allow cytoplasmic bacteria to accumulate in the presence of wortmannin. The process of AV formation could be followed when the inhibitor is removed.

While the cytosol was once considered to be permissive for the growth of bacteria, there are now ample examples to show that this is not the case. In addition to intracellular pathogens being susceptible to attack from the autophagic pathway, there is also evidence to suggest the existence of innate cytosolic host surveillance mechanisms (O'Riordan and Portnoy, 2002; O'Riordan et al., 2002). For example, the emerging function of the NOD family of pathogen-recognition receptors is to detect intracellular pathogens or danger signals in general (Inohara et al., 2005; Martinon and Tschopp, 2005). In the continuing battle between pathogen and their hosts, virulence factors produced by the bacteria have been shown to induce or repress the expression of specific host genes (Cohen et al., 2000; Goebel and Kuhn, 2000).

9.8
Cellular Fate of Metabolically Inhibited *L. monocytogenes*

It is well established that the autophagic and endocytic pathways eventually fuse. However, the sequence of events leading to this maturation is not known in detail. Previous reports show that input from the endocytic pathway can converge at different steps of the autophagic pathway (Liou et al., 1997; Lucocq and Walker, 1997; Tooze et al., 1990), but it has not been possible to determine exactly when AVs become fusion – competent with endocytic structures. In order to determine whether the vacuoles containing internalized bacteria fused with the endocytic pathway we examined intracellular bacteria to see if there was an increase in the colocalization with lysosome-associated membrane protein type 1 (LAMP-1), an established marker of late endocytic organelles. After a short incubation with chloramphenicol, few intracellular bacteria were colocalized with the LAMP-1 marker. However, after a 24-h incubation a large proportion of the intracellular bacteria had LAMP-1 associated with them. This suggests that the bacteria were being delivered to endocytic organelles during this 24-h time period.

The time course for *Listeria* to become enveloped in an AV and delivered to lysosomes, which is much slower than normal cellular physiological processes, could reflect the fact that the bacteria are actively fighting the host defensive responses. An interesting parallel is seen in the time course for maturation of *Legionella* bacteria-containing AVs. In macrophages from permissive strains of mice, maturation is much slower than in those from nonpermissive strains (Amer and Swanson, 2005). In macrophages from permissive strains, LC3 is present between 2 and 4 h after infection and the vacuoles do not become LAMP-1-positive until 8 h after infection.

The internalization of *L. monocytogenes* from the cell cytoplasm is a process with characteristics that strongly link it to autophagy. The reason that metabolic inhibition of the bacteria by treatment of the cells with chloramphenicol results in the bacteria becoming targets for autophagy is unknown at present, but understanding the mechanism may shed some light on the general process by which a cell distinguishes cargo destined for the autophagic pathway. The selectivity of the sequestration event has been the subject of some debate as there is evidence for both selective and nonselective uptake of cellular components into AVs (Kopitz et al., 1990; Lardeux and Mortimore, 1987). For example, peroxisomes are selectively removed from the cell by autophagy after cessation of drug treatment (Kondo and Makita, 1997; Luiken et al., 1992) as are mitochondria during apoptosis (Elmore et al., 2001; Xue et al., 2001). Perhaps metabolically inhibited cytoplasmic bacteria are targeted for degradation by the cell in a similar manner to nonfunctional or damaged cellular organelles.

In conclusion, we suggest that carefully designed studies using *L. monocytogenes* as a target for autophagy will eventually assist in revealing the origin of the membranes of AVs and provide a novel approach for directly examining the cellular events of AV assembly. In addition to assisting in the understanding of AV biogenesis, such studies will also help understand how cells are able to clear pathogens from their cytoplasm. These studies may also reveal how some pathogens are able to avoid host cytoplasmic factors that identify the invaders as targets for autophagy.

References

Abe, A., Matsuzawa, T., Kuwae, A. **2005**. Type-III effectors: sophisticated bacterial virulence factors. *C R Biol 328*, 413–428.

Amer, A. O., Swanson, M. S. **2005**. Autophagy is an immediate macrophage response to *Legionella pneumophila*. *Cell Microbiol 7*, 765–778.

Amer, A. O., Byrne, B. G., Swanson, M. S. **2005**. Macrophages rapidly transfer pathogens from lipid raft vacuoles to autophagosomes. *Autophagy 1*, 53–58.

Andrews, N. W. **2002**. Lysosomes and the plasma membrane: trypanosomes reveal a secret relationship. *J Cell Biol 158*, 389–394.

Andrews, N. W., Webster, P. **1991**. Phagolysosomal escape by intracellular pathogens. *Parasitol Today 7*, 335–340.

Aplin, A., Jasionowski, T., Tuttle, D. L., Lenk, S. E., Dunn, W. A., Jr. **1992**. Cytoskeletal elements are required for the formation and maturation of autophagic vacuoles. *J Cell Physiol 152*, 458–466.

Beauregard, K. E., Lee, K. D., Collier, R. J., Swanson, J. A. **1997**. pH-dependent perforation of macrophage phagosomes by listeriolysin O from *Listeria monocytogenes. J Exp Med 186*, 1159–1163.

Bielecki, J., Youngman, P., Connelly, P., Portnoy, D. A. **1990**. *Bacillus subtilis* expressing a haemolysin gene from *Listeria monocytogenes* can grow in mammalian cells. *Nature 345*, 175–176.

Blommaart, E. F., Luiken, J. J., Meijer, A. J. **1997 a**. Autophagic proteolysis: control and specificity. *Histochem J 29*, 365–385.

Blommaart, E. F., Krause, U., Schellens, J. P., Vreeling-Sindelarova, H., Meijer, A. J. **1997 b**. The phosphatidylinositol 3-kinase inhibitors wortmannin and LY294002 inhibit autophagy in isolated rat hepatocytes. *Eur J Biochem 243*, 240–246.

Brundage, R. A., Smith, G. A., Camilli, A., Theriot, J. A., Portnoy, D. A. **1993**. Expression and phosphorylation of the *Listeria monocytogenes* ActA protein in mammalian cells. *Proc Natl Acad Sci USA 90*, 11890–11894.

Bursch, W. **2001**. The autophagosomal-lysosomal compartment in programmed cell death. *Cell Death Differ 8*, 569–581.

Christie, P. J., Atmakuri, K., Krishnamoorthy, V., Jakubowski, S., Cascales, E. **2005**. Biogenesis, architecture, and function of bacterial type IV secretion systems. *Annu Rev Microbiol 59*, 451–485.

Cohen, P., Bouaboula, M., Bellis, M., Baron, V., Jbilo, O., Poinot-Chazel, C., et al. **2000**. Monitoring cellular responses to *Listeria monocytogenes* with oligonucleotide arrays. *J Biol Chem 275*, 11181–11190.

Colombo, M. I. **2005**. Pathogens and autophagy: subverting to survive. *Cell Death Differ 12*, 1481–1483.

Cossart, P., Mengaud, J. **1989**. *Listeria monocytogenes*. A model system for the molecular study of intracellular parasitism. *Mol Biol Med 6*, 463–474.

Cuervo, A. M. **2004**. Autophagy: in sickness and in health. *Trends Cell Biol 14*, 70–77.

Defacque, H., Bos, E., Garvalov, B., Barret, C., Roy, C., Mangeat, P., et al. **2002**. Phosphoinositides regulate membrane-dependent actin assembly by latex bead phagosomes. *Mol Biol Cell 13*, 1190–1202.

Desjardins, M., Huber, L. A., Parton, R. G., Griffiths, G. **1994**. Biogenesis of phagolysosomes proceeds through a sequential series of interactions with the endocytic apparatus. *J Cell Biol 124*, 677–688.

Doms, R. W., Lamb, R. A., Rose, J. K., Helenius, A. **1993**. Folding and assembly of viral membrane proteins. *Virology 193*, 545–562.

Dorn, B. R., Dunn, W. A., Jr., Progulske-Fox, A. **2001**. *Porphyromonas gingivalis* traffics to autophagosomes in human coronary artery endothelial cells. *Infect Immun 69*, 5698–5708.

Dunn, W. A., Jr. **1990**. Studies on the mechanisms of autophagy: formation of the autophagic vacuole. *J Cell Biol 110*, 1923–1933.

Dunn, W. J., Jr. **1994**. Autophagy and related mechanisms of lysosome-mediated protein degradation. *Trends Cell Biol 4*, 139–143.

Dussurget, O., Pizarro-Cerda, J., Cossart, P. **2004**. Molecular determinants of *Listeria monocytogenes* virulence. *Annu Rev Microbiol 58*, 587–610.

Elmore, S. P., Qian, T., Grissom, S. F., Lemasters, J. J. **2001**. The mitochondrial permeability transition initiates autophagy in rat hepatocytes. *FASEB J 15*, 2286–2287.

Furuno, K., Ishikawa, T., Akasaki, K., Lee, S., Nishimura, Y., Tsuji, H., et al. **1990**. Immunocytochemical study of the surrounding envelope of autophagic vacuoles in cultured rat hepatocytes. *Exp Cell Res 189*, 261–268.

Gagnon, E., Duclos, S., Rondeau, C., Chevet, E., Cameron, P. H., Steele-Mortimer, O., et al. **2002**. Endoplasmic reticulum-mediated phagocytosis is a mechanism of entry into macrophages. *Cell 110*, 119–131.

Gedde, M. M., Higgins, D. E., Tilney, L. G., Portnoy, D. A. **2000**. Role of listeriolysin O in cell-to-cell spread of *Listeria monocytogenes*. *Infect Immun 68*, 999–1003.

Goebel, W., Kuhn, M. **2000**. Bacterial replication in the host cell cytosol. *Curr Opin Microbiol 3*, 49–53.

Goetz, M., Bubert, A., Wang, G., Chico-Calero, I., Vazquez-Boland, J. A., Beck, M., et al. **2001**. Microinjection and growth of bacteria in the cytosol of mammalian host cells. *Proc Natl Acad Sci USA 98*, 12221–12226.

Gutierrez, M. G., Master, S. S., Singh, S. B., Taylor, G. A., Colombo, M. I., Deretic, V. **2004**. Autophagy is a defense mechanism inhibiting BCG and *Mycobacterium tuberculosis* survival in infected macrophages. *Cell 119*, 753–766.

Gutierrez, M. G., Vazquez, C. L., Munafo, D. B., Zoppino, F. C., Beron, W., Rabinovitch, M., et al. **2005**. Autophagy induction favours the generation and maturation of the *Coxiella*-replicative vacuoles. *Cell Microbiol 7*, 981–993.

Inohara, N., Chamaillard, M., McDonald, C., Nunez, G. **2005**. NOD-LRR proteins: role in host-microbial interactions and inflammatory disease. *Annu Rev Biochem 74*, 355–383.

Kagan, J. C., Roy, C. R. **2002**. Legionella phagosomes intercept vesicular traffic from endoplasmic reticulum exit sites. *Nat Cell Biol 4*, 945–954.

Kim, J., Huang, W. P., Stromhaug, P. E., Klionsky, D. J. **2002**. Convergence of multiple autophagy and cytoplasm to vacuole targeting components to a perivacuolar membrane compartment prior to *de novo* vesicle formation. *J Biol Chem 277*, 763–773.

Kirisako, T., Baba, M., Ishihara, N., Miyazawa, K., Ohsumi, M., Yoshimori, T., et al. **1999**. Formation process of autophagosome is traced with Apg8p/Aut7p in yeast. *J Cell Biol 147*, 435–446.

Kirkegaard, K., Taylor, M. P., Jackson, W. T. **2004**. Cellular autophagy: surrender, avoidance and subversion by microorganisms. *Nat Rev Microbiol 2*, 301–314.

Kissova, I., Deffieu, M., Manon, S., Camougrand, N. **2004**. Uth1p is involved in the autophagic degradation of mitochondria. *J Biol Chem 279*, 39068–39074.

Kondo, K., Makita, T. **1997**. Inhibition of peroxisomal degradation by 3-methyladenine (3MA) in primary cultures of rat hepatocytes. *Anat Rec 247*, 449–454.

Kopitz, J., Kisen, G. O., Gordon, P. B., Bohley, P., Seglen, P. O. **1990**. Nonselective autophagy of cytosolic enzymes by isolated rat hepatocytes. *J Cell Biol 111*, 941–953.

Lardeux, B. R., Mortimore, G. E. **1987**. Amino acid and hormonal control of macromolecular turnover in perfused rat liver. Evidence for selective autophagy. *J Biol Chem 262*, 365–368.

Lemasters, J. J. **2005**. Selective mitochondrial autophagy, or mitophagy, as a targeted defense against oxidative stress, mitochondrial dysfunction, and aging. *Rejuvenation Res 8*, 3–5.

Levine, B. **2005**. Eating oneself and uninvited guests; autophagy-related pathways in cellular defense. *Cell 120*, 159–162.

Levine, B., Klionsky, D. J. **2004**. Development by self-digestion: molecular mechanisms and biological functions of autophagy. *Dev Cell 6*, 463–477.

Liou, W., Geuze, H. J., Geelen, M. J., Slot, J. W. **1997**. The autophagic and endocytic pathways converge at the nascent autophagic vacuoles. *J Cell Biol 136*, 61–70.

Lucocq, J., Walker, D. **1997**. Evidence for fusion between multilamellar endosomes and autophagosomes in HeLa cells. *Eur J Cell Biol 72*, 307–313.

Luiken, J. J. F. P., Van Den Berg, M., Heikoop, J. C., Meijer, A. J. **1992**. Autophagic degradation of peroxisomes in isolated hepatocytes. *FEBS Lett 304*, 93–97.

Martinon, F., Tschopp, J. **2005**. NLRs join TLRs as innate sensors of pathogens. *Trends Immunol 26*, 447–454.

Meresse, S., Steele-Mortimer, O., Moreno, E., Desjardins, M., Finlay, B., Gorvel, J. P. **1999**. Controlling the maturation of pathogen-containing vacuoles: a matter of life and death. *Nat Cell Biol 1*, E183–188.

Mizushima, N. **2004**. Methods for monitoring autophagy. *Int J Biochem Cell Biol 36*, 2491–2502.

Mizushima, N. **2005**. The pleiotropic role of autophagy: from protein metabolism to bactericide. *Cell Death Differ 12*, 1535–1541.

Mizushima, N., Yamamoto, A., Hatano, M., Kobayashi, Y., Kabeya, Y., Suzuki, K., et al. **2001**. Dissection of autophagosome formation using Apg5-deficient mouse embryonic stem cells. *J Cell Biol 152*, 657–668.

Mizushima, N., Yamamoto, A., Yoshimori, T., Ohsumi, Y. **2002**. Mice with a fluorescent marker for autophagy. *Mol Biol Cell 13*, S133a.

Mortimore, G. E., Poso, A. R., Lardeux, B. R. 1989. Mechanism and regulation of protein degradation in liver. *Diabetes Metab Rev 5*, 49–70.

Mota, L. J., Cornelis, G. R. 2005. The bacterial injection kit: type III secretion systems. *Ann Med 37*, 234–249.

Nakagawa, I., Amano, A., Mizushima, N., Yamamoto, A., Yamaguchi, H., Kamimoto, T., et al. 2004. Autophagy defends cells against invading group A *Streptococcus*. *Science 306*, 1037–1040.

Noda, T., Suzuki, K., Ohsumi, Y. 2002. Yeast autophagosomes: *de novo* formation of a membrane structure. *Trends Cell Biol 12*, 231–235.

Ogawa, M., Yoshimori, T., Suzuki, T., Sagara, H., Mizushima, N., Sasakawa, C. 2005. Escape of intracellular *Shigella* from autophagy. *Science 307*, 727–731.

O'Riordan, M., Portnoy, D. A. 2002. The host cytosol: front-line or home front? *Trends Microbiol 10*, 361–364.

O'Riordan, M., Yi, C. H., Gonzales, R., Lee, K. D., Portnoy, D. A. 2002. Innate recognition of bacteria by a macrophage cytosolic surveillance pathway. *Proc Natl Acad Sci USA 99*, 13861–13866.

Petiot, A., Ogier-Denis, E., Blommaart, E. F., Meijer, A. J., Codogno, P. 2000. Distinct classes of phosphatidylinositol 3'-kinases are involved in signaling pathways that control macroautophagy in HT-29 cells. *J Biol Chem 275*, 992–998.

Pizarro-Cerda, J., Moreno, E., Sanguedolce, V., Mege, J. L., Gorvel, J. P. 1998a. Virulent *Brucella abortus* prevents lysosome fusion and is distributed within autophagosome-like compartments. *Infect Immun 66*, 2387–2392.

Pizarro-Cerda, J., Meresse, S., Parton, R. G., van der Goot, G., Sola-Landa, A., Lopez-Goni, I., et al. 1998b. *Brucella abortus* transits through the autophagic pathway and replicates in the endoplasmic reticulum of nonprofessional phagocytes. *Infect Immun 66*, 5711–5724.

Portnoy, D. A., Tweten, R. K., Kehoe, M., Bielecki, J. 1992. Capacity of listeriolysin O, streptolysin O, and perfringolysin O to mediate growth of *Bacillus subtilis* within mammalian cells. *Infect Immun 60*, 2710–2717.

Punnonen, E. L., Autio, S., Kaija, H., Reunanen, H. 1993. Autophagic vacuoles fuse with the prelysosomal compartment in cultured rat fibroblasts. *Eur J Cell Biol 61*, 54–66.

Rabouille, C., Strous, G. J., Crapo, J. D., Geuze, H. J., Slot, J. W. 1993. The differential degradation of two cytosolic proteins as a tool to monitor autophagy in hepatocytes by immunocytochemistry. *J Cell Biol 120*, 897–908.

Reggiori, F., Klionsky, D. J. 2005. Autophagosomes: biogenesis from scratch? *Curr Opin Cell Biol 17*, 415–422.

Reggiori, F., Monastyrska, I., Shintani, T., Klionsky, D. J. 2005. The actin cytoskeleton is required for selective types of autophagy, but not nonspecific autophagy, in the yeast *Saccharomyces cerevisiae*. *Mol Biol Cell 12*, 12.

Reunanen, H., Hirsimaki, P. 1983. Studies on vinblastin-induced autophagocytosis in mounse liver. IV. Origin of membranes. *Histochemistry 79*, 59–67.

Rich, K. A., Burkett, C., Webster, P. 2003. Cytoplasmic bacteria can be targets for autophagy. *Cell Microbiol 5*, 455–468.

Segal, G., Feldman, M., Zusman, T. 2005. The Icm/Dot type-IV secretion systems of *Legionella pneumophila* and *Coxiella burnetii*. *FEMS Microbiol Rev 29*, 65–81.

Seglen, P. O., Gordon, P. B. 1982. 3-Methyladenine: specific inhibitor of autophagic/lysosomal protein degradation in isolated rat hepatocytes. *Proc Natl Acad Sci USA 79*, 1889–1892.

Seglen, P. O., Bohley, P. 1992. Autophagy and other vacuolar protein degradation mechanisms. *Experientia 48*, 158–172.

Seglen, P. O., Gordon, P. B., Hoyvik, H. 1986. Radiolabelled sugars as probes of hepatocytic autophagy. *Biomed Biochim Acta 45*, 1647–1656.

Seglen, P. O., Berg, T. O., Blankson, H., Fengsrud, M., Holen, I., Stromhaug, P. E. 1996. Structural aspects of autophagy. *Adv Exp Med Biol 389*, 103–111.

Shintani, T., Klionsky, D. J. 2004. Autophagy in health and disease: a double-edged sword. *Science 306*, 990–995.

Sinai, A. P., Joiner, K. A. 1997. Safe haven: the cell biology of nonfusogenic pathogen vacuoles. *Annu Rev Microbiol 51*, 415–462.

Sodeik, B., Krijnse-Locker, J. **2002**. Assembly of vaccinia virus revisited: *de novo* membrane synthesis or acquisition from the host? *Trends Microbiol 10*, 15–24.

Stegmann, T. **2000**. Membrane fusion mechanisms: the influenza hemagglutinin paradigm and its implications for intracellular fusion. *Traffic 1*, 598–604.

Stromhaug, P. E., Berg, T. O., Fengsrud, M., Seglen, P. O. **1998**. Purification and characterization of autophagosomes from rat hepatocytes. *Biochem J 335*, 217–224.

Susan, P. P., Dunn, W. A., Jr. **2001**. Starvation-induced lysosomal degradation of aldolase B requires glutamine 111 in a signal sequence for chaperone-mediated transport. *J Cell Physiol 187*, 48–58.

Suzuki, K., Kirisako, T., Kamada, Y., Mizushima, N., Noda, T., Ohsumi, Y. **2001**. The pre-autophagosomal structure organized by concerted functions of APG genes is essential for autophagosome formation. *EMBO J 20*, 5971–5981.

Swanson, M. S., Isberg, R. R. **1995**. Association of *Legionella pneumophila* with the macrophage endoplasmic reticulum. *Infect Immun 63*, 3609–3620.

Tilney, L. G., Portnoy, D. A. **1989**. Actin filaments and the growth, movement, and spread of the intracellular bacterial parasite, *Listeria monocytogenes*. *J Cell Biol 109*, 1597–1608.

Tilney, L. G., Harb, O. S., Connelly, P. S., Robinson, C. G., Roy, C. R. **2001**. How the parasitic bacterium *Legionella pneumophila* modifies its phagosome and transforms it into rough ER: implications for conversion of plasma membrane to the ER membrane. *J Cell Sci 114*, 4637–4650.

Tooze, J., Hollinshead, M., Ludwig, T., Howell, K., Hoflack, B., Kern, H. **1990**. In exocrine pancreas, the basolateral endocytic pathway converges with the autophagic pathway immediately after the early endosome. *J Cell Biol 111*, 329–345.

Ungermann, C., Langosch, D. **2005**. Functions of SNAREs in intracellular membrane fusion and lipid bilayer mixing. *J Cell Sci 118*, 3819–3828.

Xue, L., Fletcher, G. C., Tolkovsky, A. M. **2001**. Mitochondria are selectively eliminated from eukaryotic cells after blockade of caspases during apoptosis. *Curr Biol 11*, 361–365.

Yorimitsu, T., Klionsky, D. J. **2005**. Autophagy: molecular machinery for self-eating. *Cell Death Differ 12*, 1542–1552.

Yoshimori, T. **2004**. Autophagy: a regulated bulk degradation process inside cells. *Biochem Biophys Res Commun 313*, 453–458.

Yu, W. H., Cuervo, A. M., Kumar, A., Peterhoff, C. M., Schmidt, S. D., Lee, J. H., et al. **2005**. Macroautophagy – a novel beta-amyloid peptide-generating pathway activated in Alzheimer's disease. *J Cell Biol 171*, 87–98.

10

Coxiella burnetii Hijacks the Autophagy Pathway to Survive

Maximiliano G. Gutierrez and María I. Colombo

10.1
Introduction

Phagosomes are transient membrane-bound compartments which result from the internalization of microorganisms or inert particles. The newly formed phagosome interacts with different compartments from the endocytic pathway leading to its maturation into a phagolysosome in which the incorporated material is normally degraded [1]. This maturation process, which involves not only fusion, but also fission events, is in general controlled by the phagocytic cell. However, several pathogens have developed different strategies to hijack host cell functions, hampering this maturation process or avoiding degradation by other means [2].

10.2
Coxiella burnetii

C. burnetii is an obligate intracellular pathogen that in humans causes Q fever [3, 4] characterized by flu-like symptoms and high fever in the acute phase of this disease. The chronic disease eventually leads to endocarditis and hepatitis. Q fever endocarditis is a potentially severe infection, with 24% of fatal cases [5]. The infection occurs mainly by inhalation of contaminated aerosols from natural environments [6]. Due to its low infectious dose and high environmental resistance, the Centers for Disease Control and Prevention considers this pathogen a class B agent of bioterrorism [7]. Alveolar macrophages are the initial targets, but the bacterium subsequently disseminates and replicates in a wide variety of tissues.

A broad range of animals, including most mammals, fish, birds and reptiles, can be infected by *C. burnetii*. Rodents and livestock usually host *Coxiella*, and pathogen infections are especially common in goats and sheep [8]. *In vitro*, the bacteria only replicate after phagocytosis in a wide variety of fibroblast, epithelial

Autophagy in Immunity and Infection. Edited by Vojo Deretic
Copyright © 2006 WILEY-VCH Verlag GmbH & Co. KGaA, Weinheim
ISBN: 3-527-31450-4

and macrophage-like cell lines [9], generating a bacteria-customized compartment which displays certain characteristics of a phagolysosome [9, 10]. This resistance to lysosomal degradation makes *C. burnetii* unique among other intracellular bacteria.

10.3
Bacterium Morphology and Phylogeny

C. burnetii is a nonmotile pleomorphic rod, between 0.3 and 1.0 μm in size, that possesses a membrane similar to that of Gram-negative bacteria. This microorganism is an obligate intracellular acidophil pathogen with a slow doubling time of 8–12 h [8].

C. burnetii was originally included together with *Rickettsia* (Rickettsia-like) in the α-protobacterial subdivision, but phylogenetic studies now consider *C. burnetii* γ-proteobacteria. *C. burnetii* 16S rRNA sequence studies have shown that the bacterium is distantly related to the genus *Legionella* [11]. This is supported by the homologies found between a number of *Coxiella* and *Legionella* genes [12, 13]. Due to this connection, the family *Coxiellaceae*, which includes *C. burnetii*, has been transferred to the order Legionellales, close to the family *Legionellaceae*, which cover the genus *Legionella*. However, *C. burnetii* still shares characteristics with bacteria included in the genus *Rickettsia* such as a small genome, staining by the Gimenez method [14], strict intracellular growth in eukaryotic cells and association with arthropods (a tick host, *Dermacentor andersonii*, has been implicated).

10.4
Lipopolysaccharide (LPS) and Phases

Two phases of the bacteria have been described – a highly virulent phase I and a nonvirulent phase II. This phenomenon, similar to the rough–smooth variation in enterobacteria, is due to a truncated LPS [15, 16]. The *C. burnetii* phase II Nine Mile strain contains a 26-kb deletion in the genome and, although it is avirulent in animals, it effectively infects several cultured cells [17]. This particular antigenic variation of *C. burnetii* is called phase variation. It has been extensively shown that the virulence of *C. burnetii* decreases with the transition from the phase I to the phase II during passaging in chicken embryo yolk sacs, which changes the organism LPS phenotype and causes the truncation of the O polysaccharide chain in their cell wall [18]. Structurally, phase II LPS possesses 2-keto-3-deoxyoctulonosic acid, D-mannose, D-*glycero*-D-*manno*-heptose, lipid A or a lipid A analog and a very complex mixture of fatty acids, many of which are branched [19–23]. Phase I LPS also possesses these components, but, in addition, has virenose, dihydrohydroxystreptose [21, 24, 25] and galactosaminuronyl-(1,6)-glucosamine [24].

LPS represents a major virulence determinant of *C. burnetii* [26]. When isolated from animals or humans, *C. burnetii* expresses on its surface the phase I carbohydrate and is highly infectious. It has been proposed that phase I LPS, with its full carbohydrate structure, blocks the access of antibody to surface proteins [26]. This may explain, at least in part, both the persistence of the bacterium at unknown sites recovered from acute cases of Q fever and lifelong seropositivity. LPS seems to be the only antigen and immunogen differing between phase I and II in *C. burnetii* [19]. This antigenic peculiarity is extremely valuable for the serological differentiation between acute and chronic Q fever. However, some variation in the composition of LPS has been demonstrated [27]. Moreover, 20 different genotypes have been delineated by pulsed-field gel electrophoresis [28] and/or restriction fragment length polymorphism analysis [29].

Phenotypically, phase I Nine Mile differs from phase II Nine Mile *C. burnetii* in that only phase I causes disease and persists in a guinea pig model of infection [30]. Uptake of phase I by monocytes seems to be mediated by the $a_v\beta_3$ integrin, whereas uptake of phase II is mediated both by $a_v\beta_3$ and complement receptor 3. This differential uptake has been implicated in a change of intracellular transport in the host cell [31].

10.5
Genome and Genetics

Although The Institute for Genomic Research has sequenced the *C. burnetii* Nine Mile strain complete genome [32], determining gene function in this organism has been slow, essentially due to the lack of suitable conditions for the axenic growth of the organism, making isolation of mutants impossible. The 1.995×10^6-bp genome comprises a chromosome and four copies of a plasmid of 37.393 bp called QpH1. Four different plasmids have been described for the Nine Mile strain and they vary in their size. Apart from host response, which probably plays a major role in the clinical appearance of the disease, the role of LPS and plasmid type, and the route of infection are not clear. However, some researchers have associated the presence of certain plasmids with virulence, e.g. the plasmid QpH1 was found in strains of *C. burnetii* associated with acute infections, i.e. genome groups I, II and III, while plasmid QpRS was found only in strains belonging to genome group IV, which has been related to strains that tend to produce chronic disease [33–35].

10.6
Host Response and Immunity

Using DNA microarray analysis in THP-1 cells, Ren et al. [36] identified the host transcriptional response to infection by *C. burnetii*. The analysis revealed that the number of upregulated genes and the degree to which their expression is altered during *C. burnetii* infection was relatively low (105 genes). Interestingly, *C. burnetii* infection failed to induce the expression of a cytokine response and the chemokine expression was limited to macrophage-inflammatory protein (MIP)-1 (and MIP-1β). However, *C. burnetii* infection induced the expression of the vacuolar-type (H$^+$)-ATPase (ATP6V1H), and several lysosomal glycosidases including β-hexosaminidase (HEXB) and sialidase (NEU1).

Phase I *C. burnetii* infection causes an increase in transferrin receptor expression in J774A.1 macrophages. Iron is essential for *C. burnetii* replication and it has been proposed that upregulation of transferrin receptor may facilitate the sequestration of this nutrient. Another important consequence of increasing intracellular levels of iron is the enhanced survival, because iron is known to decrease the killing of intracellular pathogens [37].

Interaction of *C. burnetii* with the host immune system is complex and still poorly understood. When infection proceeds through the respiratory route, alveolar macrophages are the primary cells to be infected during acute Q fever. The acute form of Q fever appears to trigger an efficient immune response that eventually limits bacterial replication, but fails, in many cases, to completely clear the pathogen. This failure to eradicate the organism is based on both its ability to grow and multiply within phagolysosomes, and its tendency to establish persistent infections.

Chronic infections are believed to be a result of both immunological reactions and defects [38, 39]. There is evidence suggesting that *C. burnetii* persists in fixed macrophages, and that its intracellular survival is due to the subversion of some macrophage functions and the presence of an unknown mechanism of suppression of cell mediated immunity [40, 41].

Cellular immunity plays a central role in the clearance of the organisms from infected animals [42]. Intact mice clear infection within 14 days, but the organism persists for at least 60 days in athymic mice, despite the production of antibodies [43].

The expression of antigens specific to *C. burnetii* in the membranes of infected host cells [44] supports the hypothesis that infected cells are detected by the immune system and lysed by antibody-dependent cellular cytotoxicity by monocytes and other effector cells [45]. Released organisms are then vulnerable to attack by activated macrophages. As there is greater expression of *C. burnetii* antigens by cells infected with acute isolates of the pathogen, the weaker and less immunologically visible expression of antigens by chronic strains may be a factor in persistent infections [46].

The differences in the length of the LPS molecule are also involved in the dendritic cell (DC) response to *C. burnetii* infection. Phase I *C. burnetii* infect and grow within DCs without inducing the maturation and activation of these cells. In contrast, phase II bacteria induce maturation and activation of DCs. This outcome has been attributed to the presence of full-length LPS, which acts as a shielding molecule masking other bacterial surface proteins critical for induction of dendritic cell maturation. Moreover, these studies suggest that lack of DC maturation after infection by phase I *C. burnetii* results in an immune response that lacks the potency required for complete clearance of the organism, thereby allowing persistence [47].

It has been shown that interferon (IFN)-γ limits the *in vitro* multiplication of *C. burnetii* and enhances the killing of the organism through tumor necrosis factor (TNF)-mediated apoptosis [41, 48]. Recent work indicates that IFN-γ induces apoptosis in infected monocytes [48]. Nitric oxide, whose generation is upregulated by IFN-γ and TNF-a, inhibits the multiplication of the organism *in vitro* [49, 50]. At the cellular level, IFN-γ is thought to mediate the killing of *C. burnetii* through the alteration of conditions within the phagosomes [51].

10.7
Developmental Cycle of *C. burnetii*:
Small Cell Variant (SCV) and Large Cell Variant (LCV)

Coxiella presents a pleomorphic nature during its replication in the large intracellular parasitophorous vacuole (Fig. 10.1 A) [52]. Two major morphological cell types have been described, i.e. the SCV and the LCV, which can be separated by differential centrifugation [3]. By electron microscopy, the SCV presents a compact rod-shaped aspect, with a very dense central region corresponding to condensed nucleoid filaments surrounded by ribonucleoproteins (Fig. 10.1 B–D). In contrast, the LCVs are larger and elongated less electron-dense bacteria, with the nucleoid dense filaments radiating from the central region into the more transparent cytoplasm (Fig. 10.1 E–G). These two forms are not only morphologically, but also functionally distinct – the SCV is considered the metabolically dormant, less replicating, but environmentally stable cell variant, whereas LCVs are the metabolically active and replicating large bacteria [3]. The relationship between the *Coxiella* cell variants has not been fully elucidated. SCVs are now thought to be the infective forms and the form responsible for surviving through prolonged periods in the environment [53], although spore-like particles have been described ultrastructurally.

A few proteins that are differentially expressed by the SCV and LCV have been identified. Elongation factors EF-Tu and EF-Ts, the stationary-phase sigma factor RpoS, and a protein with porin activity, termed P1, are preferentially expressed by LCVs [54–56]. In contrast, two highly basic proteins specific to the SCV, ScvA and Hq1, DNA-binding proteins that likely play roles in chromatin condensation, have been found [57]. When *C. burnetii* variants are separated on a density gradient, only the denser SCVs are labeled by an antibody to ScvA, indicating that detection of this protein can be used to distinguish the SCVs from the LCVs [58].

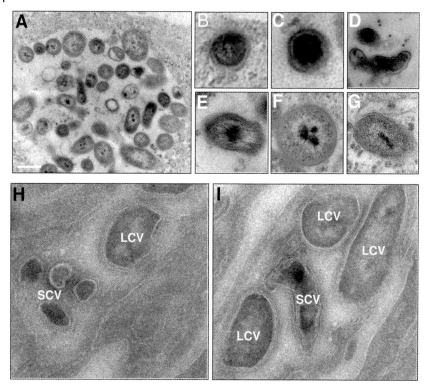

Fig. 10.1 *C. burnetii* is a pleomorphic bacterium that replicates in a large intracellular parasitophorous vacuole. (A) HeLa cells were infected with *C. burnetii* phase II for 12 h and processed for transmission electron microscopy following standard procedures. A typical large *Coxiella*-containing vacuole showing the pleomorphic bacteria is depicted. (B–D) The small cell variant (SCV) is shown. (E–G) Images show the large cell variant (LCV). (H and I) Cryosectioning images showing the SCV and LCV *Coxiella* forms in cells pre-incubated for 2 h before the infection in full nutrient media (H) or in starvation media (I). Panel (I) shows a larger proportion of LCV than in panel (H).

10.8
C. burnetii Type IV Secretion System

Many bacterial microorganisms use secretion systems as a mechanism of pathogenesis. One of the best studied systems is the type III secretion system, a device used by bacterial pathogens such as *Salmonella enterica* and *Shigella flexneri* [59]. An increasing number of bacterial pathogens are being found to have type IV secretion systems ([13], reviewed in Ref. [60]) which are used to subvert host defenses. The type IV systems have homology to plasmid transfer systems but they have been adapted as a protein export apparatus. Several intracellular pathogens such as *Legionella pneumophila*, *C. burnetii*, *Brucella abortus* and *Helicobacter pylori* possess this specialized type of secretion apparatus. Since a growing number of diverse sets of type IV genes are being reported, the type IV secretion system has been divided in two – type IVA and type IVB [61]. Type IVA systems are highly homologous to the *virB* operon of the plant pathogen *Agrobacterium tumefaciens* [62]. Type IVB has extensive homology to the transfer system of Incl plasmids. Interestingly, *L. pneumophila* possesses both types of secretion systems: the Lvh complex and the Dot/Icm (defective for organelle transport/intracellular multiplication) system that is absolutely required for virulence [63]. These genes assembly as a translocation channel (reviewed in Ref. [64]) that delivers proteins and single-stranded DNA from the bacterial cell into the host cell. Therefore, these devices are believed to be key elements in determining the intracellular fate and the replication capability of the pathogens via the secretion of effector molecules. However, for many of these systems the actual proteins exported by the apparatus have not been identified.

As indicated above, genomic sequence data indicate that *C. burnetii* has many gene products with strong homology to *L. pneumophila* genes [12]. Among them 21 genes are similar to components of the *Legionella* Dot/Icm type IV secretion system. In contrast to *L. pneumophila*, where the Dot/Icm proteins are located on two distinct pathogenicity islands, most of the *C. burnetii* Dot/Icm genes are located in a contiguous DNA fragment. It has been shown that some of these genes can actually restore the growth of *Legionella* mutants deficient for *dot/icm* gene products [65, 66]. These results clearly indicate that the *Coxiella dot/icm* related genes encode a functional type IV secretion system. Interestingly, four *C. burnetii* Dot/Icm proteins can substitute for their homologs in *L. pneumophila*, but six other proteins cannot [65, 66].

By an undefined mechanism *Coxiella* generates large, spacious vacuoles in which the bacteria replicate [3]. It is believed that the *Coxiella* type IV secretion system plays a critical role in the biogenesis of the *Coxiella* replicative niche (see below). However, since *C. burnetii* is an obligate intracellular pathogen, genetic manipulations are very difficult to perform and thus there is very limited information available about its pathogenesis systems.

10.9
Interaction with the Endocytic and Autophagic Pathways

C. burnetii can be grown *in vitro* in numerous cell types including macrophage-like cells and fibroblast-like cells. This obligate intracellular pathogen is sheltered in a replicative vacuole (RV) which manifests characteristics of mature phagolysosomes (see Tab. 10.1) [9, 57]. After entry into the host cell, *C. burnetii* is localized in early phagosomes that fuse homo- and heterotypically with other vesicles to form the large RV where this pathogen multiplies. In contrast to other pathogens, these RVs appear not to interact with Golgi-derived vesicles [57]. However, vacuoles containing *C. burnetii* are able to fuse with other vesicles of the phagocytic–endocytic system [67, 68]. The fusogenic properties of the *Coxiella* RV are likely one of the major features that may contribute to its spa-

Table 10.1 Intracellular markers associated to *C. burnetii* RVs generated at least 24 h post-infection

Markers	RVs	References
Early endosomes		
Rab5	yes	73
EEA-1	no	81
Late endosomes		
Rab7	yes	73
Lysosomal		
LAMP-1	yes	3, 81
LAMP-2	yes	3
V H$^+$-ATPase	yes	51, 57
5′-nucleotidase	yes	103
cathepsin D	yes	51, 57
acid phosphatase	yes	40, 92
CD63	yes	51
Acridine Orange	yes	57
CI-M6PR	no	57
Autophagic		
MAP-LC3	yes	73, 77
Rab24	yes	77
MDC	yes	73
Golgi		
sphingolipid	no	57

ciousness. Inert particles such as zymosan particles and latex beads phagocytosed by infected Chinese hamster ovary (CHO) cells are transferred to the lumen of vacuoles that shelter *C. burnetii*. The first report of colonization of the vacuoles of *C. burnetii* by *L. amazonensis* in a mammalian cell supports the high fusogenicity of the RVs. This study opened up numerous coinfection experiments in the search of pathogens or host-derived signals regulating intracellular transport. From this point of view, RVs can be useful to analyze the behavior of different intracellular pathogens within a lysosomal milieu. Interestingly, in

Fig. 10.2 *C. burnetii* RVs interact with the autophagic pathway. (A) Stably transfected CHO cells overexpressing GFP–LC3 (green) were infected with *C. burnetii* phase II for 48 h. Cells were fixed and the *Coxiella* (red) were detected by indirect immunofluorescence with a specific antibody. (B) CHO cells were infected as indicated in (A), fixed and subjected to indirect immunofluorescence to detect endogenous LC3 (red) using a specific antibody. (C) Stably transfected CHO cells overexpressing GFP–Rab7 (green) were infected with *Coxiella* and processed as indicated in (A). (D) HeLa cells were infected with *C. burnetii* phase II for 72 h, fixed and subjected to indirect immunofluorescence to detect endogenous Rab24 (red) using a mouse polyclonal antibody against Rab24. Confocal images are depicted. (E) A model showing the interaction between the *Coxiella*-containing vacuole and the autophagic pathway. (This figure also appears with the color plates).

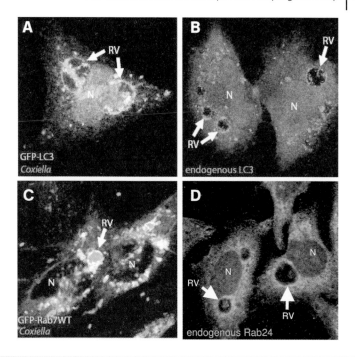

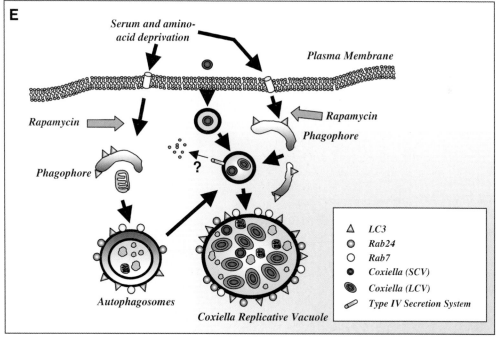

cells coinfected with *C. burnetii* and *M. avium* or *M. tuberculosis* H37Rv, the majority of mycobacteria colocalize with the lysosomal-like RVs harboring *C. burnetii* [69, 70]. This study shows that whereas *M. avium* can growth in the *C. burnetii* RVs, *M. tuberculosis* H37Rv growth appears to be inhibited. Thus, the *Coxiella* RV appears to override the regulatory mechanisms that normally stall maturation of parasitophorous vacuoles harboring *M. avium* or *M. tuberculosis* at an early endosome stage [71, 72].

We have shown that this bacterial customized RV also has the hallmarks of an autophagosomal compartment [73]. Cells infected for 48 h with *C. burnetii* phase II generate *Coxiella*-rich large vacuoles which can be easily distinguished by phase contrast microscopy. These parasitophorous vacuoles were found to accumulate the autophagosome marker monodansylcadaverine (MDC) [74, 75]. Interestingly, clusters of MDC-labeled vesicles were also observed in close proximity, suggesting that these vesicles fuse with *Coxiella*-containing compartments and contribute to the generation of the large RV [73]. Furthermore, the *Coxiella* RVs were decorated with the specific autophagosomal protein microtubule-associated protein 1 light chain 3 (LC3). Figure 10.2(A) shows a CHO cell overexpressing Green Fluorescent Protein (GFP)–LC3 with a typical *Coxiella* RV clearly labeled by the protein. Interestingly, not only GFP–LC3, but also the endogenous protein decorates the membrane of *Coxiella* RVs (Fig. 10.2 B), indicating that this localization is not due to protein overexpression. This observation clearly confirms the autophagic nature of the RV.

Work from our laboratory has shown that the small GTPase Rab24 changes its distribution from a reticular to a vesicular pattern after shifting to conditions that induce autophagy, such as amino acid deprivation [76]. Furthermore, GFP–Rab24 decorated vesicles were labeled with the autophagosome marker MDC and partially colocalized with the specific autophagosomal protein LC3, suggesting that under starvation conditions Rab24 is targeted to autophagic vacuoles (AVs).

Since LC3 colocalizes significantly with the *Coxiella* RVs, we wondered if Rab24 would also label this compartment. We showed that in CHO cells stably transfected with GFP–Rab24 and infected with *C. burnetii* some of the *Coxiella*-containing vacuoles were decorated by GFP–Rab24, indicating that Rab24 is also recruited to *Coxiella* RVs [77]. However, the degree of colocalization of Rab24 with RVs was not as evident as in the case of LC3, although colocalization was consistently observed at different post-infection times. Furthermore, we have also shown that both LC3 and Rab24 decorate the same *Coxiella* RVs. These results clearly indicate that the limiting membrane of the *C. burnetii* RVs is decorated by proteins normally localized to AVs, confirming the link with this pathway.

Endocytosis and autophagocytosis are not separated pathways. Indeed, endocytosed material can be localized within AVs in mammalian cells [78, 78]. In our laboratory we have produced evidence that Rab7, a late endosomal–lysosomal marker, is a component of the autophagic pathway since this protein was associated with AVs [79]. Furthermore, overexpression of a Rab7 dominant-negative mutant led to the accumulation of large autophagosomes, likely because this mutant hampered fusion with lysosomes, indicating that Rab7 is involved in the maturation of autophagosomes [79, 80]. When cells overexpressing Rab7wt were infected with *C. burnetii*, the generated vacuoles harboring *Coxiella* were labeled by this Rab protein [73] (Fig. 10.2 C). Thus, we have extended the number of markers that label the *Coxiella* RV as indicated in Tab. 10.1, pointing out that *Coxiella* resides in a unique pathogen customized compartment.

10.10
Contribution of Autophagy to RV Development

At 24 h post-infection *Coxiella* is contained in small vesicles, whereas at 48 h the bacteria are found in very large vacuoles, although only a few bacteria may be observed per vacuole. In contrast, by 96 h the *C. burnetii* RV is almost completely filled with bacteria, indicating a very fast replication rate between 2 and 3 days [50, 81]. This indicates that *C. burnetii* exhibits a growth cycle typical of a closed bacterial system with lag, exponential, and stationary phases. Lag phase extended to approximately 2 days post-infection and was composed primarily of SCVs differentiating to LCVs. Exponential phase occurred over the next 4 days, with parasitophorous vacuoles harboring replicating LCVs almost exclusively [82]. The synchronous infection model described in this study also allowed an initial analysis of developmentally regulated gene expression in *C. burnetii*. All genes tested, with the exception of *scvA*, demonstrated their highest expression levels during midexponential phase (3 days post-infection). *De novo* expression of *scvA* was evident at 3 days post-infection, with expression levels increasing throughout the stationary phase.

As indicated above, the vacuole that shelters *Coxiella* displays some of the features of an autophagic compartment. In order to begin to study the role of autophagy in the generation of the *Coxiella* parasitophorous vacuole, we have analyzed the development of the RVs under different experimental conditions. It is known that the activity of a member of the class III phosphatidylinositol-3-kinase (PI3K) family is required for the initial stages in the autophagic pathway [85]. Wortmannin and 3-methyladenine (3-MA) inhibit autophagy by blocking PI3K activity [83]. Therefore, to investigate the involvement of autophagy in *C. burnetii* growth, we tested the effect of wortmannin and 3-MA on the development of the large vacuoles. Cells were first infected with *C. burnetii* for 12 h and subsequently incubated with the PI3K inhibitors. Both compounds blocked *Coxiella* vacuole formation, suggesting that the autophagic pathway has a critical role in RV development and bacterial replication [73].

We next analyzed the overall effect of inducing autophagy on the generation of the *Coxiella* RV and bacterial growth. Autophagy can be regulated by the level of specific amino acids ([84], for a review see Ref. [85]), thus we tested the effect of amino acid deprivation on infection. Our results indicate that incubation of cells under starvation conditions (amino acid- and serum-free medium) for different periods of time prior infection increased the percentage of infected cells and the size and development of RVs [77]. The enhancing effect of starvation on infection was blocked by the autophagy inhibitors wortmannin or 3-MA, consistent with the results previously observed [73].

Autophagy can be activated not only physiologically (e.g. by starvation), but also pharmacologically by treatment with rapamycin, an inhibitor of target of rapamycin (TOR), a critical kinase involved in the autophagic pathway ([86], reviewed in Refs. [87, 88]). Rapamycin inhibition of TOR markedly enhanced the percentage of cells containing RVs at 12 h post-infection. Interestingly, *Coxiella* survival was also affected by conditions which increase autophagy. Both starvation and rapamycin markedly enhanced *C. burnetii* replication and viability as determined by an infectious focus-forming units assay [77], and this effect was overridden by treatment with wortmannin and 3-MA. Taken together these results indicate that inducing autophagy in the host cell favors *C. burnetii* growth.

To identify the autophagy factors that modulate the generation of *Coxiella* RVs at the molecular level, cells overexpressing proteins involved in the autophagy pathway were infected with *C. burnetii* phase II. The autophagosomal protein LC3 is recruited to the sequestering membrane, and is located in the inner and outer membranes of autophagosomes [89]. LC3 binding to the autophagosomal membrane depends on post-translational modifications which involves C-terminal cleavage of the protein and conjugation with phosphatidylethanolamine at the free Gly residue generated [89]. Replacement of the amino acid Gly120 by Ala yields a $LC3^{G120A}$ mutant that cannot bind to autophagosomes and remains cytosolic. Therefore, we studied the behavior of *C. burnetii* in stably transfected CHO cells overexpressing wild-type GFP–LC3 or the mutant $LC3^{G120A}$. Similarly, we studied pathogen fate in cells overexpressing GFP–Rab24wt or GFP–$Rab24^{S67L}$, a mutant that also remains cytosolic. As mentioned above, during the first 12 h post-infection *Coxiella* resides in small vesicles and the large RVs are generated only after 48 h. Interestingly, in cells overexpressing either GFP–LC3wt or Rab24wt, large *Coxiella* RVs were already formed at 12 h after infection, whereas in cells overexpressing the mutants both the infection and the size of RVs were drastically reduced [77]. These results clearly indicate that overexpression of proteins involved in the autophagic pathway remarkably accelerates the development of the compartment where *C. burnetii* replicates, suggesting that components of the cellular machinery of autophagosome formation are subverted to promote *C. burnetii* replication and differentiation.

As indicated above, Rab7 also decorates the *Coxiella* vacuole [73]. Interestingly, overexpression of a dominant-negative Rab7 mutant altered the development of the RVs ([73]; Romano et al., submitted). These results suggest that Rab7, another protein connected to the autophagic pathway, participates in the biogenesis of the *C. burnetii*-containing vacuoles.

Another interesting observation is that infection of CHO cells overexpressing GFP–LC3 or GFP–Rab24 with *Coxiella* causes a remarkable redistribution of these chimeric proteins, with the generation of a large number of small GFP decorated vesicles similar to those found in the course of autophagy. This observation suggests that *Coxiella* induces autophagy and, as a consequence, LC3 and Rab24 are targeted to autophagy vacuoles distributed throughout the cytoplasm [76, 89]. This redistribution of autophagic proteins has also been previously observed in cells infected with *Porphyromonas gingivalis* where the infection induced a change in the distribution of the autophagic protein Atg7 [90]. This observation raises the question of whether *Coxiella* itself activates autophagy or if this is a generalized cell response upon pathogen infection. Indeed, evidences indicate that, like other intracellular pathogens, *C. burnetii* protein synthesis is required for maturation of its RV [81]. Treatment of cells containing large and spacious RVs with chloramphenicol, an antibiotic that inhibits bacterial protein synthesis, causes the collapse of the vacuole. Thus, it is possible that *Coxiella*, through the injection of bacterial effector proteins via the type IV secretion system (see model), may activate autophagy generating the large replicative niche and a more permissive environment for bacterial replication.

All this evidence suggests that *Coxiella* benefits from the autophagic pathway [91]. However, many questions remain unanswered. Does *C. burnetii* follow the classical phagocytic pathway acquiring the early endosomal markers Rab5 and EEA1 at early times after infection? If that is the case, when are the autophagosomal markers acquired?

We have begun to address these questions by carefully analyzing the development of the *Coxiella*-containing phagosomes at different times post-infection. Results from our laboratory using CHO cells infected with *C. burnetii* Phase II indicate that indeed the *Coxiella* phagosome sequentially acquires both Rab5 and Rab7 and also the early endosomal protein EEA1 between 20 and 40 min post-infection (Romano et al., submitted). By later times (60 min post-infection), the majority of the *Coxiella* phagosomes have acquired the lysosomal enzyme cathepsin D, indicating that *Coxiella* transits the phagocytic pathway and recruits maturation markers. However, our kinetics studies indicate that the *Coxiella*-containing phagosomes do not acquire the lysosomal enzymes immediately after internalization, suggesting that there is a delay in the fusion with the lysosomal compartment (Romano et al., submitted). This is consistent with the observation by Howe and Mallavia, indicating that viable Phase I *C. burnetii* delay phagosome–lysosome fusion early during infection in J774 macrophages [92]. In this report the authors have shown that at 1 h post-incubation, 31% of the *Coxiella*-containing vacuoles were labeled with the lysosomal enzyme acid phosphatase, while inactivation of *Coxiella* resulted in an increase in the percentage of colocalization with lysosomal markers, indicating that the bacteria actively delays lysosomal fusion. It has also been shown that virulent *Coxiella* are able to survive in monocytes by altering phagosome maturation and preventing the acquisition of cathepsin D [51].

10.11
Autophagy and Bacterial Differentiation from SCVs to LCVs

Another important question is why induction of autophagy accelerates the development of RVs. What are the factors released by the bacteria that modulate host cell function, hijacking the autophagic pathway? What signals ultimately lead to bacterial differentiation and replication inside the cells? As described above, *C. burnetii* is a pleomorphic bacterium that presents two morphologically distinct forms known as SCVs and LCVs. We have tested the hypothesis that modulation of autophagy affects the intravacuolar differentiation of *C. burnetii*. For this purpose we analyzed infected HeLa cells by transmission electron microscopy to differentiate between the LCV and SCV forms, based on the morphological criteria described above. Interestingly, cells subjected to a condition that stimulates autophagy prior to the infection showed a higher number of LCVs at 24 h post-infection compared to cells pre-incubated in full medium (Fig. 10.1 H and I). A quantitative analysis of SCV and LCV forms present in the RV indicated that, in cells incubated under starvation conditions, the percentage of LCVs increased significantly from 50 to 70% [93]. Studies of the morphological differentiation under different physiological and pharmacological conditions that modulate autophagy are currently under way in our laboratory. We postulate that *C. burnetii*, after its internalization, injects effector molecules across the phagosomal membrane via type IV secretion, which stimulates the autophagic pathway in the host cell. The availability of nutrients provided by fusion with the newly generated autophagosomes may trigger specific signals required to induce the differentiation from SCVs to LCVs (Fig. 10.2 D).

As mentioned previously, *C. burnetii* survives and replicates within large, acidified, phagolysosome-like vacuoles. This bacterium has efficiently adapted to survive in this harsh environment, but the mechanism of resistance to acid hydrolases is largely unknown. Indeed, *C. burnetii* requires the low pH found within the phagolysosome to activate its metabolism [94]. *Coxiella* replication can be blocked by raising the phagolysosomal pH with chloroquine [40]. Moreover, when Bafilomycin A, the specific inhibitor of the V-ATPase, responsible for acidifying intracellular compartments, is added to the medium of *C. burnetii*-infected cells the growth of *C. burnetii* was completely inhibited [57]. These features argue that *C. burnetii* is unique among intracellular bacteria in requiring an acidic pH for development. It is noteworthy that *in vitro* incubation of isolated *C. burnetii* samples in an acidic buffer with a pH 5.5 caused a 42% decrease in the labeling of ScvA, a protein that decreases when the bacteria transform to the LCV form [92]. However, this decrease was less marked than the decrease in ScvA label seen *in vivo* during the same time period, suggesting that other factors beside the pH are responsible for the bacterial differentiation. Interestingly, the decrease in ScvA protein labeling did not occur at a lysosomal pH of 4.5, suggesting that *Coxiella* differentiation requires a moderate acidic pH and, more importantly, that the transformation is initiated in a compartment prior to fusion with lysosomes. Also protein synthesis increases in *C. burnetii* incubated at pH 5.5 (late endosomal pH) compared to that observed at pH 4.5. These results suggest that *C. burnetii* may initially prefer a slightly less acidic compartment, perhaps with characteristics of a late endosomal/autophagosomal compartment. Indeed, this is consistent with the observation that phase I *C. burnetii* delays the fusion with lysosomes at least during the first 2 h of infection [92]. This is also in agreement with our observations that *Coxiella* phagosomes do not colocalize with cathepsin D during the first 20–40 min after internalization (Romano et al., submitted).

In summary, we hypothesize that the bacterial morphological changes and its replication would depend not only on pH, as indicated above [92, 94], but also on the accessibility to nutrients [82] which are likely provided by interaction with the autophagic pathway. It is possible that *Coxiella* is diverted to the autophagic pathway to enable differentiation in a variant form that can tolerate the harsh environment of the lysosomal compartment as proposed by others [95]. Consistent with this idea, it has been suggested that the *C. burnetii* transition from SCVs to LCVs begins in a phagosomal compartment prior to fusion with the lysosomes [92]. However, it might be beneficial for the bacteria to reside in an autophagic-like vacuole that represents a continuous source of metabolites for bacteria intracellular growth and development [93].

In our current model (Fig. 10.2 D) we postulate that the development of the *Coxiella* replicative compartment is modulated by circumstances that regulate autophagy, favoring *Coxiella* replication. Our evidence also suggests that autophagic proteins play an important role in the generation and maturation of the highly specialized niche where *C. burnetii* replicates and survives. Hence, *C. burnetii* can be added to the group of intracellular bacteria that use the autophagic pathway as a means of surviving.

When cells are grown in a full nutrient medium only basal levels of autophagy are active and the number of autophagic structures is low, whereas induction of autophagy increases the number of preformed autophagosomes rich in cellular components such us lipids and proteins. After *C. burnetii* internalization into host cells, the nascent *Coxiella*-containing phagosomes fuse rapidly with these autophagic compartments. Whether *C. burnetii* injects effector molecules, which modulates the host cell autophagy via the type IV secretion system, is an open question. In any case, the nutrients provided by fusion with autophagosomes trigger signals required to induce the differentiation from SCVs to LCVs and allow bacterial replication. Thus, the morphogenesis and multiplication of the organism would depend not only on pH, but also on accessibility to nutrients as indicated above. From this point of view, fusion with autophagic compartments is likely required for the biogenesis of the RV and triggers the appearance of the vegetative forms of *C. burnetii*. It is important to mention that, normally, the RV occupies most of the cytoplasm. This observation implies that at some point when the infection proceeds and the host nutrients become scarce, LCV switches back to the infectious and more resistant form SCV. However, it is not known how the equilibrium between host and pathogen is generated at the end of infection since persistent infected cultures and minimal cytopathic effect of the infection have been reported [96].

10.12
Unanswered Questions and Future Perspectives

Results from our laboratory indicate that induction of autophagy or overexpression of proteins (e.g. LC3, Rab24) involved in the autophagic pathway favors the development and maturation of the RV during the first 12 h post-infection [77]. In contrast, overexpression of mutants of LC3 and Rab24 that cannot interact with the autophagosomal membrane delayed the formation of the RV. However, by 48 h no differences in the development of the RV were observed [77], suggesting that autophagy is involved in the early steps of vacuole development. This observation raises an important question: is autophagy actually required for *C. burnetii* replication or is it dispensable? Experiments using "knockout" cells for specific autophagy genes are underway in our laboratory to address this question.

Programmed cell death is a strictly regulated genetic and biochemical program that plays a critical physiological role during development and tissue homeostasis in multicellular organisms. Accumulating evidence suggests that programmed cell death is not confined to apoptosis. Cells use different pathways for this suicide program: programmed cell death type I or apoptosis, characterized by a prominent nuclear condensation, and programmed cell death type II or autophagic cell death, characterized by the development of notorious AVs (reviewed in Refs. [97–99]). Even though these programmed cell death types are morphologically distinct, both pathways have certain similarities and they may represent different forms by which the cell responds to physiological or pathological conditions. It has been suggested that there is abundant overlap between apoptotic and autophagic cell death. It is important to take into account that apoptosis and autophagic cell death are not mutually exclusive, and they may occur simultaneously in tissues or even conjointly or consequently in the same cell [99, 100].

There is growing evidence that apoptosis plays important roles in influencing the pathogenesis of a variety of infectious diseases. It has been shown that intracellular bacterial pathogens induce or hamper apoptosis (for reviews, see Refs. [101, 102]) and this modulation of the host cell response adds another perspective to our understanding of the exploitation of host cell biology by intracellular parasites. Induction of apoptosis in the host cell might facilitate the escape and spreading of certain intracellular pathogens. Bacteria might also induce the expression of antiapoptotic factors that will prevent apoptosis until the microorganism has completed its replication in the host cell. In this context, it is possible that *Coxiella* infection may initially activate autophagy not only to generate a more permissive niche for intracellular replication, but also as a prosurvival mechanism. It is also likely that at later stages during infection autophagic cell death and/or apoptosis might be suppressed to allow pathogen survival in the host cell. Future experiments should be designed to address these issues and their contribution to pathogenesis.

Acknowledgments

We dedicate this chapter to Michel Rabinovitch in thanks for introducing us to the fascinating world of *C. burnetii*. The work discussed in this chapter was supported in part by grants from Agencia Nacional de Promoción Científica y Tecnológica (PICT99 1-6058 and PICT2002 01-11004), Secyt (Universidad Nacional de Cuyo) and UJAM (Universidad Juan A. Maza) to M. I. C.

References

1 S. Meresse, O. Steele-Mortimer, E. Moreno, M. Desjardins, B. Finlay, J. P. Gorvel, *Nat Cell Biol 1*, 1999, E183–E188.

2 A. Alonso, F. Garcia-del Portillo, *Int Microbiol 7*, 2004, 181–191.

3 R. A. Heinzen, T. Hackstadt, J. E. Samuel, *Trends Microbiol 7*, 1999, 149–154.

4 D. Raoult, T. Marrie, J. Mege, *Lancet Infect Dis 5*, 2005, 219–226.

5 P. Brouqui, H. T. Dupont, M. Drancourt, Y. Berland, J. Etienne, C. Leport, F. Goldstein, P. Massip, M. Micoud, A. Bertrand, *Arch Intern Med 153*, 1993, 642–648.

6 J. C. Williams, M. G. Peacock, D. M. Waag, G. Kent, M. J. England, G. Nelson, E. H. Stephenson, *Ann NY Acad Sci 653*, 1992, 88–111.

7 M. G. Madariaga, K. Rezai, G. M. Trenholme, R. A. Weinstein, *Lancet Infect Dis 3*, 2003, 709–721.

8 O. G. Baca, D. Paretsky, *Microbiol Rev 47*, 1983, 127–149.

9 O. G. Baca, Y. P. Li, H. Kumar, *Trends Microbiol 2*, 1994, 476–480.

10 M. Maurin, A. M. Benoliel, P. Bongrand, D. Raoult, *Infect Immun 60*, 1992, 5013–5016.

11 W. G. Weisburg, M. E. Dobson, J. E. Samuel, G. A. Dasch, L. P. Mallavia, O. Baca, L. Mandelco, J. E. Sechrest, E. Weiss, C. R. Woese, *J Bacteriol 171*, 1989, 4202–4206.

12 A. Macellaro, E. Tujulin, K. Hjalmarsson, L. Norlander, *Infect Immun 66*, 1998, 5882–5888.

13 G. Segal, H. A. Shuman, *Mol Microbiol 33*, 1999, 669–670.

14 D. F. Gimenez, *Stain Technol 39*, 1964, 135–140.

15 T. Hackstadt, M. G. Peacock, P. J. Hitchcock, R. L. Cole, *Infect Immun 48*, 1985, 359–365.

16 R. Toman, L. Skultety, *Carbohydr Res 283*, 1996, 175–185.

17 M. Maurin, D. Raoult, *Clin Microbiol Rev 12*, 1999, 518–553.

18 M. A. Quevedo Diaz, M. Lukacova, *Acta Virol 42*, 1998, 181–185.

19 K. Amano, J. C. Williams, *J Bacteriol 160*, 1984, 994–1002.

20 S. Schramek, H. Mayer, *Infect Immun 38*, 1982, 53–57.

21 R. Toman, L. Skultety, P. Ftacek, M. Hricovini, *Carbohydr Res 306*, 1998, 291–296.

22 R. Toman, L. Skultety, J. Kazar, *Acta Virol 37*, 1993, 196–198.

23 H. W. Wollenweber, S. Schramek, H. Moll, E. T. Rietschel, *Arch Microbiol 142*, 1985, 6–11.

24 K. Amano, J. C. Williams, S. R. Missler, V. N. Reinhold, *J Biol Chem 262*, 1987, 4740–4747.

25 S. Schramek, J. Radziejewska-Lebrecht, H. Mayer, *Eur J Biochem 148*, 1985, 455–461.

26 T. Hackstadt, *Ann NY Acad Sci 590:27-32.*, 1990, 27–32.

27 R. Toman, J. Kazar, *Acta Virol 35*, 1991, 531–537.

28 R. A. Heinzen, M. E. Frazier, L. P. Mallavia, *Nucleic Acids Res 18*, 1990, 6437.

29 D. Thiele, H. Willems, G. Kopf, H. Krauss, *Eur J Epidemiol 9*, 1993, 419–425.

30 A. Moos, T. Hackstadt, *Infect Immun 55*, **1987**, 1144–1150.

31 C. Capo, F. P. Lindberg, S. Meconi, Y. Zaffran, G. Tardei, E. J. Brown, D. Raoult, J. L. Mege, *J Immunol 163*, **1999**, 6078–6085.

32 R. Seshadri, I. T. Paulsen, J. A. Eisen, T. D. Read, K. E. Nelson, W. C. Nelson, N. L. Ward, H. Tettelin, T. M. Davidsen, M. J. Beanan, R. T. Deboy, S. C. Daugherty, L. M. Brinkac, R. Madupu, R. J. Dodson, H. M. Khouri, K. H. Lee, H. A. Carty, D. Scanlan, R. A. Heinzen, H. A. Thompson, J. E. Samuel, C. M. Fraser, J. F. Heidelberg, *Proc Natl Acad Sci USA 100*, **2003**, 5455–5460.

33 J. E. Samuel, M. E. Frazier, M. L. Kahn, L. S. Thomashow, L. P. Mallavia, *Infect Immun 41*, **1983**, 488–493.

34 J. E. Samuel, M. E. Frazier, L. P. Mallavia, *Infect Immun 49*, **1985**, 775–779.

35 L. P. Mallavia, *Eur J Epidemiol 7*, **1991**, 213–221.

36 Q. Ren, S. J. Robertson, D. Howe, L. F. Barrows, R. A. Heinzen, *Ann NY Acad Sci 990*, **2003**, 701–713.

37 D. Howe, L. P. Mallavia, *Infect Immun 67*, **1999**, 3236–3241.

38 F. T. Koster, J. C. Williams, J. S. Goodwin, *J Infect Dis 152*, **1985**, 1283–1289.

39 A. Honstettre, G. Imbert, E. Ghigo, F. Gouriet, C. Capo, D. Raoult, J. L. Mege, *J Infect Dis 187*, **2003**, 956–962.

40 E. T. Akporiaye, J. D. Rowatt, A. A. Aragon, O. G. Baca, *Infect Immun 40*, **1983**, 1155–1162.

41 J. L. Mege, M. Maurin, C. Capo, D. Raoult, *FEMS Microbiol Rev 19*, **1997**, 209–217.

42 J. Turco, H. A. Thompson, H. H. Winkler, *Infect Immun 45*, **1984**, 781–783.

43 R. A. Kishimoto, H. Rozmiarek, E. W. Larson, *Infect Immun 22*, **1978**, 69–71.

44 T. R. Jerrells, H. Li, D. H. Walker, *Adv Exp Med Biol 239*, **1988**, 193–200.

45 F. T. Koster, T. L. Kirkpatrick, J. D. Rowatt, O. G. Baca, *Infect Immun 43*, **1984**, 253–256.

46 O. G. Baca, *Eur J Epidemiol 7*, **1991**, 222–228.

47 J. G. Shannon, D. Howe, R. A. Heinzen, *Proc Natl Acad Sci USA 102*, **2005**, 8722–8727.

48 J. Dellacasagrande, E. Ghigo, D. Raoult, C. Capo, J. L. Mege, *J Immunol 169*, **2002**, 6309–6315.

49 D. Howe, L. F. Barrows, N. M. Lindstrom, R. A. Heinzen, *Infect Immun 70*, **2002**, 5140–5147.

50 D. S. Zamboni, M. Rabinovitch, *Infect Immun 71*, **2003**, 1225–1233.

51 E. Ghigo, C. Capo, C. H. Tung, D. Raoult, J. P. Gorvel, J. L. Mege, *J Immunol 169*, **2002**, 4488–4495.

52 T. F. McCaul, J. C. Williams, *J Bacteriol 147*, **1981**, 1063–1076.

53 T. Hackstadt, *Infect Agents Dis 5*, **1996**, 127–143.

54 R. Seshadri, J. E. Samuel, *Infect Immun 69*, **2001**, 4874–4883.

55 R. Seshadri, L. R. Hendrix, J. E. Samuel, *Infect Immun 67*, **1999**, 6026–6033.

56 S. Varghees, K. Kiss, G. Frans, O. Braha, J. E. Samuel, *Infect Immun 70*, **2002**, 6741–6750.

57 R. A. Heinzen, M. A. Scidmore, D. D. Rockey, T. Hackstadt, *Infect Immun 64*, **1996**, 796–809.

58 R. A. Heinzen, D. Howe, L. P. Mallavia, D. D. Rockey, T. Hackstadt, *Mol Microbiol 22*, **1996**, 9–19.

59 G. R. Cornelis, G. F. Van, *Annu Rev Microbiol 54*, **2000**, 735–774.

60 J. A. Sexton, J. P. Vogel, *Traffic 3*, **2002**, 178–185.

61 P. J. Christie, J. P. Vogel, *Trends Microbiol 8*, **2000**, 354–360.

62 P. J. Christie, *Mol Microbiol 40*, **2001**, 294–305.

63 G. Segal, M. Feldman, T. Zusman, *FEMS Microbiol Rev 29*, **2005**, 65–81.

64 P. J. Christie, E. Cascales, *Mol Membr Biol 22*, **2005**, 51–61.

65 T. Zusman, G. Yerushalmi, G. Segal, *Infect Immun 71*, **2003**, 3714–3723.

66 D. S. Zamboni, S. McGrath, M. Rabinovitch, C. R. Roy, *Mol Microbiol 49*, **2003**, 965–976.

67 P. S. Veras, C. de Chastellier, M. F. Moreau, V. Villiers, M. Thibon, D. Mattei, M. Rabinovitch, *J Cell Sci 107*, **1994**, 3065–3076.

68 P. S. Veras, C. Moulia, C. Dauguet, C. T. Tunis, M. Thibon, M. Rabinovitch, *Infect Immun 63*, **1995**, 3502–3506.

69 C. C. de, M. Thibon, M. Rabinovitch, *Eur J Cell Biol 78*, **1999**, 580–592.

70 M. S. Gomes, S. Paul, A. L. Moreira, R. Appelberg, M. Rabinovitch, G. Kaplan, *Infect Immun 67*, **1999**, 3199–3206.

71 D. L. Clemens, M. A. Horwitz, *J Exp Med 184*, **1996**, 1349–1355.

72 S. Sturgill-Koszycki, P. H. Schlesinger, P. Chakraborty, P. L. Haddix, H. L. Collins, A. K. Fok, R. D. Allen, S. L. Gluck, J. Heuser, D. G. Russell, *Science 263*, **1994**, 678–681.

73 W. Beron, M. G. Gutierrez, M. Rabinovitch, M. I. Colombo, *Infect Immun 70*, **2002**, 5816–5821.

74 A. Biederbick, H. F. Kern, H. P. Elsasser, *Eur J Cell Biol 66*, **1995**, 3–14.

75 D. B. Munafo, M. I. Colombo, *J Cell Sci 114*, **2001**, 3619–3629.

76 D. B. Munafo, M. I. Colombo, *Traffic 3*, **2002**, 472–482.

77 M. G. Gutierrez, C. L. Vazquez, D. B. Munafo, F. C. Zoppino, W. Beron, M. Rabinovitch, M. I. Colombo, *Cell Microbiol 7*, **2005**, 981–993.

78 P. B. Gordon, P. O. Seglen, *Biochem Biophys Res Commun 151*, **1988**, 40–47.

79 M. G. Gutierrez, D. B. Munafo, W. Beron, M. I. Colombo, *J Cell Sci 117*, **2004**, 2687–2697.

80 S. Jager, C. Bucci, I. Tanida, T. Ueno, E. Kominami, P. Saftig, E. L. Eskelinen, *J Cell Sci 117*, **2004**, 4837–4848.

81 D. Howe, J. Melnicakova, I. Barak, R. A. Heinzen, *Cell Microbiol 5*, **2003**, 469–480.

82 S. A. Coleman, E. R. Fischer, D. Howe, D. J. Mead, R. A. Heinzen, *J Bacteriol 186*, **2004**, 7344–7352.

83 E. F. Blommaart, U. Krause, J. P. Schellens, H. Vreeling-Sindelarova, A. J. Meijer, *Eur J Biochem 243*, **1997**, 240–246.

84 G. E. Mortimore, A. R. Poso, *Adv Enzyme Regul 25*, **1986**, 257–276.

85 A. J. Meijer, P. Codogno, *Int J Biochem Cell Biol 36*, **2004**, 2445–2462.

86 N. S. Cutler, J. Heitman, M. E. Cardenas, *Mol Cell Endocrinol 155*, **1999**, 135–142.

87 D. A. van Sluijters, P. F. Dubbelhuis, E. F. Blommaart, A. J. Meijer, *Biochem J 351*, **2000**, 545–550.

88 D. J. Klionsky, S. D. Emr, *Science 290*, **2000**, 1717–1721.

89 Y. Kabeya, N. Mizushima, T. Ueno, A. Yamamoto, T. Kirisako, T. Noda, E. Kominami, Y. Ohsumi, T. Yoshimori, *EMBO J 19*, **2000**, 5720–5728.

90 B. R. Dorn, W. A. Dunn, Jr., A. Progulske-Fox, *Infect Immun 69*, **2001**, 5698–5708.

91 M. I. Colombo, *Cell Death Differ 12 (Suppl 2)*, **2005**, 1481–1483.

92 D. Howe, L. P. Mallavia, *Infect Immun 68*, **2000**, 3815–3821.

93 M. G. Gutierrez, M. I. Colombo, *Autophagy 1*, **2005**, 179–181.

94 T. Hackstadt, J. C. Williams, *Proc Natl Acad Sci USA 78*, **1981**, 3240–3244.

95 M. S. Swanson, E. Fernandez-Moreira, E. Fernandez-Moreia, *Traffic 3*, **2002**, 170–177.

96 M. J. Roman, P. D. Coriz, O. G. Baca, *J Gen Microbiol 132*, **1986**, 1415–1422.

97 W. Bursch, *Cell Death Differ 8*, **2001**, 569–581.

98 W. Bursch, A. Ellinger, C. Gerner, U. Frohwein, R. Schulte-Hermann, *Ann NY Acad Sci 926*, **2000**, 1–12.

99 R. A. Lockshin, Z. Zakeri, *Int J Biochem Cell Biol 36*, **2004**, 2405–2419.

100 R. A. Gonzalez-Polo, P. Boya, A. L. Pauleau, A. Jalil, N. Larochette, S. Souquere, E. L. Eskelinen, G. Pierron, P. Saftig, G. Kroemer, *J Cell Sci 118*, **2005**, 3091–3102.

101 L. Gao, K. Y. Abu, *Microbes Infect 2*, **2000**, 1705–1719.

102 L. Y. Gao, Y. A. Kwaik, *Trends Microbiol 8*, **2000**, 306–313.

103 P. R. Burton, N. Kordova, D. Paretsky, *Can J Microbiol 17*, **1971**, 143–150.

11
Utilization of Endoplasmic Reticulum Membranes to Establish a Vacuole that Supports Replication of *Legionella pneumophila*

Mary-Pat Stein and Craig R. Roy

11.1
Introduction

Legionella pneumophila are Gram-negative bacteria that live in fresh water environments. *Legionella* normally replicate within protozoan host cells, but can infect humans when contaminated water sources are aerosolized. Inhaled *Legionella* are internalized by human alveolar macrophages, replicate and spread within the lung, causing inflammation and tissue damage. Healthy individuals are usually able to clear *Legionella* from the lung; however, immunocompromised and elderly individuals sometimes develop Legionnaires' disease – a severe pneumonia that can be life threatening.

In amoebae, the natural reservoir of *Legionella* [1], and in alveolar macrophages, intracellular survival and replication relies on the ability of *Legionella* to modulate transport of the vacuole in which they reside. Normally, pathogens internalized by macrophages are transported to lysosomes where they are killed and degraded. In contrast, after uptake by host cells, *Legionella* inhibit phagosome transport down this degradative pathway and subsequently remodel the compartment in which they reside to a replication-competent endoplasmic reticulum (ER)-like vacuole [2–4]. Modulation of vacuole transport is dependent on the expression of a functional type IV secretion system called the Dot/Icm (defective for organelle transport/intracellular multiplication) apparatus [5, 6]. The Dot/Icm apparatus directs the translocation of *Legionella* effector proteins across the bacterial cell and into the host cell cytoplasm. Within the host cell cytosol, effector proteins have activities that allow internalized bacteria to avoid phagosome–lysosome fusion and direct the remodeling of the vacuolar membrane to generate a replicative niche.

Significant progress has been made in understanding the cell biology of *Legionella* infection in host cells. Upon contact with host cells, *Legionella* are able to immediately translocate bacterial effector proteins into the host cell cytosol using the Dot/Icm Type IV secretion apparatus [7]. Within 5 min of infection, *Legionella*-containing vacuoles (LCVs) are surrounded by host cell vesicles and mitochondria [8]. Vesicles recruited to the LCVs are derived from the ER [3, 9]. These vesicles flatten

Autophagy in Immunity and Infection. Edited by Vojo Deretic
Copyright © 2006 WILEY-VCH Verlag GmbH & Co. KGaA, Weinheim
ISBN: 3-527-31450-4

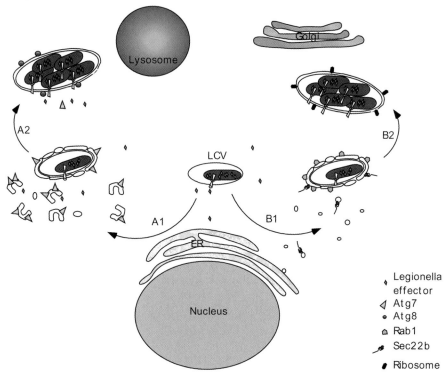

Fig. 11.1 Model depicting two alternative pathways for the intracellular trafficking of *Legionella*. (A1) *Legionella* induces autophagy by injecting effector proteins into the host cell cytoplasm using the type IV Dot/Icm secretion apparatus. Autophagic membranes labeled with Atg7 and Atg8 are recruited to the LCV, remodeling the limiting membrane into an autophagosome-like compartment where *Legionella* replication takes place. (A2) After a long delay of 16–24 h after infection, autophagosomes containing *Legionella* fuse with lysosomes. (B1) *Legionella* inject effec-tor proteins into the host cell cytoplasm using the type IV Dot/Icm secretion appara-tus to inhibit transport to lysosomes and di-rect the recruitment of ER-derived vesicles to the LCV limiting membrane. Remodeling of the LCV involves the recruitment of Rab1- and Sec22b-containing ER-derived vesicles to the LCV. (B2) Remodeling of the LCV mem-brane to resemble the ER, including thinning of the membrane and attachment of ribo-somes, facilitates bacterial growth and repli-cation. (This figure also appears with the color plates).

and fuse together with the limiting membrane of the vacuole, encapsulating the LCV within a membrane compartment that resembles the ER [10]. At 4 h post-in-fection, ribosomes decorate the LCV membrane [8, 10, 11] and within these ribo-some-studded ER-like compartments, *Legionella* replicate. Large replicative va-cuoles (RVs) crowded with *Legionella* are clearly visible 10–12 h post-infection. The recruitment and fusion of host vesicles derived from the ER indicates that *Le-gionella* has the ability to subvert the transport of host ER membranes.

The process of autophagy, which in mammalian cells has been linked to cel-lular homeostasis, differentiation, tissue remodeling, development, cancer, neurodegeneration and myopathies, involves the sequestration of proteins, organelles or organisms into a membrane-bounded compartment derived from host cell membranes [12]. Autophagy can be divided into several distinct steps [13–15]. First, cells recognize intracellular and/or extracellular signals that induce autophagy. Following induction, autophagosome formation proceeds through the action of a variety of autophagy (Atg) proteins [16, 17]. These proteins direct the recruitment and fusion of host vesicles to generate a sequestering organelle. Subsequently, autophagosomes fuse with lysosomes and the breakdown of autophagosomal content completes the process.

The role of autophagy in the establishment of the *Legionella* RV was first hypothesized based on the ultrastructural similarities between the LCV and autophagosomes [11]. In this chapter we summarize recent studies that suggest a role for the autophagic pathway in the generation of the *Legionella* RV and provide an alternative model (Fig 11.1) that does not invoke an essential role for autophagy in the intracellular growth of *Legionella*.

11.2
Evidence that *Legionella* Utilizes the Autophagy Machinery for Biogenesis of a Replicative Organelle

11.2.1
Induction of Autophagy

The first step in the formation of autophagosomes is the induction of the autophagy machinery [14]. Induction occurs nonselectively when cells sense extracellular cues, such as nutrient starvation or changes in local cytokine concentrations. In response to these signals, double membrane-bound autophagosomes form and envelop intracellular constituents including cytosol and organelles. Selective autophagy involves activation of the autophagy machinery and incorporation of specific intracellular cargo into the autophagosome. Selective autophagy has been described for specific organelles, such as peroxisomes (pexophagy) [18], and for a number of intracellular bacteria, including *Legionella*, group A *Streptococcus* and *Shigella flexneri*. Although a role for a bacterial protein in the induction of autophagy has been described for *Shigella* [19], the mechanisms by which the other intracellular organisms induce autophagy remain to be determined.

It is attractive to speculate that *Legionella* induces autophagy by injection of effector substrates into host cells by the Dot/Icm secretion system. *Legionella* mutants defective in the Dot/Icm system are transported directly from early endosome-like phagosomes to lysosomes through the endocytic pathway [20, 21]. Interestingly, recent experimental evidence suggests that proteins secreted by *Legionella* may be capable of inducing autophagy. It was shown that cell-free supernatants from virulent, but not from a type IV secretion-defective mutant, *dotA*, *Legionella*, were capable of inducing the formation of autophagosomes in murine macrophages [22]. Vesicle formation was not induced by supernatants prepared from *Escherichia coli* or from noninfectious exponential-phase *Legionella*. Furthermore, the formation of autophagosomes was limited to bone-marrow-derived macrophages treated with supernatant fractions containing the 10- to 30-kDa protein species, but not fractions containing protein species of greater than 30 kDa or less than 10 kDa. *Legionella* supernatants treated with proteinase K lost their ability to stimulate autophagosome formation, indicating that the autophagosome-inducing component in the supernatants was proteinacious in nature. These data suggest that an unidentified bacterial protein produced by *Legionella* has the capacity to induce autophagy in murine macrophages. In future work, it will be important to determine if the 10- to 30-kDa *Legionella* proteins that induce autophagy are proteins secreted via the Dot/Icm apparatus into the host cell cytosol where they could exert their action in host cells during infection.

While the data suggest that partially purified proteins from *Legionella* can induce autophagy in murine macrophages, the induction of autophagy following infection with *Legionella* has not been demonstrated. Recruitment of autophagy-related proteins to internalized *Legionella* has only been observed when macrophages were starved or treated with pharmacological agents to induce autophagy. The observation that purified proteins from *Legionella* may be capable of inducing autophagy suggests that following infection, the induction of autophagy may be below the level of detection or it may not be required to establish a replicative niche *in vivo*. Identification of the specific *Legionella* proteins capable of inducing autophagy in host cells will improve our understanding of the role of autophagy in *Legionella*'s intracellular survival and propagation.

11.2.2
Formation of Autophagosomes

After induction, autophagosome formation proceeds through a series of molecular conjugation reactions leading to the formation of a double membrane-bound structure. In yeast, the molecular cascade of events leading to the formation of these macromolecular complexes has been described [15–17, 23]. Two of the main reactions in autophagosome formation in yeast are the conjugation of an Atg5–Atg12 complex and the lipidation of Atg8 [microtubule-associated protein 1 light chain 3 (LC3)] to phosphatidylethanolamine. Both these reactions require the action of Atg7, which acts as an E1-ubiquitin ligase. In mammalian cells, homologs of many of the yeast autophagy proteins have been identified and the molecular details of autophagosome assembly closely parallel those described for yeast [15]. Localization of Atg5, Atg7 and Atg8 is routinely used to analyze the formation and trafficking of autophagosomes [24–26].

The initiation of autophagy involves the recruitment of the Atg5–Atg12 complex to a membrane-bound compartment referred to as autophagosome precursors, sequestration crescents, phagophores or pre-autophagosome structures (PAS) [14, 15, 27]. The source of the membranes constituting these autophagosomal precursors is a continued topic of dispute. Early morphological studies suggested that autophagosomal membranes were derived from the ER [28]. Immunological labeling experiments demonstrated the absence of plasma membrane, Golgi and endosomal proteins on these membrane structures. In contrast, proteins derived from the ER, such as ribophorin II, serum albumin and α2a-globulin and several ER-specific proteins were localized to autophagosome membranes [28]. Recently, however, the formation of autophagosomes from non-ER derived pre-existing autophagosome precursor membranes has been postulated [26]. Ultrastructural analyses demonstrated that upon the induction of autophagy, small crescent-shaped membranes labeled with Atg5 expand and elongate forming first curved, then semi-spherical structures [26]. While the nature of these small precursor membranes has not been determined, it remains entirely possible that these autophagosomal precursors are ultimately derived from the ER.

The initial observation suggesting that *Legionella* utilizes autophagy in the formation of its replicative niche comes from morphological observations of LCVs early after infection. Ultrastructural and immunofluorescence analyses demonstrated that smooth vesicles are recruited to the limiting membrane of the LCV, and flatten and fuse with the vacuole [8, 10]. *Legionella* has been observed in a double-membrane-bound compartment [8, 10, 11] morphologically similar to those surrounding autophagosomes. Furthermore, localization of ER markers including BIP, calnexin and KDEL to the LCVs suggests that vesicles recruited to the limiting membrane of the LCVs are derived from the ER [3, 11]. Since enclosure in an ER-derived double-membrane-bound compartment is the hallmark of autophagosome formation, these data indicate that LCV are remodeled by a process that is strikingly similar to autophagy.

To further investigate whether autophagy might play a significant role in establishment of the *Legionella* replicative organelle, the effects of demonstrated activators and inhibitors of autophagy on *Legionella* intracellular growth were analyzed. Stimulation or inhibition of autophagy in murine macrophage cells was followed by infection with *Legionella*. RV formation and bacterial growth were measured. In macrophages starved of amino acids to induce autophagy, the kinetics of *Legionella* RV formation was more rapid than in non-induced macrophages. Increased association of LCV with the ER marker, BIP, 3 h postinfection and increased yields of intracellular bacteria recovered 3 days following infection were observed in macrophages in which autophagy was induced compared to macrophages cultured in non-inducing conditions [11]. The increased association of LCVs with ER and increased bacterial loads recovered from starvation-induced cells were modest with 2.5- and 3.5-fold increases, respectively. In contrast, cells treated with bafilomycin A or 3-methyladenine to inhibit autophagy exhibited decreased *Legionella* RV formation and increased association of LCVs with lysosomes compared to non-treated cells [22]. These results suggest that in the absence of autophagy, the ability of *Legionella* to create an ER-derived vacuole is diminished, leading to decreased bacterial survival. Together, experiments using pharmacological agents to manipulate cellular autophagy provide additional evidence in support of the theory that *Legionella* are capable of creating a RV using the autophagy machinery of the host cell.

11.2.3
Maturation of Autophagosomes and Fusion with Lysosomes

The final stage of autophagosomal trafficking in mammalian cells involves docking and fusion with late endosomes and lysosomes. The docking and fusion of membrane compartments in mammalian cells utilizes specific Rab and *N*-ethylmalemide-sensitive fusion protein attachment receptor (SNARE) proteins to direct the specificity between compartments. Rab7 and Rab24 have been shown to play a role in mammalian autophagy [29, 30]. In yeast, several SNARE proteins, including Vam3, Vam7 and Vti1, have been proposed to facilitate fusion of autophagosomes with the vacuole [31–33]. As additional components required for fusion are identified, more specific details of the fusion mechanism between autophagosomes and lysosomes will likely be revealed.

Fusion of autophagosomes with late endosomes and lysosomes results in the acquisition of hydrolytic enzymes and vacuolar acidification. Both of these activities are required for the breakdown of the interior vesicles and contents of autophagosomes in the autolysosome [13]. Inhibition of autophagosome–lysosome fusion by expression of a dominant-negative mutant of SDK1, a AAA ATPase protein important in late endosomal transport to the lysosomes, demonstrated that autophagosomes interact with late endosomes prior to fusion with lysosomes [34]. Additionally, treatment of cells with Bafilomycin A1, an inhibitor of the V-ATPase, resulted in the accumulation of autophagosomes and decreased protein degradation [35]. Microtubule-based vesicular transport also plays a role in fusion of autophagosomes with both endosomes and lysosomes, since accumulation of cellular proteins and acidified autophagosomes was observed following treatment of cells with the microtubule depolymerizing agent nocodozole [36]. The accumulation of acidified autophagosomes following treatment with nocodozole suggested that the recruitment of the V-ATPase to autophagosomes precedes fusion with lysosomes. Taken together, these results suggest that interactions between autophagosomes and late endosomal compartments may provide required molecules for autophagosome fusion with lysosomes.

Following this paradigm, the LCVs should fuse with late endosomes and lysosomes resulting in acidification of the *Legionella* RV if these vacuoles are utilizing the autophagosome trafficking pathway. Consistent with this hypothesis, it was demonstrated that numerous vacuoles containing replicating *Legionella* had a pH of approximately 5.6 at 16–24 h post-infection. In addition, the accumulation of monodansylcadaverine (MDC), a marker of acidic autophagosomes, was observed in the LCV as early as 6 h post-infection (30% MDC-positive) [37]. Finally, the acidification of the LCV was inhibited by treatment of cells with Bafilomycin A1. These results demonstrate that the LCV acidifies slowly from 6 to 24 h after infection, similar to the acidification of autophagosomes, albeit with significantly slower kinetics.

The fusion of the LCVs with lysosomes was delayed compared to starvation-induced autophagy. Accumulation of the lysosomal markers, lysosome-associated membrane protein type 1 (LAMP-1) and cathepsin D, was observed on LCVs at 16 h post-infection [38]. Confirmation that LAMP-1- and cathepsin D-positive LCVs were lysosomal compartments was demonstrated by chasing replicating LCVs into lysosomal compartments labeled with Texas Red-labeled ovalbumin or dextran. These data demonstrated that at 6 h post-infection, acidification of the LCVs could be detected, followed by fusion of LCVs with lysosomes between 16 and 20 h after infection [38]. One interpretation of these data is that the LCV is acting as an autophagosome, acidifying and fusing with lysosomes at very late time points after infection.

11.3
Evidence that the Host Autophagy Machinery is not Essential for Transport or Growth of *Legionella*

11.3.1
Dictyostelium discoideum Autophagy Mutants Support *Legionella* Growth

A remarkable aspect of *Legionella* biology is the ability of these bacteria to grow in different eukaryotic hosts by exploiting cellular processes that are conserved evolutionarily. Although in a human infection *Legionella* grow within macrophages, in nature *Legionella* grow in a diverse number of different protozoan hosts. Importantly, it was shown that the ER-derived vacuole in which *Legionella* replicates in macrophages is indistinguishable from that observed in a natural host, such as the amoeba, *Dictyostelium discoideum* [39–41]. Thus, it was possible to directly test whether host autophagy machinery is required for *Legionella* replication by analyzing bacterial growth in autophagy mutants that were constructed in *D. discoideum* [42]. These data revealed that loss-of-function mutations in *atg1*, *atg5*, *atg6*, *atg7* and *atg8* had no effect on growth of *Legionella*, but as predicted these mutations severely reduced the turnover of host proteins by the autophagy pathway. In addition, the morphology of the LCV was unchanged in *D. discoideum* autophagy mutants when compared to wild-type cells. Additionally, it was shown that Atg8 was not recruited to the LCVs at detectable levels in *D. discoideum*. These results indicate that autophagy does not play a significant role in the pathogenesis of *Legionella* in *D. discoideum*, suggesting that the autophagy machinery is not essential for biogenesis of the *Legionella* RV.

11.3.2
Morphological Differences between LCVs and Autophagosomes

Morphologically, the LCV and autophagosome membranes display some notable differences. First, mitochondria, nuclear membranes and ribosomes have been observed at various time points following infection attached to the membranes encircling internalized *Legionella* [8, 10]. In contrast, autophagosomes containing engulfed organelles and host cell cytoplasm have only smooth vesicles localized to the limiting membranes. Second, ultrastructural analyses of LCV membranes and their associated vesicles demonstrated that small osmiophilic hair-like projections exist between the inner LCV membrane and the outer vesicular membranes [10]. These projections are hypothesized to represent connections between these two compartments. In autophagosomes, no such osmiophilic projections between membranes have been observed. Third, over time, the thickness of the LCV limiting membrane decreased, thinning from the width of the plasma membrane to the width of the typical ER membrane [10], suggesting fusion between the attached ER-derived vacuoles and the LCV. In contrast, the thickness of the isolation membrane of autophagosomes remains constant and fusion between pre-autophagosomal membranes and membranes of the organelle being engulfed has not been documented. These results suggest that active remodeling of the LCV membrane continues over time by fusion of the attached vesicles with the vacuole, reducing the thickness of the LCV membrane to be consistent with ER membranes. In contrast, autophagosomal membranes elongate to enclose cytosolic content without any alterations in membrane thickness. These distinctions demonstrate that significant differences between these membrane-bound compartments exist, in particular with regard to the thickness of the final limiting membranes.

The time course for the remodeling of the LCV and for trafficking to lysosomes is significantly longer than that observed for cellular autophagy induced by rapamycin or amino acid starvation. Induction of autophagy results in the formation of Atg7-positive compartments as early as 5 min after induction. Atg7-positive compartments are transient and Atg8-positive compartments become visible within 1 h. In contrast, LCVs have been reported to stain positive for Atg7 for the first 3 h post-infection [22] and Atg8 accumulation occurred only after Atg7 diminished approximately 4 h after infection. It has been argued that this delay represents a "pregnant pause" that may be caused by *Legionella* effector proteins that delay the maturation of autophagosomes, thereby allowing *Legionella* to replicate prior to delivery to lysosomes [43].

11.3.3
Differences Between *Legionella* and Other Bacteria that Utilize Autophagy for Intracellular Survival

Brucella abortus, *Porphyromonas gingivalis* and *Coxiella burnetii* are bacterial pathogens that have been proposed to enter the autophagic pathway at some point during their intracellular lifecycle. Similar to *Legionella*, increased *Brucella* and *Coxiella* growth were observed in cells where autophagy was induced and decreased bacterial loads were detected following inhibition of autophagy [44–46]. However, in contrast to *Legionella*, virulent *Brucella*-containing autophagosome-like structures did not fuse with lysosomes as determined by the lack of cathepsin D colocalization with replicating bacteria [44]. Similarly, when *Por-*

phyromonas infection of human coronary artery endothelial cells was examined, clear changes in both the occurrence of autophagosomes and in Atg7 (HsGsa7p) localization were observed [47]. However, no colocalization of cathepsin L with *Porphyromonas* RVs was noted, indicating inhibition of autophagosome–lysosome fusion. Experimental evidence from the intracellular trafficking of another bacterial pathogen, *Coxiella*, demonstrated that although some bacteria are able to utilize the autophagy pathway to create a replicative niche, autophagy is not an absolute requirement for intracellular survival and replication [46]. Transient expression of dominant-negative mutants of Atg8 and Rab24, proteins associated with autophagosome formation [30], decreased targeting of *Coxiella* to autophagosome-like structures. At late time points, however, infection recovered to levels similar to those observed in nontransfected control cells. Results suggested that while autophagy may aid in the establishment of the *Coxiella* RV, the ability of *Coxiella* to survive and multiply is not dependent on transit through an autophagosome-like compartment.

11.4
Creation of an ER-derived Vacuole that Supports *Legionella* Replication by Subversion of the Host Secretory Pathway

An alternative model for how *Legionella* creates a replicative organelle that does not rely upon the autophagy machinery of host cells proposes that the LCV acquires membranes and vesicles exiting the host secretory pathway. Acquisition of secretory vesicles allows *Legionella* to modify the vacuole into an organelle that resembles the ER. In this ER-like compartment, *Legionella* both avoids fusion with lysosomes and obtains the nutrients required for bacterial growth and replication.

In addition to early electron microscopy studies showing the recruitment of ribosomes to the *Legionella* replicative organelle, evidence in support of the LCV intercepting secretory vesicles exiting the ER comes from studies demonstrating that host proteins contained in early secretory vesicles, such as BIP and proteins with the KDEL motif, localized to the LCVs within 30 min of infection [3, 11]. At 10 h post-infection, KDEL staining of the LCVs was lost and vacuoles stained positive for the resident ER protein calnexin [3]. These results suggested that ER-derived vesicles containing secretory cargo, such as KDEL-containing proteins, are recruited to the LCVs early after internalization. The appearance of ribosomes on the LCVs [8], which are never observed on autophagosomes, and the accumulation of the ER-resident protein calnexin at later time points signified the remodeling of the LCV to an ER-like compartment.

Initial insight into the mechanism by which ER-derived vesicles were recruited to the LCV came from experiments using inhibitors of the secretory pathway [3, 9]. Prior to infection, pretreatment of cells with Brefeldin A, a fungal metabolite known to inhibit ARF-dependent secretory transport [48], resulted in decreased RV formation 12 h post-infection and decreased bacterial loads at

24 and 48 h post-infection [3]. Furthermore, expression of Sar1H79G and Arf1T31N, mutant proteins previously shown to block release of secretory vesicles from the ER [49], also inhibited *Legionella* RVs formation [3]. These results established that early secretory vesicles from the ER are important in the formation of *Legionella*'s RVs.

The discovery that early secretory transport was important for the establishment of the LCV prompted the investigation into the roles other secretory proteins play in LCV biogenesis. First, the importance of Rab1, the small GTPase that mediates transport of secretory vesicles from the ER to the Golgi, was examined. Immunofluorescence analyses of cells infected with wild-type *Legionella* demonstrated that Rab1, but not Rab2 or Rab6, was recruited to the LCV early after infection [9]. Furthermore, expression of a dominant-negative mutant of Rab1 inhibited intracellular growth of *Legionella*, whereas the expression of mutant Rab2 and mutant Rab6 had no such effect [9]. Sec22b, a SNARE protein that normally participates in the fusion of ER-derived vesicles to ERGIC (ER Golgi intermediate compartment) and Golgi membranes, was also observed on the LCVs early after infection [9]. Intriguingly, Membrin, rBet and Syntaxin5, the cognate SNARE binding partners for Sec22b, were not observed with Sec22b on LCVs, suggesting that other host SNARE proteins or *Legionella* effector proteins localized to LCVs could provide binding partners for Sec22b. The recruitment of Sec22b to LCVs was inhibited by overexpression of Membrin, which also inhibited the intracellular growth of *Legionella*. Membrin-dependent growth inhibition was reversed by overexpression of Sec22b, and recruitment of Sec22b to LCVs was demonstrated to be dependent on the relative ratio of Membrin to Sec22b expressed in transfected cells [9]. These results suggested that titration of Sec22b off LCVs was associated with the inability of *Legionella* to establish a productive RV. Roles for Rab1 and Sec22b in autophagosome formation have not been reported, and, as such, the importance of Rab1 and Sec22b in the biogenesis of *Legionella*'s RV demonstrates that *Legionella* modulates the host cell secretory machinery to form a specialized compartment for replication.

11.5
Conclusion

Although similarities exist between autophagosomes and LCVs, the data to suggest that the trafficking of *Legionella* utilizes the autophagy pathway in mammalian cells to generate a RV remains inconclusive. The data suggest that when cells are induced to initiate autophagy by artificial means, LCVs can utilize newly formed autophagosomal membranes to increase the limiting membrane of their vacuolar compartments. In both *Legionella* and *Coxiella*, it is clear that induction of autophagy can enhance bacterial growth, but in neither case is autophagy strictly required for growth. Future experiments to identify *Legionella* effector proteins that are translocated into host cells to induce autophagy would provide support for an autophagy model of *Legionella* growth. If instead autophagy is a normal cellular response to infection by *Legionella*, it would be beneficial to demonstrate increased autophagy gene transcription and/or increased mRNA for conserved autophagy genes following infection. In this paradigm, additional *Legionella* effector proteins that inhibit or delay the maturation of autophagosomes should be isolated and identified. With the increasing availability of RNAi, it will also be possible to determine which autophagy proteins play a role in *Legionella* intracellular trafficking by directly inhibiting protein production. Thus, recently developed methodologies should allow researchers to definitively address the importance of autophagy in *Legionella* trafficking and intracellular growth in mammalian cells.

References

1 T. J. Rowbotham, *J Clin Pathol* **1980**, *33*, 1179–1183.

2 M. A. Horwitz, *J Exp Med* **1983**, *158*, 2108–2126.

3 J. C. Kagan, C. R. Roy, *Nat Cell Biol* **2002**, *4*, 945–954.

4 C. R. Roy, *Trends Microbiol* **2002**, *10*, 418–424.

5 G. Segal, M. Purcell, H. A. Shuman, *Proc Natl Acad Sci USA* **1998**, *95*, 1669–1674.

6 J. P. Vogel, H. L. Andrews, S. K. Wong, R. R. Isberg, *Science* **1998**, *279*, 873–876.

7 H. Nagai, E. D. Cambronne, J. C. Kagan, J. C. Amor, R. A. Kahn, C. R. Roy, *Proc Natl Acad Sci USA* **2005**, *102*, 826–831.

8 M. A. Horwitz, *J Exp Med* **1983**, *158*, 1319–1331.

9 J. C. Kagan, M.-P. Stein, M. Papaert, C. Roy, *J Exp Med* **2004**, *199*, 1201–1211.

10 L. G. Tilney, O. S. Harb, P. S. Connelly, C. G. Robinson, C. R. Roy, *J Cell Sci* **2001**, *114*, 4637–4650.

11 M. S. Swanson, R. R. Isberg, *Infect Immun* **1995**, *63*, 3609–3620.

12 B. Levine, D. J. Klionsky, *Dev Cell* **2004**, *6*, 463–477.

13 F. Reggiori, D. J. Klionsky, *Eukaryot Cell* **2002**, *1*, 11–21.

14 F. Reggiori, D. J. Klionsky, *Curr Opin Cell Biol* **2005**, *17*, 415–422.

15 N. Mizushima, Y. Ohsumi, T. Yoshimori, *Cell Struct Funct* **2002**, *27*, 421–429.

16 P. E. Stromhaug, D. J. Klionsky, *Traffic* **2001**, *2*, 524–531.

17 D. J. Klionsky, J. M. Cregg, W. A. Dunn, Jr., S. D. Emr, Y. Sakai, I. V. Sandoval, A. Sibirny, S. Subramani, M. Thumm, M.

Veenhuis, Y. Ohsumi, *Dev Cell* **2003**, *5*, 539–545.

18 J. C. Farre, S. Subramani, *Trends Cell Biol* **2004**, *14*, 515–523.

19 M. Ogawa, T. Yoshimori, T. Suzuki, H. Sagara, N. Mizushima, C. Sasakawa, *Science* **2005**, *307*, 727–731.

20 A. Marra, H. A. Shuman, *J Bacteriol* **1989**, *171*, 2238–2240.

21 K. H. Berger, R. R. Isberg, *Mol Microbiol* **1993**, *7*, 7–19.

22 A. O. Amer, M. S. Swanson, *Cell Microbiol* **2005**, *7*, 765–778.

23 T. Noda, K. Suzuki, Y. Ohsumi, *Trends Cell Biol* **2002**, *12*, 231–235.

24 T. Kirisako, M. Baba, N. Ishihara, K. Miyazawa, M. Ohsumi, T. Yoshimori, T. Noda, Y. Ohsumi, *J Cell Biol* **1999**, *147*, 435–446.

25 Y. Kabeya, N. Mizushima, T. Ueno, A. Yamamoto, T. Kirisako, T. Noda, E. Kominami, Y. Ohsumi, T. Yoshimori, *EMBO J* **2000**, *19*, 5720–5728.

26 N. Mizushima, A. Yamamoto, M. Hatano, Y. Kobayashi, Y. Kabeya, K. Suzuki, T. Tokuhisa, Y. Ohsumi, T. Yoshimori, *J Cell Biol* **2001**, *152*, 657–668.

27 K. Kirkegaard, M. P. Taylor, W. T. Jackson, *Nat Rev Microbiol* **2004**, *2*, 301–314.

28 W. A. Dunn, Jr., *J Cell Biol* **1990**, *110*, 1923–1933.

29 M. G. Gutierrez, D. B. Munafo, W. Beron, M. I. Colombo, *J Cell Sci* **2004**, *117*, 2687–2697.

30 D. B. Munafo, M. I. Colombo, *Traffic* **2002**, *3*, 472–482.

31 T. Darsow, S. E. Rieder, S. D. Emr, *J Cell Biol* **1997**, *138*, 517–529.

32 T. K. Sato, T. Darsow, S. D. Emr, *Mol Cell Biol* **1998**, *18*, 5308–5319.

33 N. Ishihara, M. Hamasaki, S. Yokota, K. Suzuki, Y. Kamada, A. Kihara, T. Yoshimori, T. Noda, Y. Ohsumi, *Mol Biol Cell* **2001**, *12*, 3690–3702.

34 A. Nara, N. Mizushima, A. Yamamoto, Y. Kabeya, Y. Ohsumi, T. Yoshimori, *Cell Struct Funct* **2002**, *27*, 29–37.

35 A. Yamamoto, Y. Tagawa, T. Yoshimori, Y. Moriyama, R. Masaki, Y. Tashiro, *Cell Struct Funct* **1998**, *23*, 33–42.

36 A. Aplin, T. Jasionowski, D. L. Tuttle, S. E. Lenk, W. A. Dunn, Jr., *J Cell Physiol* **1992**, *152*, 458–466.

37 A. O. Amer, M. S. Swanson, *Curr Opin Microbiol* **2002**, *5*, 56–61.

38 S. Sturgill-Koszycki, M. S. Swanson, *J Exp Med* **2000**, *192*, 1261–1272.

39 S. Hagele, R. Kohler, H. Merkert, M. Schleicher, J. Hacker, M. Steinert, *Cell Microbiol* **2000**, *2*, 165–171.

40 J. M. Solomon, R. R. Isberg, *Trends Microbiol* **2000**, *8*, 478–480.

41 J. M. Solomon, A. Rupper, J. A. Cardelli, R. R. Isberg, *Infect Immun* **2000**, *68*, 2939–2947.

42 G. P. Otto, M. Y. Wu, M. Clarke, H. Lu, O. R. Anderson, H. Hilbi, H. A. Shuman, R. H. Kessin, *Mol Microbiol* **2004**, *51*, 63–72.

43 M. S. Swanson, E. Fernandez-Moreira, E. Fernandez-Moreia, *Traffic* **2002**, *3*, 170–177.

44 J. Pizarro-Cerda, E. Moreno, V. Sanguedolce, J. L. Mege, J. P. Gorvel, *Infect Immun* **1998**, *66*, 2387–2392.

45 W. Beron, M. G. Gutierrez, M. Rabinovitch, M. I. Colombo, *Infect Immun* **2002**, *70*, 5816–5821.

46 M. G. Gutierrez, C. L. Vazquez, D. B. Munafo, F. C. Zoppino, W. Beron, M. Rabinovitch, M. I. Colombo, *Cell Microbiol* **2005**, *7*, 981–993.

47 B. R. Dorn, W. A. Dunn, Jr., A. Progulske-Fox, *Cell Microbiol* **2002**, *4*, 1–10.

48 C. L. Jackson, J. E. Casanova, *Trends Cell Biol* **2000**, *10*, 60–67.

49 T. H. Ward, R. S. Polishchuk, S. Caplan, K. Hirschberg, J. Lippincott-Schwartz, *J Cell Biol* **2001**, *155*, 557–570.

Part III
Autophagy and Viruses

Autophagy in Immunity and Infection. Edited by Vojo Deretic
Copyright © 2006 WILEY-VCH Verlag GmbH & Co. KGaA, Weinheim
ISBN: 3-527-31450-4

12
Endogenous Major Histocompatibility Complex Class II Antigen Processing of Viral Antigens

Dorothee Schmid and Christian Münz

12.1
Introduction

Immune control of intracellular pathogens like viruses relies to a large degree on T cells of the adaptive immune system. In contrast to B cells and their antibodies, T cells scan with their T cell receptor the cell surface for fragments, which are generated within all body tissues and displayed on major histocompatibility complex (MHC) molecules. Detection of a pathogen-derived constituent leads either to direct destruction of the infected cell or to activation of other components of the immune system. Classical T cells come in two main flavors – CD8$^+$ and CD4$^+$ T cells. Both recognize peptides from intracellular or endocytosed proteins after presentation on MHC class I and II molecules, respectively. However, these peptides are generated by distinct proteolytic systems within cells. MHC class I ligands are primarily produced by proteasomes, whereas MHC class II ligands originate mainly from lysosomal degradation. Access to these proteolytic compartments determines how efficiently T cells detect viral antigens and, therefore, an understanding of these antigen-processing pathways is crucial to our understanding of successful versus insufficient immune responses as well as vaccine development.

In this chapter we summarize recent evidence that autophagy contributes to CD4$^+$ T cell *immunity* via delivery of antigens for MHC class II presentation. We will outline the canonic pathways of MHC class I and II antigen processing. Furthermore, we discuss viral antigens that gain access to MHC class II loading within cells and which pathways have been implicated in their processing. Moreover, we highlight autophagy substrates that get presented on MHC class II, and point out similarities between source proteins of MHC class II ligands and autophagy substrates. In addition, we discuss where MHC class II loading of autophagy substrates might occur in cells. Finally, we speculate how this pathway might be generally involved in the immune control of viral infections and how it might be harnessed for enhanced MHC class II presentation of viral antigens in vaccination strategies.

Autophagy in Immunity and Infection. Edited by Vojo Deretic
Copyright © 2006 WILEY-VCH Verlag GmbH & Co. KGaA, Weinheim
ISBN: 3-527-31450-4

12.2
Classical Pathways of Antigen Processing for MHC Presentation

MHC class I and II molecules, the two main categories of T cell restriction elements, are loaded with their peptide cargo in different cytoplasmic locations [1]. Furthermore, MHC class I and II ligands are generated by distinct proteolytic events. Antigens for MHC class I presentation are primarily degraded by proteasomes – large multicatalytic proteases found in cytosol and nucleus [2]. Access to proteasomal degradation is primarily regulated via ubiquitinylation of its substrates and this degradation pathway regulates mainly short-lived proteins [3]. One group of short-lived proteasomal substrates that has recently received a lot of attention are the so-called defective ribosomal products (DRiPs), which are degraded by the ubiquitin–proteasome system immediately after misfolding or premature termination of their translation [4]. Degradation of short-lived pathogenic products for MHC class I presentation enables the immune system to detect pathogenic determinants rapidly after infection. Proteasomal products, primarily peptides of below 10 amino acids length, are then translocated via the transporter associated with antigen processing (TAP) into the endoplasmic reticulum (ER), where they meet newly synthesized MHC class I molecules, which are cotranslationally inserted into the ER membrane. Within the MHC class I loading complex in the ER, which contains chaperones, aminopeptidases and thiol oxidoreductases, primarily octamer or nonamer peptides are loaded onto MHC class I molecules [5]. Stably associated MHC class I/peptide complexes are then exported to the cell membrane via the Golgi apparatus. As MHC class I ligands are mainly generated in a TAP- and proteasome-dependent fashion, they are thought to mainly originate from cytosolic and nuclear source proteins.

MHC class II ligands, however, are thought to primarily originate from extracellular antigens after endocytosis and degradation in lysosomes [1]. Newly synthesized MHC class II molecules associate in the ER with a chaperone called invariant chain (Ii), which blocks the MHC class II peptide-binding groove and with its cytosolic targeting signal directs MHC class II molecules to the MHC class II loading compartment, which consists of late endosomes [6]. In these late endosomes Ii is degraded by lysosomal proteases and the remaining peptide [class II-associated peptide (CLIP)] is replaced by lysosomal products under the influence of the nonclassical MHC class II molecule and chaperone HLA-DM/H2-M [7]. Assembled MHC class II/peptide complexes then migrate to the cell surface for surveillance by CD4$^+$ T cells.

These classical MHC antigen-processing pathways suggest that intracellular antigens get preferentially presented on MHC class I, while extracellular antigens are the main source of MHC class II ligands. However, both the elution of natural ligands from MHC class II molecules and the analysis of the processing requirements for several CD4$^+$ T cell epitopes revealed that intracellular proteins, including important viral antigens, gain access to MHC class II presentation.

Table 12.1 Viral proteins processed endogenously onto MHC class II

Antigen	Localization	Cell type	References
Measles virus matrix protein	cytosol	HLA-DR-transfected fibroblasts	12
Measles virus nucleocapsid protein	cytosol	HLA-DR-transfected fibroblasts	12
Influenza A virus matrix protein 1	cytosol, nucleus	B cells	13, 14, 26
Influenza A virus hemagglutinin	cytosol, ER	HLA-DR-transfected HeLa cells, B cells	15, 16
Hepatitis C virus core protein	cytosol	B cells	17
Vesicular stomatitis virus ER-restricted protein Gpt	ER	B cells	18
EBNA1	nucleus	B cells	27

12.3
Endogenous MHC Class II Antigen Processing of Viral Antigens

Initial evidence that intracellular antigens gain access to MHC class II presentation came from acid extractions of natural MHC class II ligands. The analysis of eluted peptides from immunoaffinity purified MHC class II molecules of mouse and human macrophages and B cells, including Epstein–Barr virus (EBV)-transformed human B cell lines, revealed that the majority of MHC class II ligands were derived from intracellular proteins [8]. While most ligands were found to originate from membrane and secretory proteins, around 20% were found to have nuclear and cytosolic sources [9–11]. These studies suggested that intracellular proteins gain access to MHC class II presentation by one or more nonclassical endogenous antigen-processing pathways.

These findings were supported when CD4$^+$ T cell recognition after endogenous MHC class II antigen processing was found for several viral antigens (Tab. 12.1). Initially measles virus matrix and nucleocapsid proteins were determined to be presented on HLA-DR molecules to CD4$^+$ T cells after transfection of human fibroblasts [12]. Furthermore, endogenous MHC class II antigen processing was characterized for influenza matrix protein (M1) [13, 14] and hemagglutinin [15, 16] in human HLA-DR-transfected HeLa cervical carcinoma cells or EBV-transformed B cells. In addition, hepatitis C virus core protein was found to be presented to CD4$^+$ T cells after intracellular antigen processing in human EBV-transformed B cells [17]. Moreover, the ER-restricted glycoprotein Gpt of vesicular stomatitis virus followed endogenous MHC class II antigen processing in the murine A20 B cell lymphoma cell line [18]. Finally, the nuclear antigen 1 of EBV (EBNA1) was found to sensitize human EBV-transformed B cell lines for CD4$^+$ T cell recognition [19, 20]. From these examples it becomes clear that important viral antigens are presented on MHC class II molecules to CD4$^+$ T cells after endogenous processing.

The analysis of characteristic processing steps revealed that at least four distinct endogenous MHC class II antigen-processing pathways exist [21]. First, newly synthesized MHC class II molecules can associate with secreted or membrane proteins, such as influenza hemagglutinin [16], and travel together to MHC class II loading compartments, where the respective antigens are processed and loaded onto MHC class II ligands. Second, MHC class II ligands can be generated by the classical MHC class I antigen-processing pathway [15]. Exogenous influenza hemagglutinin and neuraminidase follow this route in dendritic cells after escape from endosomes [22]. Third, cytosolic antigens can gain access to MHC class II loading in a TAP-independent, but proteasome-dependent pathway, thought to involve peptide transport directly into lysosomes [23, 24]. Recently, lysosome-associated membrane protein type 2A (LAMP-2A), the transporter of chaperone-mediated autophagy, was suggested to mediate this antigenic peptide import into lysosomes [25]. However, no viral- or pathogen-derived protein has been found so far to follow this pathway. As a fourth endogenous MHC class II antigen-processing pathway, TAP- and proteasome-independent processing of endogenous antigens in lysosomes was found to lead to MHC class II antigen processing. Two viral antigens have so far been described to follow this pathway – M1 [14, 26] and EBNA1 [27]. For at least one of them, EBNA1, macroautophagy contributes to the delivery of the antigen to lysosomal degradation, followed by MHC class II presentation to CD4$^+$ T cells. It is conceivable that the three latter pathways (proteasome- and TAP-dependent, proteasome-dependent/TAP-independent, and proteasome- and TAP-independent) generate the 20% of natural MHC class II ligands of cytosolic and nuclear origin [10].

12.4
Autophagic Delivery of Antigens for Lysosomal Degradation and MHC Class II Presentation

Six antigens have so far been shown to use autophagy pathways for their presentation on MHC class II. Two autoantigens, glutamate decarboxylase and mutated human immunoglobulin κ chain (SMA), were more efficiently presented to CD4$^+$ T cells upon LAMP-2A overexpression [25]. Since LAMP-2A is the transporter involved in chaperone-mediated autophagy [28, 29], this autophagic pathway was suggested to deliver proteasomal products of autoantigens to MHC class II loading compartments for presentation to CD4$^+$ T cells (Fig. 12.1). The other autophagic pathway implicated in endogenous MHC class II antigen processing is macroautophagy. Neomycin phosphotransferase II [30], complement C5 [31], MUC1 [32] and EBNA1 [27] have been described to be presented on

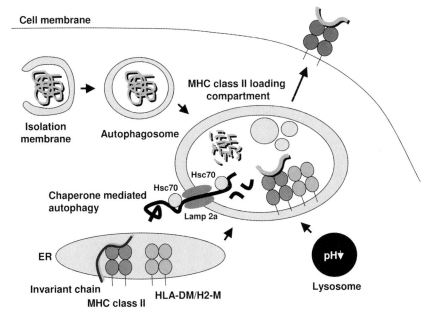

Fig. 12.1 Macroautophagy and chaperone-mediated autophagy target antigens to MHC class II loading compartments. During macroautophagy an isolation membrane engulfs cytoplasmic constituents and delivers them to late endosomes, which are equipped with the MHC class II antigen-loading machinery. MHC class II is transported to this compartment from the ER under the guidance of the Ii. HLA-DM/H2-M, a nonclassical MHC class II molecule and chaperone, reaches this compartment also from the ER and replaces the proteolytic peptide product of Ii (CLIP) with antigenic peptides that are generated from autophagy substrates by lysosomal proteases in the MHC class II loading compartment. Chaperone-mediated autophagy probably directly imports antigens into the same MHC class II loading vesicles for processing onto MHC class II. Cytosolic and endosomal forms of Hsc70 chaperones assist LAMP-2A during the peptide or protein transport in this form of autophagy. MHC class II molecules associated with high-affinity peptide ligands then travel to the cell surface for immune surveillance by CD4+ T cells.

MHC class II after macroautophagy (Fig. 12.1). In all of these studies, inhibitors of class III phosphatidylinositol-3-kinases (PI3Ks) like 3-methyladenine (3-MA) and wortmannin have been used to inhibit autophagosome formation and to downregulate CD4+ T cell recognition of the respective antigen. For EBNA1, localization of the antigen in autophagosomes was demonstrated in deconvolution immunofluorescence and immune electron microcopy. CD4+ T cell recognition of EBNA1 could in addition to inhibition by 3-MA also be downregulated by RNA interference against *atg12*, an essential autophagy gene. The studies described above implicate two autophagic pathways, i.e. macroautophagy and chaperone-mediated autophagy, in the endogenous MHC class II antigen processing and CD4+ T cell recognition of B cells, dendritic cells and macrophages.

For EBV-transformed B cell lines (LCLs) autophagy has also been demonstrated to generate natural MHC class II ligands. In biochemical studies HLA-

DR ligands were characterized with and without autophagy induction by starvation [11]. After 24 h of autophagy induction peptide presentation on MHC class II from intracellular and lysosomal proteins rose by 50%, while ligands derived from membrane and secreted proteins remained constant. Three cytosolic/nuclear proteins (eukaryotic translation elongation factor 1α, ubiquitin protein ligase NEDD4La and RAD23 homolog B nucleotide excision repair protein) and one lysosomal protein (cathepsin D) were most dramatically upregulated in their presentation on HLA-DR. In addition to MHC class II ligand induction upon starvation, the same study found two peptides of the autophagosome-associated protein Atg8/microtubule-associated protein 1 light chain 3 (LC3) presented on HLA-DR of LCLs cultured under nutrient-rich conditions. Therefore, Atg8/LC3, which is coupled to autophagosome membranes and partially degraded with autophagy substrates, is processed onto MHC class II under steady-state conditions. This study suggested that autophagy delivers cytoplasmic antigens for MHC class II presentation constitutively and in an enhanced manner upon starvation.

12.5
Similarities Between Sources of MHC Class II Ligands and Autophagy Substrates

Autophagic delivery seems to allow cytosolic and nuclear antigens to gain access to MHC class II presentation. Therefore, immune surveillance of intracellular antigens seems not to be exclusively the domain of CD8$^+$ T cells. Since antigen localization is not the decisive factor for MHC class I versus class II presentation anymore, the question arises which protein features target antigens primarily for CD8$^+$ versus CD4$^+$ T cell recognition. One possibility is that the characteristics of proteasome and lysosome substrates overlap with the qualities targeting preferentially for MHC class I and class II loading, respectively. Proteasome substrates are primarily short-lived proteins [3] and one class of short-lived proteins, i.e. cyclins, was frequently found to be the source of natural MHC class I, but so far not of MHC class II ligands [33]. Conversely, autophagy substrates destined for lysosomal proteolysis have been reported to be mainly long-lived ($t_{1/2} > 30$ min) [34], e.g. glyceraldehyde-3-phosphate dehydrogenase (GAPDH) with its estimated half-life of 130 h is a representative of this protein category [35]. GAPDH was found to be both a substrate of chaperone-mediated autophagy [36] and to be contained in autophagosomes [37]. Consistent with long-lived autophagy substrates being presented on MHC class II are the reports of natural HLA class II ligands of GAPDH that have been isolated from four different HLA-DR and -DQ alleles by different investigators [8, 38, 39]. So far no natural MHC class I ligands of GAPDH have been found. Supporting this analysis, half-life modification of the M1 influenced its MHC class I versus class II presentation [26]. Long-lived wild-type M1 ($t_{1/2} = 5$ h) gets endogenously processed both for MHC class I and II presentation. Shortening the half-life to 10 min abolished endogenous MHC class II presentation to CD4$^+$ T cells of M1, but maintained CD8$^+$ T cell recognition. Along these lines EBNA1 is recognized by CD4$^+$ T cells after endogenous MHC class II processing involving autophagy [27]. EBNA1 is a very long-lived protein ($t_{1/2} > 20$ h in EBV transformed B cells) due to its glycine–alanine repeat (GA) domain, which inhibits rapid degradation of EBNA1 by proteasomes [40]. When the GA domain is removed, EBNA1's half-life shortens and CD8$^+$ T cell recognition of EBV-transformed B cells is increased 4-fold [41]. Finally, DRiPs are rapidly turned over by proteasomes and contribute substantially to MHC class I ligand pools [42, 43]. Taken together these studies suggest that long-lived cytosolic and nuclear antigens are preferentially targeted for MHC class II presentation via autophagy and short-lived primarily for loading onto MHC class I.

How might protein half-life influence the degradation pathways of proteins? One possible mechanism could be that long-lived proteins preferentially gain access to protein aggregates in cells, which then in turn serve as autophagy substrates. There is so far little evidence for this hypothesis. However, in the autophagy-related cytosol-to-vacuole (Cvt) pathway in yeast, aminopeptidase I is transported to the vacuole, the lysosome-equivalent in yeast, in a double-membrane-coated vesicle only after its proenzyme aggregates to dodecamers in the cytosol [44]. Furthermore, disruption of the essential autophagy gene Atg5 leads to the accumulation of protein aggregates in mouse embryonic fibroblasts [45]. Such protein aggregates containing polyglutamine and polyalanine harboring proteins get degraded by autophagy *in vitro* in COS-7 cells and *in vivo* in mice and flies [46, 47]. Finally, the autophagy-linked FYVE protein (Alfy) localizes to ubiquitin-containing protein aggregates, which are partially targeted to autophagosomes [48]. All these studies suggest that protein aggregates might be preferentially cleared by autophagy. However, why long-lived proteins might be selectively incorporated into protein aggregates and how these are recognized by the autophagy machinery remains unclear, requiring additional studies in the future.

12.6
Overlap Between the Vesicular Transport Pathways of Autophagosomes and MHC Class II Loading Compartments

Two lines of cell biological evidence suggest that autophagosomes deliver their cargo to MHC class II loading compartments. First, autophagosomes morphologically resemble MHC class II loading vesicles. These cellular compartments, in which the Ii is degraded and MHC class II is loaded with high-affinity antigenic peptides under the influence of the nonclassical MHC class II molecule HLA-DM/H2-M [1], have been described to be multivesicular or multilamellar in morphology (Fig. 12.1) [49]. Isolated autophagosomes display a similar phenotype with multiple intravesicular membranes [50, 51]. In addition, multilamellar body development was found to depend on autophagy in lung epithelial cells [52]. Therefore, autophagosomes might contribute intravesicular membranes to MHC class II loading compartments.

Autophagosomes have also been described to fuse with late endosomes [53, 54], which in professional antigen-presenting cells are equipped with the MHC class II antigen loading machinery [55]. Fusion products of autophagosomes and late endosomes, so-called amphisomes, were found to contain colloidal gold-loaded endosomes and gold-negative, double-membrane-coated autophagosomes. These amphisomes could be stabilized by inhibition of lysosomal degradation. These studies suggest that autophagosomes frequently target their cargo to late endosomes, which in B cells, macrophages and dendritic cells are equipped with the machinery for MHC class II loading.

12.7
Possible Functions of MHC Class II Presentation after Autophagy in the Immune Control of Viral Infections

The above studies on autophagy in endogenous MHC class II antigen processing have so far primarily focused on classical antigen-presenting cells like B cells [11, 25, 27, 30], macrophages [31] and dendritic cells [32]. However, this pathway might be even more important for cells with limited endocytic capacity. In support of this hypothesis, thymic cortical epithelial cells have been found to show a high level of constitutive macroautophagy in mice, in which autophagosomes were selectively visualized via a transgene encoding for Green Fluorescent Protein (GFP)-tagged Atg8/LC3 [56]. Cortical epithelial cells are capable of mediating positive selection of T lymphocytes in the thymus, including CD4[+] T cells that have to be educated on MHC class II [57]. Since thymic cortical epithelial cells are only weakly phagocytic, endogenous MHC class II antigen processing might play a major role in loading MHC class II molecules for positive selection of CD4[+] T cells and autophagy might contribute self-antigens for this positive T cell selection.

In addition to constitutive MHC class II expression on some cell types with low endocytic capacity, such as thymic cortical epithelial cells, it is also upregulated on many tissues after inflammation [58]. Mediated primarily by interferon-γ, endothelial and epithelial tissues upregulate MHC class II after inflammation. In addition, nearly all activated human lymphocytes display HLA class II on their cell surface. Since endocytosis is not increased by these treatments, a substantial proportion of MHC class II molecules on these tissues might be loaded with intracellular antigens after endogenous antigen processing. Immune surveillance of intracellular pathogens by this pathway might provide T cell help for adaptive immune responses and evoke direct effector functions of CD4[+] T cells against infected cells. Specifically, CD4[+] T cell help has been reported to be essential for the maintenance of protective CD8[+] T cell responses [59, 60]. In various viral infections, CD4[+] T cells have also been found to directly lyse infected cells [20, 61] and mediate immune control on their own [62–66]. Therefore, endogenous MHC class II processing after autophagy at sites of inflammation might enhance CD4[+] T cell help and elicit antiviral effector mechanisms.

Long-lived autophagy substrates could complement immune surveillance by CD8$^+$ T cells, primarily focusing on short-lived antigens degraded by proteasomes.

12.8
Future Directions of Research into Endogenous MHC Class II Antigen Processing

The novel endogenous MHC class II processing pathways via autophagy promise to allow a better understanding of immune surveillance by CD4$^+$ T cells and should be targeted for enhanced presentation of antigens on MHC class II. For both purposes the autophagic delivery of antigens has to be characterized in more detail. Currently, there are two main challenges in our understanding of substrate selection for autophagy. One is the characterization of autophagy substrates. The preferential degradation of long-lived proteins and damaged cell organelles by lysosomes after autophagy suggests that autophagosomes not just unspecifically engulf cytosolic material. The characterization of more substrates for this degradation pathway should allow us to define features of proteins that render them susceptible to autophagic degradation. The second challenge is how constituents of other cellular compartments gain access to this pathway. In this respect, nuclear antigens like EBNA1 might either be exported from the nucleus prior to MHC class II antigen processing via autophagy or engulfed into autophagosomes before they can reach the nucleus. In addition, *Mycobacterium tuberculosis* (Mtb) containing phagosomes have been found to fuse with autophagosomes [67] and apart from hydrolytic destruction this might lead to MHC class II presentation of Mtb constituents. Characterization of the mechanisms that make noncytosolic antigens accessible to autophagy will allow us to understand which antigens can be surveilled by CD4$^+$ T cells after endogenous MHC class II antigen processing via autophagy.

Furthermore, autophagy can be targeted for enhanced MHC class II presentation of antigens. Two autophagy pathways, i.e. macroautophagy [27] and chaperone-mediated autophagy [25], have been implicated in endogenous MHC class II antigen processing and should be harnessed for efficient antigen display by MHC class II. Known autophagy substrates or, once the import characteristics into autophagosomes have been defined, domains of these can be fused to the antigen of choice to enhance endogenous MHC class II antigen processing and increase CD4$^+$ T cell recognition. Hybrid antigens constructed in this fashion can then be delivered to antigen-presenting cells with viral vectors for immunizations. One can envision that the same antigen, targeted for MHC class I and class II presentation with two hybrid proteins could be incorporated into a vaccine for optimal CD8$^+$ and CD4$^+$ T cell stimulations to achieve comprehensive immune control and long-lasting immune memory.

12.9
Summary

Endogenous MHC class II antigen processing has been demonstrated for some antigens of influenza, measles, hepatitis C, vesicular stomatitis and EBV. For EBNA1, macroautophagy has been shown to contribute to antigen delivery for HLA class II presentation. These findings raise the possibility that long-lived proteins, which are preferentially degraded by autophagy, give rise to MHC class II ligands, including viral antigens relevant to disease. CD4$^+$ T cell recognition of antigens displayed by this pathway can contribute to antiviral immune control by assisting adaptive *immunity* and target infected cells directly. The antigen characteristics targeting this pathway should be defined in more detail in order to harness them for targeting of viral and tumor proteins for enhanced MHC class II presentation.

Acknowledgments

We thank the National Cancer Institute (R01CA108609), the Arnold and Mabel Beckman Foundation, and the Alexandrine and Alexander Sinsheimer Foundation for supporting our research (to C.M.). D.S. is recipient of a Predoctoral Fellowship from the Schering Foundation.

References

1 Trombetta ES, Mellman I. **2005**. Cell biology of antigen processing *in vitro* and *in vivo*. *Annu Rev Immunol 23*, 975–1028.

2 Kloetzel PM. **2001**. Antigen processing by the proteasome. *Nat Rev Mol Cell Biol 2*, 179–187.

3 Ciechanover A, Finley D, Varshavsky A. **1984**. Ubiquitin dependence of selective protein degradation demonstrated in the mammalian cell cycle mutant ts85. *Cell 37*, 57–66.

4 Yewdell JW, Reits E, Neefjes J. **2003**. Making sense of mass destruction: quantitating MHC class I antigen presentation. *Nat Rev Immunol 3*, 952–961.

5 Shastri N, Schwab S, Serwold T. **2002**. Producing nature's gene-chips: the generation of peptides for display by MHC class I molecules. *Annu Rev Immunol 20*, 463–493.

6 Bryant P, Ploegh H. **2004**. Class II MHC peptide loading by the professionals. *Curr Opin Immunol 16*, 96–102.

7 Busch R, Doebele RC, Patil NS, Pashine A, Mellins ED. **2000**. Accessory molecules for MHC class II peptide loading. *Curr Opin Immunol 12*, 99–106.

8 Rammensee H-G, Bachman J, Stevanovic S. **1997**. *MHC Ligands and Peptide Motifs*. Austin, TX: Springer/Landes Bioscience.

9 Chicz RM, Urban RG, Gorga JC, Vignali DA, Lane WS, Strominger JL. **1993**. Specificity and promiscuity among naturally processed peptides bound to HLA-DR alleles. *J Exp Med 178*, 27–47.

10 Dongre AR, Kovats S, deRoos P, McCormack AL, Nakagawa T, Paharkova-Vatchkova V, Eng J, Caldwell H, Yates JR, 3rd., Rudensky AY. **2001**. *In vivo* MHC class II presentation of cytosolic proteins revealed by rapid automated tandem mass spectrometry and functional analyses. *Eur J Immunol 31*, 1485–1494.

11 Dengjel J, Schoor O, Fischer R, Reich M, Kraus M, Muller M, Kreymborg K, Al-

tenberend F, Brandenburg J, Kalbacher H, Brock R, Driessen C, Rammensee HG, Stevanovic S. **2005**. Autophagy promotes MHC class II presentation of peptides from intracellular source proteins. *Proc Natl Acad Sci USA 102*, 7922–7927.

12 Jacobson S, Sekaly RP, Jacobson CL, McFarland HF, Long EO. **1989**. HLA class II-restricted presentation of cytoplasmic measles virus antigens to cytotoxic T cells. *J Virol 63*, 1756–1762.

13 Nuchtern JG, Biddison WE, Klausner RD. **1990**. Class II MHC molecules can use the endogenous pathway of antigen presentation. *Nature 343*, 74–76.

14 Jaraquemada D, Marti M, Long EO. **1990**. An endogenous processing pathway in vaccinia virus-infected cells for presentation of cytoplasmic antigens to class II-restricted T cells. *J Exp Med 172*, 947–954.

15 Malnati MS, Marti M, LaVaute T, Jaraquemada D, Biddison W, DeMars R, Long EO. **1992**. Processing pathways for presentation of cytosolic antigen to MHC class II-restricted T cells. *Nature 357*, 702–704.

16 Aichinger G, Karlsson L, Jackson MR, Vestberg M, Vaughan JH, Teyton L, Lechler RI, Peterson PA. **1997**. Major histocompatibility complex class II-dependent unfolding, transport, and degradation of endogenous proteins. *J Biol Chem 272*, 29127–29136.

17 Chen M, Shirai M, Liu Z, Arichi T, Takahashi H, Nishioka M. **1998**. Efficient class II major histocompatibility complex presentation of endogenously synthesized hepatitis C virus core protein by Epstein–Barr virus-transformed B-lymphoblastoid cell lines to CD4[+] T cells. *J Virol 72*, 8301–8308.

18 Bartido SM, Diment S, Reiss CS. **1995**. Processing of a viral glycoprotein in the endoplasmic reticulum for class II presentation. *Eur J Immunol 25*, 2211–2219.

19 Münz C, Bickham KL, Subklewe M, Tsang ML, Chahroudi A, Kurilla MG, Zhang D, O'Donnell M, Steinman RM. **2000**. Human CD4[+] T lymphocytes consistently respond to the latent Epstein–Barr virus nuclear antigen EBNA1. *J Exp Med 191*, 1649–1660.

20 Paludan C, Bickham K, Nikiforow S, Tsang ML, Goodman K, Hanekom WA, Fonteneau JF, Stevanovic S, Münz C. **2002**. EBNA1 specific CD4[+] T$_h$1 cells kill Burkitt's lymphoma cells. *J Immunol 169*, 1593–1603.

21 Lechler R, Aichinger G, Lightstone L. **1996**. The endogenous pathway of MHC class II antigen presentation. *Immunol Rev 151*, 51–79.

22 Tewari MK, Sinnathamby G, Rajagopal D, Eisenlohr LC. **2005**. A cytosolic pathway for MHC class II-restricted antigen processing that is proteasome and TAP dependent. *Nat Immunol 6*, 287–294.

23 Lich JD, Elliott JF, Blum JS. **2000**. Cytoplasmic processing is a prerequisite for presentation of an endogenous antigen by major histocompatibility complex class II proteins. *J Exp Med 191*, 1513–1524.

24 Dani A, Chaudhry A, Mukherjee P, Rajagopal D, Bhatia S, George A, Bal V, Rath S, Mayor S. **2004**. The pathway for MHCII-mediated presentation of endogenous proteins involves peptide transport to the endo-lysosomal compartment. *J Cell Sci 117*, 4219–4230.

25 Zhou D, Li P, Lott JM, Hislop A, Canaday DH, Brutkiewicz RR, Blum JS. **2005**. Lamp-2a facilitates MHC class II presentation of cytoplasmic antigens. *Immunity 22*, 571–581.

26 Gueguen M, Long EO. **1996**. Presentation of a cytosolic antigen by major histocompatibility complex class II molecules requires a long-lived form of the antigen. *Proc Natl Acad Sci USA 93*, 14692–14697.

27 Paludan C, Schmid D, Landthaler M, Vockerodt M, Kube D, Tuschl T, Münz C. **2005**. Endogenous MHC class II processing of a viral nuclear antigen after autophagy. *Science 307*, 593–596.

28 Cuervo AM, Dice JF. **1996**. A receptor for the selective uptake and degradation of proteins by lysosomes. *Science 273*, 501–503.

29 Cuervo AM, Dice JF. **2000**. Unique properties of lamp2a compared to other lamp2 isoforms. *J Cell Sci 113*, 4441–4450.

30 Nimmerjahn F, Milosevic S, Behrends U, Jaffee EM, Pardoll DM, Bornkamm

GW, Mautner J. **2003**. Major histocompatibility complex class II-restricted presentation of a cytosolic antigen by autophagy. *Eur J Immunol 33*, 1250–1259.

31 Brazil MI, Weiss S, Stockinger B. **1997**. Excessive degradation of intracellular protein in macrophages prevents presentation in the context of major histocompatibility complex class II molecules. *Eur J Immunol 27*, 1506–1514.

32 Dörfel D, Appel S, Grunebach F, Weck MM, Muller MR, Heine A, Brossart P. **2005**. Processing and presentation of HLA class I and II epitopes by dendritic cells after transfection with *in vitro* transcribed MUC1 RNA. *Blood 105*, 3199–3205.

33 Rammensee H, Bachmann J, Emmerich NP, Bachor OA, Stevanovic S. **1999**. SYFPEITHI: database for MHC ligands and peptide motifs. *Immunogenetics 50*, 213–219.

34 Henell F, Berkenstam A, Ahlberg J, Glaumann H. **1987**. Degradation of short- and long-lived proteins in perfused liver and in isolated autophagic vacuoles – lysosomes. *Exp Mol Pathol 46*, 1–14.

35 Dice JF, Goldberg AL. **1975**. A statistical analysis of the relationship between degradative rates and molecular weights of proteins. *Arch Biochem Biophys 170*, 213–219.

36 Aniento F, Roche E, Cuervo AM, Knecht E. **1993**. Uptake and degradation of glyceraldehyde-3-phosphate dehydrogenase by rat liver lysosomes. *J Biol Chem 268*, 10463–10470.

37 Fengsrud M, Raiborg C, Berg TO, Stromhaug PE, Ueno T, Erichsen ES, Seglen PO. **2000**. Autophagosome-associated variant isoforms of cytosolic enzymes. *Biochem J 352*, 773–781.

38 Friede T, Gnau V, Jung G, Keilholz W, Stevanovic S, Rammensee HG. **1996**. Natural ligand motifs of closely related HLA-DR4 molecules predict features of rheumatoid arthritis associated peptides. *Biochim Biophys Acta 1316*, 85–101.

39 Harris PE, Maffei A, Colovai AI, Kinne J, Tugulea S, Suciu-Foca N. **1996**. Predominant HLA-class II bound self-peptides of a hematopoietic progenitor cell line are derived from intracellular proteins. *Blood 87*, 5104–5112.

40 Levitskaya J, Sharipo A, Leonchiks A, Ciechanover A, Masucci MG. **1997**. Inhibition of ubiquitin/proteasome-dependent protein degradation by the Gly-Ala repeat domain of the Epstein–Barr virus nuclear antigen 1. *Proc Natl Acad Sci USA 94*, 12616–12621.

41 Lee SP, Brooks JM, Al-Jarrah H, Thomas WA, Haigh TA, Taylor GS, Humme S, Schepers A, Hammerschmidt W, Yates JL, Rickinson AB, Blake NW. **2004**. CD8 T cell recognition of endogenously expressed Epstein–Barr virus nuclear antigen 1. *J Exp Med 199*, 1409–1420.

42 Reits EA, Vos JC, Gromme M, Neefjes J. **2000**. The major substrates for TAP *in vivo* are derived from newly synthesized proteins. *Nature 404*, 774–778.

43 Schubert U, Anton LC, Gibbs J, Norbury CC, Yewdell JW, Bennink JR. **2000**. Rapid degradation of a large fraction of newly synthesized proteins by proteasomes. *Nature 404*, 770–774.

44 Kim J, Klionsky DJ. **2000**. Autophagy, cytoplasm-to-vacuole targeting pathway, and pexophagy in yeast and mammalian cells. *Annu Rev Biochem 69*, 303–342.

45 Yoshimori T. **2004**. Autophagy: a regulated bulk degradation process inside cells. *Biochem Biophys Res Commun 313*, 453–458.

46 Ravikumar B, Vacher C, Berger Z, Davies JE, Luo S, Oroz LG, Scaravilli F, Easton DF, Duden R, O'Kane CJ, Rubinsztein DC. **2004**. Inhibition of mTOR induces autophagy and reduces toxicity of polyglutamine expansions in fly and mouse models of Huntington disease. *Nat Genet 36*, 585–595.

47 Ravikumar B, Duden R, Rubinsztein DC. **2002**. Aggregate-prone proteins with polyglutamine and polyalanine expansions are degraded by autophagy. *Hum Mol Genet 11*, 1107–1117.

48 Simonsen A, Birkeland HC, Gillooly DJ, Mizushima N, Kuma A, Yoshimori T, Slagsvold T, Brech A, Stenmark H. **2004**. Alfy, a novel FYVE-domain-containing protein associated with protein granules and autophagic membranes. *J Cell Sci 117*, 4239–4251.

49 Zwart W, Griekspoor A, Kuijl C, Marsman M, van Rheenen J, Janssen H, Calafat J, van Ham M, Janssen L, van Lith M, Jalink K, Neefjes J. 2005. Spatial separation of HLA-DM/HLA-DR interactions within MIIC and phagosome-induced immune escape. *Immunity* 22, 221–233.

50 Fengsrud M, Erichsen ES, Berg TO, Raiborg C, Seglen PO. 2000. Ultrastructural characterization of the delimiting membranes of isolated autophagosomes and amphisomes by freeze-fracture electron microscopy. *Eur J Cell Biol* 79, 871–882.

51 Stromhaug PE, Berg TO, Fengsrud M, Seglen PO. 1998. Purification and characterization of autophagosomes from rat hepatocytes. *Biochem J* 335, 217–224.

52 Lajoie P, Guay G, Dennis JW, Nabi IR. 2005. The lipid composition of autophagic vacuoles regulates expression of multilamellar bodies. *J Cell Sci* 118, 1991–2003.

53 Berg TO, Fengsrud M, Stromhaug PE, Berg T, Seglen PO. 1998. Isolation and characterization of rat liver amphisomes. Evidence for fusion of autophagosomes with both early and late endosomes. *J Biol Chem* 273, 21883–21892.

54 Liou W, Geuze HJ, Geelen MJ, Slot JW. 1997. The autophagic and endocytic pathways converge at the nascent autophagic vacuoles. *J Cell Biol* 136, 61–70.

55 Kleijmeer MJ, Morkowski S, Griffith JM, Rudensky AY, Geuze HJ. 1997. Major histocompatibility complex class II compartments in human and mouse B lymphoblasts represent conventional endocytic compartments. *J Cell Biol* 139, 639–649.

56 Mizushima N, Yamamoto A, Matsui M, Yoshimori T, Ohsumi Y. 2004. *In vivo* analysis of autophagy in response to nutrient starvation using transgenic mice expressing a fluorescent autophagosome marker. *Mol Biol Cell* 15, 1101–1111.

57 Starr TK, Jameson SC, Hogquist KA. 2003. Positive and negative selection of T cells. *Annu Rev Immunol* 21, 139–76.

58 Reith W, Mach B. 2001. The bare lymphocyte syndrome and the regulation of MHC expression. *Annu Rev Immunol* 19, 331–373.

59 Sun JC, Williams MA, Bevan MJ. 2004. CD4$^+$ T cells are required for the maintenance, not programming, of memory CD8$^+$ T cells after acute infection. *Nat Immunol* 5, 927–933.

60 Bevan MJ. 2004. Helping the CD8$^+$ T-cell response. *Nat Rev Immunol* 4, 595–602.

61 Jellison ER, Kim SK, Welsh RM. 2005. Cutting edge: MHC class II-restricted killing *in vivo* during viral infection. *J Immunol* 174, 614–618.

62 Nikiforow S, Bottomly K, Miller G, Münz C. 2003. Cytolytic CD4$^+$-T-cell clones reactive to EBNA1 inhibit Epstein–Barr virus-induced B-cell proliferation. *J Virol* 77, 12088–12104.

63 Stevenson PG, Cardin RD, Christensen JP, Doherty PC. 1999. Immunological control of a murine gammaherpesvirus independent of CD8$^+$ T cells. *J Gen Virol* 80, 477–483.

64 Sparks-Thissen RL, Braaten DC, Kreher S, Speck SH, Virgin HW. 2004. An optimized CD4 T-cell response can control productive and latent gammaherpesvirus infection. *J Virol* 78, 6827–6835.

65 Robertson KA, Usherwood EJ, Nash AA. 2001. Regression of a murine gammaherpesvirus 68-positive B-cell lymphoma mediated by CD4 T lymphocytes. *J Virol* 75, 3480–3482.

66 Fu T, Voo KS, Wang RF. 2004. Critical role of EBNA1-specific CD4$^+$ T cells in the control of mouse Burkitt lymphoma *in vivo*. *J Clin Invest* 114, 542–550.

67 Gutierrez MG, Master SS, Singh SB, Taylor GA, Colombo MI, Deretic V. 2004. Autophagy is a defense mechanism inhibiting BCG and *Mycobacterium tuberculosis* survival in infected macrophages. *Cell* 119, 753–766.

13
Autophagy in Antiviral Host Defense

Beth Levine

13.1
Introduction

Autophagy is emerging as an important mechanism of innate immunity against viral pathogens. This concept is supported by several lines of evidence, including (a) links between type I and type II interferon (IFN) signaling and autophagy regulation, (b) antiviral effects of plant and mammalian autophagy genes, and (c) suppression of host autophagy by viral virulence factors. The recently described role of autophagy in antigen presentation suggests that autophagy may also foster adaptive immune responses to viral infections. In this chapter we highlight recent discoveries in each of these areas. The reader is referred to the addendum after this chapter for a discussion of how some viruses (e.g. poliovirus, murine hepatitis virus) may subvert the host autophagy pathway to foster their own replication.

13.2
Role of Antiviral Signaling Pathways in Autophagy Regulation

Two major signaling molecules involved in antiviral immunity have been shown to positively regulate autophagy – the type I IFN-inducible double-stranded (ds) RNA-dependent protein kinase R (PKR) and type II (or IFN-γ) (Fig. 13.1).

13.2.1
PKR is Required for Virus-induced Autophagy

The IFN-inducible molecule, PKR, plays an important role in innate immunity against viral infections and many viruses encode gene products that block PKR function (reviewed in Refs. [1, 2]). Until recently, it was believed that PKR functions as an antiviral effector molecule entirely via its ability to shutdown host and viral protein synthesis in infected cells through phosphorylation of the α subunit of eukaryotic initiation factor 2 (eIF2α). However, PKR and its downstream

Autophagy in Immunity and Infection. Edited by Vojo Deretic
Copyright © 2006 WILEY-VCH Verlag GmbH & Co. KGaA, Weinheim
ISBN: 3-527-31450-4

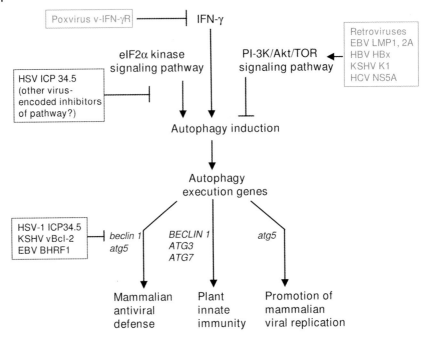

Fig. 13.1 Conceptual overview of the inter-relationships between autophagy signaling pathways, autophagy execution genes, autophagy functions in viral infections and viral inhibitors of autophagy. The viral gene products shown in gray are known to modulate the indicated autophagy signals, but have not yet been shown to inhibit autophagy. See text for a more detailed explanation of these concepts. The role of autophagy in promoting viral replication is discussed in the addendum after this chapter.

phosphorylation target, the Ser51 residue of eIF2a, are both required for autophagy induction in virally infected cells [3]. At present, the relative contributions of PKR-dependent host cell shutoff and PKR-dependent autophagy in mediating its antiviral effects are unknown. Nonetheless, since autophagy execution genes are known to inhibit viral replication (see below), the possibility that some of the antiviral effects of PKR may be mediated through autophagy is intriguing.

The regulation of virus-induced autophagy by PKR is part of an evolutionarily conserved strategy by which cells activate autophagy in response to different forms of cellular stress [3]. In yeast, disruption of the starvation-activated eIF2a kinase, Gcn2, mutation of the Ser51 phosphorylation site of eIF2a and mutation of a transcription factor downstream of Gcn2/eIF2a, Gcn4, block starvation-induced autophagy, and mammalian PKR rescues starvation-induced autophagy in *gcn2* null yeast. In mammalian cells, disruption of PKR blocks virus-induced autophagy, and mutation of the Ser51 residue of eIF2a blocks starvation and virus-induced autophagy. These findings indicate an unequivocal role for the eIF2a kinase signaling pathway in both virus- and starvation-induced autophagy. Since many viruses also activate the endoplasmic reticulum (ER) stress-induced

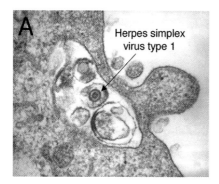

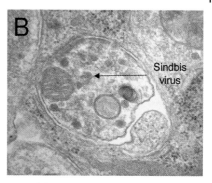

Fig. 13.2 Electron micrographs demonstrating autophagic degradation of HSV-1 (A) and Sindbis virus (B) in neurons. The HSV-1 strain is a mutant lacking the virus-encoded PKR inhibitor, ICP34.5. Similar structures are not observed in neurons infected with wild-type HSV-1 (containing ICP34.5), or in HSV-1-infected or Sindbis virus-infected *pkr*$^{-/-}$ neurons.

eIF2a kinase, PERK, the prediction is that virus-induced ER stress may represent yet another mechanism to activate host cellular autophagy.

Studies using a herpes simplex virus (HSV) type 1 model have begun to unravel a possible antiviral role of PKR-dependent autophagy [4]. HSV-1 encodes a neurovirulence protein, ICP34.5, that binds to a protein phosphatase and causes it to dephosphorylate eIF2a, thereby negating the activity of PKR [5–7]. A neuroattenuated HSV-1 mutant lacking ICP34.5 exhibits wild-type replication and virulence in mice genetically lacking *pkr*, proving that the ICP34.5 gene product mediates neurovirulence by antagonizing PKR-dependent functions [8]. Talloczy et al. found that in murine embryonic fibroblasts and neurons infected with the mutant HSV-1 virus lacking the PKR inhibitor, ICP34.5, significantly more HSV-1 virions are degraded inside autophagosomes (Fig. 13.2); moreover, the degradation of these mutant virions is blocked in cells with a null mutation in *pkr* [4]. Thus, PKR stimulates the autophagic degradation of HSV-1 and this function of PKR is antagonized by the ICP34.5 neurovirulence gene product. Although further studies are needed to dissect the relative contributions of PKR-dependent translational control and PKR-dependent autophagy in the regulation of HSV-1 replication, it is reasonable to speculate that PKR-dependent autophagic degradation of viruses may be involved in innate antiviral immunity.

During HSV-1 infection, PKR-dependent autophagy not only targets viruses for degradation, but also increases the autophagic degradation of long-lived cellular proteins [3]. In theory, the degradation of cellular proteins could have either a proviral role by generating building blocks for viral protein synthesis or an antiviral role by degrading host proteins necessary for viral replication. Independently of effects on viral replication, the autophagic degradation of self-constituents may benefit host cells during viral infection, either by promoting adaptations to nutrient, oxidative, ER or other forms of cellular stress (Fig. 13.3).

To date, the role of PKR in autophagy regulation has only been studied in the context of HSV-1 infection. However, it is important to note the multiplicity of

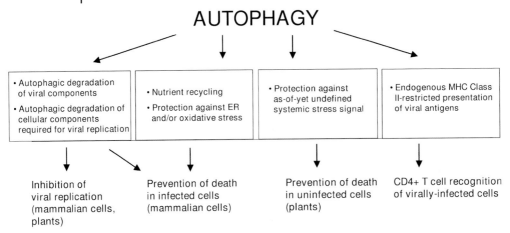

Fig. 13.3 Conceptual overview of the protective roles of autophagy in mammalian and plant viral infections. Areas in boxed regions represent potential mechanisms by which autophagy exerts each type of protective effect.

strategies that different viruses have evolved to antagonize PKR function (reviewed in Refs. [2, 9]). For example (a) vaccinia virus E3, influenza virus NS1, HSV-1 Us11, reovirus σ3 and rotavirus NSP3 encode dsRNA-binding proteins that prevent PKR activation; (b) adenovirus VAI RNAs and HIV Tar RNAs bind to dsRNA substrates and inhibit PKR; (c) hepatitis C virus NS5A protein inhibits the dimerization of PKR; (d) influenza virus recruits a cellular protein, P58^{IPK}, that directly interacts with PKR and inhibits its dimerization; and (e) vaccinia virus K3L, hepatitis C virus E2 and HIV Tat proteins act as pseudosubstrates of PKR. Given the evolutionarily conserved requirement for an intact eIF2α kinase signaling pathway in autophagy induction, the prediction is that these other viral inhibitors of PKR, like HSV-1 ICP34.5, will also function as antagonists of host autophagy. Further studies are needed to test this prediction and to study the role of viral evasion of autophagy in the pathogenesis of diseases caused by viral pathogens that encode PKR (and putative autophagy) inhibitors.

13.2.2
IFN-γ Stimulates Autophagy

Type II IFN (IFN-γ) is synthesized by cells of the immune system, including natural killer cells, CD4$^+$ T cells and CD8$^+$ T cells, and plays a role in host defense against viral as well as intracellular bacterial pathogens (reviewed in Ref. [10]). Although not yet studied in the context of viral infection, three different studies have shown that IFN-γ induces autophagy. Inbal et al. and Pyo et al. both found that IFN-γ stimulates autophagy in HeLa cells [11, 12], and Gutierrez et al found that IFN-γ activates autophagy in macrophages [13]. The latter observations are particularly relevant to antimycobacterial immunity, since IFN-γ is an important antituberculosis immune mediator and autophagy has recently been shown to be a defense mechanism inhibiting *Mycobacterium tuberculosis* survival in infected macrophages (reviewed in Ref. [14]). These findings raise the possibility that IFN-γ may, at least in part, exert antimycobacterial activity via autophagy induction.

A critical question for the field of viral immunology is whether IFN-γ acts in a similar fashion during viral infections and whether the activation of autophagy by IFN-γ contributes to antiviral host defense. Moreover, as with PKR, viruses have evolved strategies to antagonize the antiviral actions of IFN-γ (Fig. 13.1). For example, several poxviruses encode viral soluble receptors for IFN-γ (reviewed in Ref. [15]), and paramyxoviruses and herpesviruses encode inhibitors of the signal transducer and activators of transcription (STAT) family of proteins that activate gene transcription downstream of IFN-γ (reviewed in Refs. [16, 17]). It will therefore be important to examine the effects of viral antagonism of IFN-γ activity on autophagy regulation.

13.3
Role of Mammalian Autophagy Genes in Antiviral Host Defense

The role of PKR and IFN-γ in antiviral immunity is well established, but both of these molecules regulate many different cellular processes and it is not yet known exactly what role autophagy induction plays in their antiviral effects. The concept that autophagy is important in innate immunity is more directly supported by studies involving components of the mammalian and plant autophagic machinery.

The first identified mammalian autophagy gene product, Beclin 1, was isolated in a yeast two-hybrid screen, in the context of studies of the mechanism by which the antiapoptotic protein, Bcl-2, protects mice against lethal Sindbis virus encephalitis [18]. Similar to the neuroprotective effects of Bcl-2 [19], enforced neuronal expression of wild-type Beclin 1 in a recombinant chimeric Sindbis virus vector reduces Sindbis virus replication, reduces Sindbis virus-induced apoptosis and protects mice against lethal Sindbis virus encephalitis [18] (Fig. 13.4). Mutations in the Bcl-2-binding domain of Beclin 1 (that preserve autophagy function) and mutations in other regions of Beclin 1 that block autophagy function inhibit the protective effects of Beclin 1 during Sindbis virus infection [18, 20]. Thus, it appears that both the interaction with Bcl-2 and the autophagy function may be required for the antiviral effects of Beclin 1.

Preliminary studies with *beclin 1* null embryonic stem (ES) cells and *atg5* null mouse embryonic fibroblasts (MEFs) also indicate a role for these two endogenous autophagy genes in innate immunity against Sindbis virus infection [21]. Sindbis virus replicates to higher titers and results in accelerated death in *beclin 1* null ES cells as compared to wild-type ES cells and *atg5* null MEFs as compared to wild-type MEFs. It is not yet known whether the acceleration of virus-induced death

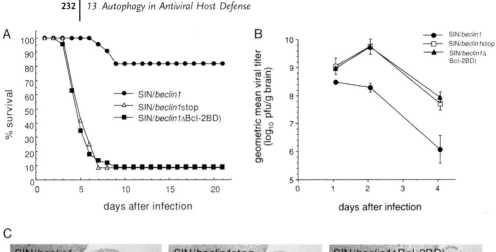

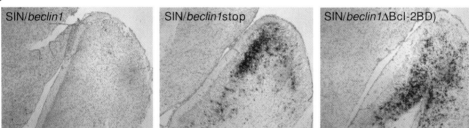

Fig. 13.4 Enforced neuronal expression of Beclin 1 protects mice against lethal Sindbis virus encephalitis (A), decreases CNS viral replication (B) and decreases CNS virus-induced apoptosis (C). Data are shown for recombinant chimeric Sindbis virus constructs that express either wild-type human Beclin 1 (SIN/*beclin1*), human Beclin 1 lacking the Bcl-2-binding domain of Beclin 1 (SIN/*beclin1*ΔBcl-2BD) or human Beclin 1 containing a premature stop codon at nucleotide position 270 (SIN/*beclin1*stop). Adapted with permission from Ref. [18].

in Sindbis virus infected *beclin 1* or *atg5* null cells is the indirect result of increased viral replication or results from more direct effects of *atg* gene deficiency on cell death. However, studies comparing HSV-1 infection in wild-type and *beclin 1* null ES cells suggest that Beclin 1 can protect against virus-induced cell death in the absence of inhibitory effects on viral replication [22].

The mechanisms by which autophagy genes exert protective effects on Sindbis virus infection are not yet known. Presumably, the autophagic breakdown of viral components, as observed by electron microscopic analyses of Sindbis virus-infected neurons (Fig. 13.2), leads to decreased viral yields. However, it as also possible that autophagy leads to the breakdown of cellular components required for viral replication. As noted above, the protective affects against cell death may be secondary to inhibitory effects on viral replication. Alternatively, the protective effects may relate to the nutrient recycling functions or "damage control" functions of autophagy. It is also possible that autophagy may protect against cell death by degrading specific viral proteins (e.g. the Sindbis virus E1 and E2 envelope glycoproteins) that are involved in triggering the apoptotic pathway [23].

In summary, the studies of Sindbis virus infection in neurons overexpressing Beclin 1 or in cultured cells lacking *beclin 1* or *atg5* demonstrate a role for mammalian autophagy genes in both restricting viral replication and in protection against virus-induced cell death. It will be important to examine whether other autophagy genes have a similar antiviral function and to examine whether autophagy genes also protect against other types of virus infections. Moreover, it will also be important to determine the role of autophagy genes in antiviral host defense *in vivo* using newly developed tissue-specific autophagy gene knockout mouse models.

13.4
Role of Plant Autophagy Genes in Antiviral Host Defense

Autophagy genes play a role not only in antiviral host defense in mammalian cells, but also in plants [24]. Similar to mammalian *beclin 1*, the plant orthologs of three autophagy genes, *beclin 1*, *atg3* and *atg7*, restrict viral replication. However, in contrast to mammalian *beclin 1*, which prevents the death of Sindbis virus- and HSV-1-infected cells, plant autophagy genes play an interesting role in preventing the death of uninfected cells.

Liu et al. used a tobacco mosaic virus (TMV) infection model to study the role of autophagy genes in plant innate immunity [24]. During virus infection or in response to expression of a virally encoded pathogen avirulence protein, the successful plant innate immune response is characterized by a hypersensitive response. This is a programmed cell death response that occurs around the infected areas, and serves to limit virus spread and confer pathogen resistance. During TMV infection, the hypersensitive response is triggered by the tobacco mosaic virus protein, TMV p50, which is the helicase domain of the viral replicase [25]. TMV p50 is recognized by a specific plant resistance (R) protein, the N protein, which is composed of a Toll/interleukin-1 receptor domain, a nucleotide-binding domain and a leucine-rich repeat domain [26]. Therefore, in tobacco plants containing the N protein, there is death localized to cells that are either infected with TMV or that express the TMV p50 protein, but no viral spread or death of uninfected cells.

RNA interference (RNAi) silencing of autophagy genes in N protein-containing tobacco plants results in a significant increase in local replication of TMV (consistent with the antiviral role of autophagy genes during mammalian virus infection) without increasing viral spread [24] (Fig. 13.5). Furthermore, it results in the uncontrolled spread of cell death that extends far beyond the sites of TMV infection until there is death of the entire inoculated leaf and other uninoculated leaves. A similar spreading cell death phenotype is seen with local expression of the TMV p50 protein in autophagy gene-silenced plants, confirming that the cell death occurs in response to a specific signal triggered by the pathogen-encoded avirulence protein and is not due to increased TMV replication or altered virus movement.

At present, it is not yet clear how autophagy genes protect uninfected cells against death during the plant hypersensitive response. One possibility is that the absence of autophagy genes in infected cells somehow modifies the N gene-

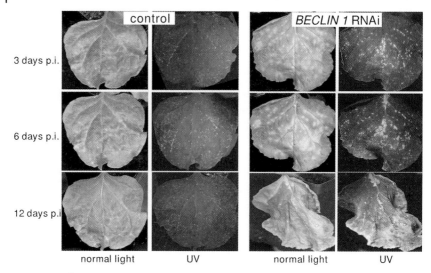

Fig. 13.5 Beclin 1 limits tobacco mosaic virus replication and limits the spread of tobacco mosaic virus-induced programmed cell death in tobacco plants. UV images show sites of replication of a GFP-expressing TMV and normal light images show areas of cell death (yellow) in nonsilenced (control) and *BECLIN 1*-silenced (*BECLIN 1* RNAi) plants. p.i. = post-infection. Adapted with permission from Ref. [24]. (This figure also appears with the color plates).

mediated signal transduction pathway in a way that instructs uninfected cells to die. An alternative, perhaps more likely, possibility is that the absence of autophagy genes in uninfected cells renders them more susceptible to pro-death signals emitted from infected cells. A third possibility is that autophagy blocks the movement of pro-death signals from infected areas to uninfected areas of the plant. Regardless of the mechanism, this newly defined role for autophagy genes in preventing the spread of cell death during plant innate immunity has significant implications for understanding the role of autophagy in systemic protection against viral infections. A critical question is whether autophagy genes play a similar role during animal virus infections.

13.5
Evasion of Autophagy by Viruses

The evolutionarily conserved function of both mammalian and plant autophagy genes in restricting viral replication and/or protection against cell death suggests an essential role for autophagy in innate immunity. This concept is further supported by recent studies indicating that viral virulence proteins possess different strategies to disarm host autophagy.

The HSV neurovirulence protein, ICP34.5, has at least two mechanisms to antagonize host autophagy. As noted above, HSV-1 ICP34.5 can antagonize the

PKR signaling pathway required for autophagy induction. In preliminary studies, we have also found that ICP34.5 can bind to and inhibit the function of a component of the autophagic machinery, Beclin 1 [27]. Using HSV-1 constructs containing different mutations in ICP34.5, our results demonstrate that the Beclin 1-binding activity of ICP34.5, rather than its ability to antagonize host cell shutoff, is essential for viral neurovirulence. Deletion of the beclin 1-binding domain of ICP34.5 from HSV-1 completely blocks the ability of HSV-1 to cause lethal encephalitis in mice and decreases its ability to inhibit autophagy in mouse brain, without affecting the ability of ICP34.5 to dephosphorylate eIF2α or block translational arrest. These results provide the first example of a viral virulence protein directly targeting an autophagy protein to elicit disease.

In addition to HSV-1 ICP34.5, other viral gene products may also interact with Beclin 1 or other mammalian autophagy proteins. Beclin 1 was originally isolated in a yeast two-hybrid screen with the cellular anti-apoptotic protein Bcl-2 [18]. Subsequently, Beclin 1 has also been shown to interact with viral Bcl-2-like protein encoded by different gammaherpesviruses, including Epstein–Barr virus (EBV)-encoded BHRF1, Kaposi's sarcoma-associated herpesvirus (KSHV)-encoded v-Bcl-2 and murine gHV68-encoded M11 [20, 28]. Like ICP34.5, these viral Bcl-2 family proteins also inhibit the autophagy function of Beclin 1 in yeast and mammalian assays. Preliminary evidence indicates that other KSHV-encoded proteins may also target other autophagy proteins [29].

Not only may viruses target the autophagy execution machinery, viruses may also modulate autophagy signaling networks. As discussed above, many viruses encode antagonists of PKR, a signaling molecule required for autophagy induction, and of IFN-γ, a cytokine that stimulates autophagy. An as-yet unexplored area is whether viruses also inhibit autophagy by activating autophagy inhibitory signaling networks. For example, the class I phosphatidylinositol-3-kinase (PI3K)/Akt signaling pathway negatively regulates autophagy in *Caenorhabditis elegans*, *Drosophila* and mammalian cells [30–33], and several different viruses activate this pathway. Certain oncogenic retroviruses encode the catalytic subunit of PI3K and Akt (reviewed in Ref. [34]). In addition, the EBV latent membrane proteins, LMP1 and 2A, the hepatitis B virus protein, Hbx, the Kaposi's sarcoma virus protein, K1, and the hepatitis C virus protein, NS5A, all activate the PI3K/Akt signaling pathway [35–40]. Presumably, such activation plays a role in autophagy inhibition, although this has not yet been formally tested.

While further studies are required to more precisely define the interactions between viral gene products and autophagy regulatory signals and autophagy proteins, there is, however, accumulating evidence that viruses do target multiple different steps of the host autophagy pathway. This observation strongly suggests an evolutionary advantage for viruses to inhibit host autophagy and, by extrapolation, a beneficial role for host autophagy in defense against viral infections.

13.6
How Might Viral Evasion of Autophagy Contribute to Viral Pathogenesis?

Viruses may have originally evolved mechanisms to evade host autophagy to defend themselves against an important arm of the innate immune response. However, in addition to innate immunity, the autophagy pathway plays an important role in many other fundamental biological processes, including tissue homeostasis, differentiation and development, cell growth control, and the prevention of aging and neurodegenerative diseases (reviewed in Refs. [41, 42]). Accordingly, the inhibition of host autophagy by viral gene products has implications not only for understanding mechanisms of immune evasion, but also for understanding novel mechanisms of viral pathogenesis. In this section, we provide a few illustrations of different ways in which viral evasion of autophagy might contribute to distinct aspects of viral pathogenesis.

Genetic studies have revealed an essential role for components of the autophagic machinery in differentiation and developmental processes in several different organisms, including sporulation in yeast, multicellular development in *Dictyostelium*, dauer development in *C. elegans* and embryonic development in mice (reviewed in Ref. [41]). Moreover, the mammalian autophagy gene, *beclin 1*, appears to play a role in epithelial cell differentiation, since the mammary glands in *beclin* 1 heterozygous-deficient mice display striking morphological abnormalities [43]. Since viral gene products inhibit the autophagy function of Beclin 1 and potentially other autophagy proteins, it is possible that autophagy blockade represents a mechanism by which viruses can affect cellular differentiation and development. As examples, the Bcl-2-like BHRF1 protein encoded by EBV binds to Beclin 1 [20], blocks it autophagy function [44] and also perturbs epithelial cell differentiation [45]. Intrauterine HSV infection leads to severe fetal anomalies [46] and HSV ICP34 inhibits the autophagy function of Beclin 1 [27], a protein required for normal embryonic development [43, 47]. Further studies are needed to determine the role of Beclin 1 binding in the perturbation of epithelial cell differentiation by BHRF1 during EBV infection or in fetal developmental defects induced by HSV infection, as well as to investigate the effects of other viral inhibitors of autophagy on cellular differentiation and multicellular development.

Another intriguing area is the role of viral evasion of autophagy in viral oncogenesis. There is increasing evidence that autophagy deregulation contributes to tumorigenesis; several different oncogenic signaling molecules negatively regulate autophagy (e.g. class I PI3K, Akt, members of the Rhos and Ras family of GTPases), the PTEN (phosphatase and tensin homolog deleted on chromosome 10) tumor suppressor negatively regulates autophagy and components of the autophagic machinery, such as *beclin 1*, function as tumor suppressors (reviewed in Refs. [48, 49]). Oncogenic viruses, including retroviruses, gammaherpesviruses, papillomaviruses and hepatitis viruses, likely possess mechanisms to block host autophagy. For example, many oncogenic viruses turn on autophagy-inhibitory signaling pathways, either by encoding viral oncoproteins that represent activated forms of the corresponding cellular proto-oncogene (reviewed in Ref. [34]) or by upregulating Rho/Ras or class I PI3K/Akt/target of rapamycin (TOR) signaling through alternative mechanisms [35–40, 50]. Further, the onco-genic gammaherpesivurses encode Bcl-2-like proteins that antagonize the au-tophagy function of Beclin 1 [28]. Given the emerging importance of autophagy in tumor suppression, it is tempting to speculate that these autophagy-inhibi-tory actions of viral oncoproteins contribute to viral oncogenesis.

It is also theoretically possible that the ability of HSV-1 to evade host autopha-gy contributes to neurodegenerative diseases. Although the topic remains con-troversial, several studies have suggested an epidemiological link between the type 4 allele of apolipoprotein E, productive HSV-1 replication and risk of Alz-heimer's disease [51–54]. Moreover, there is increasing evidence that autophagy is involved in protection against different types of neurodegenerative diseases, including Alzheimer's disease (reviewed in Refs. [42, 55, 56]). Of note, de-creased brain expression of Beclin 1 was identified in unbiased search for gene products associated with Alzheimer's disease in humans and heterozygous dele-tion of *beclin 1* promotes Alzheimer's disease pathology in mice [57]. Thus, it will be interesting to determine whether ICP34.5 antagonism of the autophagy function of Beclin 1 contributes to the putative link between HSV-1 infection and Alzheimer's disease.

In the preceding paragraphs, we have discussed a few speculative hypotheses of how viral suppression of autophagy might contribute to the pathogenesis of medical diseases. With the rapid growth of research on the role of autophagy deregulation in disease and the new identification of viral gene products that in-hibit autophagy, it should be possible to rigorously test these and other related hypotheses.

13.7
Autophagy and Antigen Presentation

The role of autophagy in antiviral host defense is not limited to its function in innate immunity. There is increasing evidence that autophagy may play a role in adaptive immunity during viral infection through its role in MHC class II-re-stricted presentation of endogenous cytosolic and nuclear antigens (see Chapter 12). Paludan et al found that nuclear antigen 1 of EBV (EBNA1) enters endoge-nous MHC class II processing through a pathway involving autophagic degrada-tion [58]. Blockade of autophagy with the pharmacological inhibitor, 3-methyla-denine (3-MA), or RNAi against the autophagy gene, *atg12*, decreased EBNA1-specific CD4$^+$ T cell recognition of EBV-transformed B cells and EBNA1-trans-fected Hodgkin's lymphoma cells. In addition, other groups have reported that the endogenous presentation of epitopes derived from cytosolic proteins in-volves the autophagy pathway [59, 60]. Based on these studies, it is reasonable to postulate that autophagy plays a role in MHC class II antigen presentation during viral infection. Moreover, the stimulation of autophagy by IFN-γ may be important in promoting autophagy-dependent endogenous MHC class II-restricted antigen presentation in virally infected cells. A wide open question is whether autophagy also plays a role in processing viral antigens for MHC class I-restricted presentation.

13.8
Concluding Remarks

Although research in this area is still in a stage of infancy, there is accumulating evidence that the lysosomal degradation pathway of autophagy plays an evolutionarily conserved role in antiviral immunity. Two antiviral signaling molecules, i.e. PKR and IFN-γ, positively regulate autophagy, and both mammalian and plant autophagy genes restrict viral replication and protect against virus-induced death. Furthermore, autophagy may foster T cell responses to viral infections through its role in MHC class II-restricted presentation of endogenous viral antigens. Given this role of autophagy in both innate and adaptive immunity, it is not surprising that viruses have evolved numerous strategies to inhibit host autophagy. Different viral gene products can either modulate autophagy regulatory signals or directly interact with components of the autophagy execution machinery. Moreover, as discussed in the addendum after this chapter, certain RNA viruses have managed to "co-apt" the autophagy pathway, selectively utilizing certain components of the dynamic membrane rearrangement system to promote their own replication inside the cytoplasm.

There are many important questions to be addressed. We do not yet understand exactly how cells sense viral infections and turn on autophagy, how autophagy restricts viral replication or how autophagy prevents virus-induced cell death. In terms of regulation, it is not yet known whether type I IFN stimulates autophagy (either through PKR or alternative pathways), and precisely which molecules are downstream in PKR- and IFN-γ—induced autophagy. At the cell biology level, it is not known whether viruses, like intracellular bacteria (reviewed in Ref. [61]), are specifically targeted to autophagosomes or whether their engulfment by autophagic membranes is a chance occurrence. In the case of HSV-1 infection, newly synthesized virions appear to be trapped inside autophagosomes during egress from the nucleus out of the cell [4]; however, it is not yet known whether viruses, like the bacteria Group A *Streptococcus* [62], are also targeted to autophagosomes upon cell entry. Moreover, nothing is understood about the molecular basis for why certain RNA viruses (e.g. Sindbis virus, tobacco mosaic virus) are attacked by autophagy [18, 24], whereas other RNA viruses (e.g. poliovirus, mouse hepatitis virus) require autophagy proteins for replication [63, 64]. In the forthcoming years, we will likely begin to unravel these and many more mysteries regarding the role of autophagy in antiviral host defense.

References

1 Williams, B.R. **1999**. PKR: a sentinel kinase for cellular stress. *Oncogene 18*, 6112–6120.

2 Tan, S.L., Katze, M.G. **2000**. HSV.com: maneuvering the internetworks of viral neuropathogenesis and evasion of the host defense. *Proc Natl Acad Sci USA 97*, 5684–5686.

3 Talloczy, Z., Jiang, W., Virgin, H.W., Leib, D.A., Scheuner, D., Kaufman, R.J., Eskelinen, E.-L., Levine, B. **2002**. Regulation of starvation- and virus-induced autophagy by the eIF2a kinase signaling pathway. *Proc Natl Acad Sci USA 99*, 190–195.

4 Talloczy, Z., Virgin, H., Levine, B. **2006**. PKR-dependent autophagic degradation of herpes simplex virus type 1. *Autophagy* in press.

5 Chou, J., Kern, E.R., Whitley, R.J., Roizman, B. **1990**. Mapping of herpes simplex virus-1 neurovirulence to $\gamma_1$34.5, a gene nonessential for growth in culture. *Science 250*, 1262–1266.

6 Chou, J., Chen, J.J., Gross, M., Roizman, B. **1995**. Association of a M_r 90,000 phosphoprotein with protein kinase PKR in cells exhibiting enhanced phosphorylation of translation initiation factor eIF-a2 and premature shutoff of protein synthesis after infection with $\gamma_1$34.5 mutants of herpes simplex virus 1. *Proc Natl Acad Sci USA 23*, 10516–10520.

7 He, B., Gross, M., Roizman, B. **1997**. The $\gamma_1$34.5 protein of herpes simplex virus 1 complexes with protein phosphatase 1a to dephosphorylate the a subunit of the eukaryotic translation initiation factor 2 and preclude the shutoff of protein synthesis by double-stranded RNA-activated protein kinase. *Proc Natl Acad Sci USA 94*, 843–848.

8 Leib, D.A., Machalek, M.A., Williams, B.R.G., Silverman, R.H., Virgin, H.W. **2000**. Specific phenotypic restoration of an attenuated virus by knockout of a host resistance gene. *Proc Natl Acad Sci USA 97*, 6097–6101.

9 Gale Jr., M., Katze, M.G. **1998**. Molecular mechanisms of interferon resistance mediated by viral-directed inhibition of PKR, the interferon-induced protein kinase. *Pharmacol Ther 78*, 29–46.

10 Shtrichman, R., Samuel, C.E. **2001**. The role of γ-interferon in antimicrobial immunity. *Curr Opin Microbiol 4*, 251–259.

11 Inbal, B., Bialik, S., Sabanay, I., Shani, G., Kimchi, A. **2002**. DAP kinase and DRP-1 mediate membrane blebbing and the formation of autophagic vesicles during programmed cell death. *J Cell Biol 157*, 455–468.

12 Pyo, J.-O., Jang, M.-H., Kwon, Y.-K., Lee, H.-J., Jun, J.-I., Woo, H.-N., Cho, D.-H., Choi, B.-Y., Lee, H., Kim, J.-H., Mizushima, N., Oshumi, Y., Jung, Y.-K. **2005**. Essential roles of Atg5 and FADD in autophagic cell death: dissection of autophagic cell death into vacuole formation and cell death. *J Biol Chem 280*, 20722–20729.

13 Gutierrez, M.G., Master, S.S., Singh, S.B., Taylor, G.A., Colombo, M.I., Deretic, V. **2004**. Autophagy is a defense mechanism inhibiting BCG and *Mycobacterium tuberculosis* survival in infected macrophages. *Cell 119*, 753–766.

14 Deretic, V. **2005**. Autophagy in innate and adaptive immunity. *Trends Immunol 26*, 523–528.

15 Alcami, A. **2003**. Viral mimicry of cytokines, chemokines and their receptors. *Nat Rev Immunol 3*, 36–50.

16 Horvath, C.M. **2004**. Weapons of STAT destruction: interferon evasion by paramyxovirus V proteins. *Eur J Biochem 271*, 4621–4628.

17 Hengel, H., Koszinowski, U.H., Conzelmann, K.-K. **2005**. Viruses know it all: new insights into IFN networks. *Trends Immunol 26*, 396–401.

18 Liang, X.H., Kleeman, L.K., Jiang, H.H., Gordon, G., Goldman, J.E., Berry, G., Herman, B., Levine, B. **1998**. Protection against fatal Sindbis virus encephalitis by Beclin, a novel Bcl-2-interacting protein. *J Virol 72*, 8586–8596.

19 Levine, B., Goldman, J.E., Jiang, H.H., Griffin, D.E., Hardwick, J.M. **1996**. Bcl-2 protects mice against fatal alphavirus encephalitis. *Proc Natl Acad Sci USA 93*, 4810–4815.

20 Liang, X. H., Levine, B. Unpublished data.

21 MacPhearson, S., Yu, J., Levine, B. Unpublished data.

22 Talloczy, Z., Levine, B. Unpublished data.

23 Joe, A., Foo, H., Kleeman, L., Levine, B. **1998**. The transmembrane domains of Sindbis virus envelope glycoproteins induce cell death. *J Virol 72*, 3935–3943.

24 Liu, Y., Schiff, M., Czymmek, K., Talloczy, Z., Levine, B., Dinesh-Kumar, S. P. **2005**. Autophagy regulates programmed cell death during the plant innate immune response. *Cell 121*, 567–577.

25 Erickson, F. L., Hozberg, S., Calderon-Urrea, A., Handley, V., Axtell, M., Corr, C., Baker, B. **1999**. The helicase domain of the TMV replicase proteins induces the N-mediated defence response in tobacco. *Plant J 18*, 67–75.

26 Whitham, S., Dinesh-Kumar, S. P., Choi, D., Hehl, R., Corr, C., Baker, B. **1994**. The product of the tobacco mosaic virus resistance gene N: similarity to toll and the interleukin-1 receptor. *Cell 78*, 1101–1115.

27 Orvedahl, A., Talloczy, Z., Alexander, D., Leib, D., Levine, B. Unpublished data.

28 Pattingre, S., Tassa, A., Qu, X., Garuti, R., Liang, X. H., Mizushima, N., Packer, M., Schneider, M. D., Levine, B. **2005**. Bcl-2 antiapoptotic proteins inhibit Beclin 1-dependent autophagy. *Cell 122*, 927–939.

29 Jung, J. **2004**. Personal Communication.

30 Melendez, A., Talloczy, Z., Seaman, M., Eskelinen, E.-L., Hall, D. H., Levine, B. **2003**. Autophagy genes are essential for dauer development and lifespan extension in *C. elegans. Science 301*, 1387–1391.

31 Rusten, T. E., Lindmo, K., Juhasz, G., Sass, M., Seglen, P. O., Brech, A., Stenmark, H. **2004**. Programmed autophagy in the *Drosophila* fat body is induced by ecdysone through regulation of the PI3K pathway. *Dev Cell 7*, 179–192.

32 Scott, R. C., Schuldiner, O., Neufeld, T. P. **2004**. Role and regulation of starvation-induced autophagy in the *Drosophila* fat body. *Dev Cell 7*, 167–178.

33 Petiot, A., Ogier-Denis, E., Blommaart, E. F., Meijer, A. J., Codogno, P. **2000**. Distinct classes of phosphatidylinositol 3′-kinases are involved in signaling pathways that control macroautophagy in HT-29 cells. *J Biol Chem 275*, 992–998.

34 Aoki, M., Vogt, P. K. **2004**. Retroviral oncogenes and TOR. *Curr Top Microbiol Immunol 279*, 321–338.

35 Scholle, F., Bendt, K. M., Raab-Traub, N. **2000**. Epstein–Barr virus LMP2A transforms epithelial cells, inhibits cell differentiation, and activates Akt. *J Virol 74*, 10681–10689.

36 Fukuda, M., Longnecker, R. **2004**. Latent membrane protein 2A inhibits transforming growth factor-beta 1-induced apoptosis through the phosphatidylinositol 3-kinase/Akt pathway. *J Virol 78*, 1697–1705.

37 Morrison, J. A., Gulley, M. L., Pathmanathan, R., Raab-Traub, N. **2004**. Differential signaling pathways are activated in the Epstein–Barr virus-associated malignancies nasopharyngeal carcinoma and Hodgkin lymphoma. *Cancer Res 64*, 5251–5260.

38 hung, T. W., Lee, Y. C., Kim, C. H. **2004**. Hepatitis B viral HBx induces matrix metalloproteinase-9 gene expression through activation of ERK and PI-3K/Akt pathways: involvement of invasive potential. *FASEB J 18*, 1123–1125.

39 Tomlinson, C. C., Damania, B. **2004**. The K1 protein of Kaposi's sarcoma-associated herpesvirus activates the Akt signaling pathway. *J Virol 78*, 1918–1927.

40 Street, A., Macdonald, A., Crowder, K., Harris, M. **2004**. The hepatitis C virus NS5A protein activates a phosphoinositide 3-kinase-dependent survival signaling cascade. *J Biol Chem 279*, 12232–12241.

41 Levine, B., Klionsky, D. J. **2004**. Development by self-digestion: molecular mechanisms and biological functions of autophagy. *Dev Cell 6*, 463–477.

42 Shintani, T., Klionsky, D. J. **2004**. Autophagy in health and disease: a double-edged sword. *Science 306*, 990–995.

43 Qu, X., Yu, J., Bhagat, G., Furuya, N., Hibshoosh, H., Troxel, A., Rosen, J., Eskelinen, E.-L., Mizushima, N., Ohsumi, Y., Cattoretti, G., Levine, B. **2003**. Promotion of tumorigenesis by heterozygous

disruption of the *beclin 1* autophagy gene. *J Clin Invest 112*, 1809–1820.

44 Pattingre, S., Liang, X. H., Levine, B. Unpublished data.

45 Dawson, C. W., Eliopoulos, A. G., Dawson, J., Young, L. S. 1995. BHRF1, a viral homologue of the Bcl-2 oncogene, disturbs epithelial cell differentiation. *Oncogene 10*, 69–77.

46 Grosse, C. 1994. Congenital infections caused by varicella zoster virus and herpes simplex virus. *Semin Pediatr Neurol 1*, 43–49.

47 Yue, Z., Jin, S., Yang, C., Levine, A. J., Heintz, N. 2003. Beclin 1, an autophagy gene essential for early embryonic development, is a haploinsufficient tumor suppressor. *Proc Natl Acad Sci USA 100*, 15077–15082.

48 Furuya, N., Liang, X. H., Levine, B. 2004. Autophagy and cancer. In: *Autophagy*. D. J. Klionsky (Ed.). Georgetown, TX: Landes Bioscience, pp. 244–253.

49 Meijer, A. J., Codogno, P. 2004. Regulation and role of autophagy in mammalian cells. *Int J Biochem Cell Biol 36*, 2445–2462.

50 Payne, E., Bowles, M. R., Don, A., Hancock, J. F., McMillan, N. A. 2001. Human papillomavirus type 6b virus-like particles are able to activate the Ras-MAP kinase pathway and induce cell proliferation. *J Virol 75*, 4150–4157.

51 Wozniak, M. A., Shipley, S. J., Combrinck, M., Wilcock, G. K., Itzhaki, R. F. 2005. Productive herpes simplex virus in brain of elderly normal subjects and Alzheimer's disease patients. *J Med Virol 75*, 300–306.

52 Itzhaki, R. 2004. Herpes simplex virus type 1, apolipoprotein E and Alzheimer' disease. *Herpes 11*, 77A–82A.

53 Mori, I., Kimura, Y., Naiki, H., Matsubara, R., Takeuchi, T., Yokochi, T., Nishiyama, Y. 2004. Reactivation of HSV-1 in the brain of patients with familial Alzheimer's disease. *J Med Virol 73*, 605–611.

54 Itzhaki, R. F., Lin, W. R., Shang, D., Wilcock, G. K., Faragher, B., Jamieson, G. A. 1997. Herpes simplex virus type 1 in brain and risk of Alzheimer's disease. *Lancet 349*, 241–244.

55 Larsen, K. E., Sulzer, D. 2002. Autophagy in neurons: a review. *Histol Histopathol 17*, 897–908.

56 Ravikumar, B., Rubinsztein, D. C. 2004. Can autophagy protect against neurodegeneration caused by aggregate-prone proteins? *Neuroreport 15*, 2443–2445.

57 Pickford, F., Masliah, E., Small, S., Mizushima, N., Levine, B., Wyss-Coray, T. 2006. Deficiency of the Beclin 1 autophagy protein promotes Alzheimer's disease pathology. Submitted.

58 Paludan, C., Schmid, D., Landthaler, M., Vockerodt, M., Kube, D., Tuschl, T., Munz, C. 2005. Endogenous MHC class II processing of a viral nuclear antigen after autophagy. *Science 307*, 593–596.

59 Nimmerjahn, F., Milosevic, S., Behrends, U., Jaffe, E. M., Pardoll, D. M., Bornkamm, G. W., Mautner, J. 2003. Major histocompatibility complex class II-restricted presentation of a cytosolic antigen by autophagy. *Eur J Immunol 33*, 1250–1259.

60 Dengjel, J., Schoor, O., Fischer, R., Reich, M., Kraus, M., Muller, M., Kreymborg, K., Altenberend, F., Brandenburg, J., Kalbacher, H., Brock, R., Driessen, C., Rammensee, H.-G., Stevanovic, S. 2005. Autophagy promotes MHC class II presentation of peptides from intracellular source proteins. *Proc Natl Acad Sci USA 102*, 7922–7927.

61 Levine, B. 2005. Eating oneself and uninvited guests: autophagy-related pathways in cellular defense. *Cell 120*, 159–162.

62 Nakagawa, I., Amano, A., Mizushima, N., Yamamoto, A., Yamaguchi, H., Kamimoto, T., Nara, A., Funao, J., Nakata, M., Tsuda, K., Hamada, S., Yoshimori, T. 2004. Autophagy defends cells against invading Group A *Streptococcus*. *Science 306*, 1037–1040.

63 Jackson, W. T., Giddings Jr., T. H., Taylor, M. P., Mulinyawe, S., Rabinovitch, M., Kopita, R. R., Kirkegaard, K. 2005. Subversion of cellular autophagosomal machinery by RNA viruses. *PLOS Biol 3*, 861–871.

64 Prentice, E., Jerome, W. G., Yoshimori, T., Mizushima, N., Denison, M. R. 2004. Coronavirus replication complex formation utilizes components of cellular autophagy. *J Biol Chem 279*, 10136–10141.

Addendum

Jennifer Sparks and Mark R. Denison

Some viruses have evolved specific strategies to subvert host autophagic pathways in order to facilitate their replication [1, 2]. The coronavirus murine hepatitis virus (MHV) induces the formation of cytoplasmic double-membrane vesicles (DMVs) that serve as sites of viral replication, and are required for that process [2, 3]. Endogenous LC3 (homolog of Atg8) and plasmid expressed LC3-GFP colocalize with Atg12 and MHV replication products on DMVs [2]. In murine embryonic stem cells that are incapable of forming autophagosomes due to lack of Atg5, MHV replication is decreased by 10^4 pfu/ml, an inhibition that is reversed by plasmid expression of Atg5. MHV induction of DMVs occurs by a mechanism resistant to the autophagy inhibitor 3-methyladenine, and MHV-induced DMVs do not label for LAMP-1 or Rab7, markers for degradative autophagolysosomes. These findings suggest that MHV may directly interact with initiation machinery from the autophagy pathway while inhibiting the maturation of DMVs into autophagolysosomes.

Poliovirus provides another example of how RNA viruses may use autophagosomal machinery for their benefit. Cells infected with poliovirus accumulate 150–400 nm diameter DMVs [1]. The induction of DMV formation is mediated by poliovirus proteins, 2BC and 3A [4], presumably by hydrophobic domains that are hypothesized to mediate interaction with host membranes. In contrast to MHV, poliovirus-induced DMVs share several hallmarks of autophagosomes, including colocalization of LAMP1 and LC3 [5]. Additionally, stimulation of cellular autophagy with either tamoxifen or rapamycin increases poliovirus yield, while inhibition of the autophagy with 3-methyladenine decreases virus yield [5]. These data suggest that the process of autophagy may be directly beneficial for poliovirus replication by providing membraneous support for viral RNA replication complexes.

1 Schlegel, A., et al., Cellular origin and ultrastructure of membranes induced during poliovirus infection. *J Virol* **1996**, *70*(10), p. 6576–6588.

2 Prentice, E., et al., Coronavirus replication complex formation utilizes components of cellular autophagy. *J Biol Chem* **2004**, *279*(11), p. 10136–10141.

3 Gosert, R., et al., RNA replication of mouse hepatitis virus takes place at double-membrane vesicles. *J Virol* **2002**, *76*(8), p. 3697–3708.

4 Suhy, D.A., Giddings, T.H. Jr., Kirkegaard, K., Remodeling the endoplasmic reticulum by poliovirus infection and by individual viral proteins: an autophagy-like origin for virus-induced vesicles. *J Virol* **2000**, *74*(19), p. 8953–8965.

5 Jackson, W.T., et al., Subversion of cellular autophagosomal machinery by RNA viruses. *PLoS Biol* **2005**, *3*(5), p. e156.

Subject Index

Autophagy in Immunity and Infection. Edited by Vojo Deretic
Copyright © 2006 WILEY-VCH Verlag GmbH & Co. KGaA, Weinheim
ISBN: 3-527-31450-4